Advanced Optics Using

Aspherical Elements

Advanced Optics Using
Aspherical Elements

Bernhard Braunecker

Rüdiger Hentschel

Hans J. Tiziani

Editors

SPIE
PRESS

Bellingham, Washington USA

Library of Congress Cataloging-in-Publication Data

Braunecker, B. (Bernhard)
 Advanced optics using aspherical elements / B. Braunecker, R. Hentschel, H. Tiziani.
 p. cm.
 Includes bibliographical references.
 ISBN 978-0-8194-6749-2
 1. Aspherical lenses. 2. Optical instruments--Design and construction. I. Hentschel, R. (Rudiger), 1949- II. Tiziani, Hans J. III. Title.

TS517.5.A86B73 2007
681'.423--dc22

2007028838

Published by

SPIE
P.O. Box 10
Bellingham, Washington 98227-0010 USA
Phone: +1 360 676 3290
Fax: +1 360 647 1445
Email: spie@spie.org
Web: SPIE.org

Printed in the United States of America.

Contents

Chapter 1

Introduction

1.1 Motivation

Modern optical systems rely on leading-edge production technologies, especially for the development of aspherical optical elements. Many activities, worldwide, are targeted to the development of production technologies for aspheres that are as accurate, reliable, and cost-attractive as those for spherical lenses. Today, aspherical lenses of reasonably good quality are still up to 10 times more expensive than the corresponding spherical ones, which indicates how far we are away from achieving this. We must understand the current technologies to identify improvement potential.

We will describe several examples of applications and technologies, which give a good understanding along the value-added chain. Our priority is to understand general principles. Therefore, we disregard high-volume applications such as handy phone cameras, which on the one hand apply these principles but on the other hand use very special solutions for mass production because of the high competitiveness of these market segments.

This compendium is primarily written to be an optical technology reference book for development and production engineers. Due to the inherent complexity of aspheres, all efforts to push the development of technologies are still risky. To minimize risk, clear decisions based on a good understanding of technology are therefore indispensable at the management level.

Decision-makers for the implementation of optical technologies need to have technological background information available in a short and compressed way. This also holds for strategy management consultants, who have to propose solutions and prepare decisions. Today it is rather difficult to get access to all the information, and very often only experts can extract and interpret the relevant part that is needed. We see a real need for a reference book, and the best way to pull together authentic information is to involve experts in the field of optics.

We therefore asked leading engineers from renowned optical companies to join as authors. Their contributions are summarized in standardized templates, which help to understand a variety of technologies in a common way. To facilitate reading, all templates have a unified structure and present their content in a brief yet comprehensive form. Links to more detailed information lead to short presentations of essentials.

The experts' contributions help to understand the fundamental features in an easily accessible way by

- Presenting the "state of the art" in optical design and production technology,
- Pointing out trends and ongoing activities, and
- Performing benefit and risk analysis of different production technologies.

The focus on aspheres includes the standard processes for producing spherical components, but it also points out the true challenges for optical technologies to achieve the same reliability and cost structure in aspheres.

Based on this concept, the compendium is divided into two parts. The first part, "Review and Summary," is an introduction to the technologies, but it also summarizes the detailed results of the templates. The second part, "Experts' Contributions," is the collection of templates, which were each created by their particular authors.

Both parts are organized into sections that reflect the typical workflow to produce optical systems with aspheres. We start with "basic considerations," comments on "design and application cases," and subsequently treat "materials," "surface processing," "metrology," "coatings," and "assembly." We finish Part I with a technology forecast, "Future Trends," where we try to predict the progress of all technologies involved. This outlook is rather speculative and reflects our own opinions and the personal opinions of our experts.

The user can profit from the dual structure in the following ways:

- It allows a focused reading of the basics but also provides important details.
- It enables understanding of the differences between sometimes competing, but mostly complementing, technology variants.
- The Internet links may be used for further reading of publications, patents, or industrial information materials.

All these features should enable the reader to navigate through the sections and find the most appropriate combination of materials and processes for a particular application.

This book relates to a fast-moving technology. Having this in mind, we asked our experts to add web links for immediate and further reading in more detail.

1.2 Guideline

We here present in condensed form the key contents of the different sections of Part I: Review and Summary.

Chapter 2: Basic Considerations

We start by explaining the physics of image generation, studying the propagation of an optical wavefront through a single lens. We find the equivalence of the optical material parameters and the geometry parameters of the lens surfaces, concerning wavefront errors.

We consider design criteria for good images and what advantages we expect from aspheres. Aspheres are then defined in more detail. We present their mathematical description and how technical drawings have to look according to the ISO Norm. We warn of potential communication errors between designer and factories due to misunderstandings in the ISO Norm. To illustrate this point, we present an example with a single aspheric lens used for laser collimation to guide a machine for line-marking in sports fields. Here we analyse how residual manufacturing errors of different surface shaping methods would degrade the application.

Chapter 3: Applications

In Chapter 3 we describe several application cases using aspherical elements. Our industrial authors will explain in Part II their individual systems and their motivation to insert or at least consider inserting aspherical elements. They will comment on the performance improvements they gain but also what requirements must be fulfilled by the manufacturer. They address the choice of suitable materials, typical features of tolerancing, and how they ensure that image quality is maintained over the whole life-cycle of the instrument. The application spectrum extends from mass production for illumination purposes to single space telescopes.

Chapter 4: Materials of Aspheres

At first glance, most optical materials seem to be suited either for spherical or aspherical components. However, special surface-shaping technologies such as "precise pressing" need low-melting, low T_g glasses, which therefore are presented in more detail. Other materials such as ceramics, crystals, and polymers are of growing importance for aspheric systems. We also address the following questions. What are the optical, mechanical, and thermal parameters? How do these parameter vary during production and what are typical cost structures? Is there a need for new materials or at least for new material features?

Chapter 5: Processing Technologies

A variety of surface-forming technologies are described, as practised by leading manufacturing companies. Common to all of them is the understanding of mass removal physics, its mathematical modeling, and its numerically controlled realization. We outline the major differences between the technologies, and their dependance on materials, point out their limits, and comment the cost structure. Are other materials needed to exceed the current limits? What residual surface errors (profile amplitudes and spatial frequencies: long, medium, or short) are to be expected? How precise and stable is the orientation of an aspherical surface to another body axis? Do larger lot sizes favor the cost structure?

Chapter 6: Metrology

Optical methods are well suited to verifying the image performance of complete systems. The optical transfer function (OTF) or the modulation transfer function (MTF) and interferometric methods are frequently used to measure the performance of an optical system. Interferometry is mainly used to measure the shape deviation of polished surfaces. Due to the higher production complexity of aspheres, one needs more elaborate metrology methods. For the measurement of shape deviations using interferometry, Null lenses such as computer generated holograms (CGH) are needed, making the measurement more difficult and expensive. Alternative methods are discussed. The ultimate choice would be a complete integration of metrology in the production process. This is unfortunately not yet available and is perhaps one of the reasons why aspheres are still too expensive. We present several methods to measure the shape, texture, and microroughness of a surface. We will analyse the application range, the limits, and the constraints of metrology methods.

Chapter 7: Coating Technologies

Similar to our comments made above about "materials," there should also be no difference between the coating of spherical and aspherical elements. However, when describing several alternative technologies in more detail, we will see differences. We mention here only two points. Low-melting, low T_g glasses are favored for precision glass molding methods for producing aspheres. A coating process operating at lower temperatures is then preferred to avoid a thermally induced change of the refractive glass index. Another point to watch might be the evaporating geometry. To achieve a thin-layer structure within some sub-micrometer tolerance, the geometry and the rate of evaporation must be optimized to the surface shape to be coated, whether it is a spherical or a (strong) aspherical surface.

Chapter 8: Assembly Technologies

It will be established in this chapter that the assembly of aspheres is quite different to that of spherical elements. Because aspheres carry more surface shape parameters than spherical elements, they are consequently more sensitive to "decentring" errors. This should make the assembly more complicated. However, it is shown in Chapter 5 that the centring of aspheric lenses is a necessary part of the surface-shaping process. Thus, aspherical lenses are already well centred when being delivered to assembly. It will be shown what assembly strategies exist today and what technical possibilities are available to compensate residual material and form errors of single components, compounds, and of the assembly process itself.

Chapter 9: Future Trends

Here, we will summarize our own opinions and the opinions of our coauthors, what progress they expect in single disciplines, what efforts must be undertaken, and what system benefits will be the most prominent drivers of the technology. Furthermore, we give an outlook on future developments.

Chapter 10: Mathematical Formulation

This chapter presents in more detail the general mathematical description of aspheres. It starts with surface function of second order and extends then to higher orders. Problems with the definition as used in the ISO Norm are indicated.

Part II: Chapters 11–16

All experts' contributions in Part II describe the actual technologies. These are presented in a standardized format, which is given in a template used by all contributors of a chapter. Each chapter in Part II describing the value added chain corresponds to a chapter in Part I: Review and Summary.

Part I

Review and Summary

B. Braunecker, R. Hentschel, H. J. Tiziani

with contributions from

R. Litschel	**(Definition of Aspherical Lenses)**
B. Schreder, J. Zimmer	**(Materials)**
K. Beckstette, R. Börret	**(Processing)**
N. Kaiser	**(Coating)**
C. Gunkel	**(Assembly)**
U. Tippner	**(Mathematics)**

Chapter 2

Basic Considerations

2.1 Preliminary Remarks

We will start with a single lens to illustrate the physics of imaging and will find the equivalence of material and geometry parameters for the image quality.

2.1.1 Optical element and wavefront propagation

Optical imaging is generally performed by lenses, that is, by pieces of glass of thickness d and two properly shaped glass–air surfaces. Such a component can be considered as a "black box," hopefully transparent, which transfers an optical input wave of amplitude $A_{\text{In}}(r)$ and phase $\Phi_{\text{In}}(r)$,

$$U_{\text{In}}(r) = A_{\text{In}}(r) * \exp[i * \Phi_{\text{In}}(r)],$$

into an outgoing wave,

$$U_{\text{Out}}(r) = A_{\text{Out}}(r) * \exp[i * \Phi_{\text{Out}}(r)],$$

where i is the imaginary unit, and r is the radial coordinate normal to the light propagation direction z. We describe here the most simple case of a monochromatic wave originating from a far distant point source.

Such a lens alters the amplitude and the phase. We observe the phase change in Fig. 2.1, where a plane wave is transformed into a spherical wave. The spherical wave converges to an intensity spot at a distance f, called the focal length of the lens. Although an amplitude change results in a light intensity loss, this is ignorable here, and the phase term is much more important. It determines the optical quality, that is, the sharpness and contrast of an image. We see in Fig. 2.1 that a wavefront error of the perfect spherical output wave would result in an undesirable spot broadening.

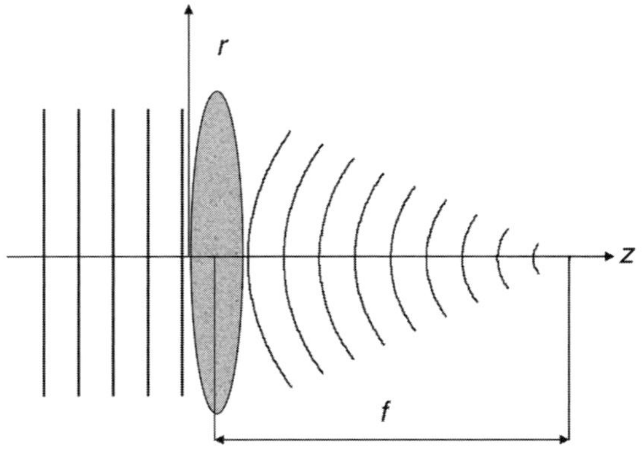

Figure 2.1 Lens action, showing an incident plane wave transformed into a spherical wave.

Let us further assume in our example an input plane wave with $\Phi_{\mathrm{In}}(r) = 0$. Then the phase change between "in" and "out" is given by

$$\Delta\Phi_{\mathrm{Out}}(r) = \Phi_{\mathrm{Out}}(r) - \Phi_{\mathrm{In}}(r) = \Phi_{\mathrm{Out}}(r).$$

If light travels a distance z, which is exactly one wavelength λ or an integer multiple of it, the phase change is 2π. For arbitrary z in air, we obtain $\Phi_{\mathrm{Out}}(r) = 2\pi/\lambda * z(r)$, while inside glass, with refractive index n, the wavelength is shorter, $\lambda' = \lambda/n$, and we obtain

$$\Phi_{\mathrm{Out}}(r) = 2\pi/\lambda' * z(r) = 2\pi/\lambda * [n(r) * z(r)] = 2\pi/\lambda * \mathrm{OP}(r),$$

where OP is the abbreviation for "optical path," $n * z$.

We are interested in how the incoming plane wave is deformed by the element, so we consider in the output plane the phase difference or, equivalently, the optical path difference (OPD) between position r and the optical axis $r = 0$,

$$\Phi_{\mathrm{Out}}(r)/(2\pi/\lambda) = \mathrm{OP}(r) - \mathrm{OP}(0) = \mathrm{OPD}(r),$$

and we also call this phase error the wavefront error.

A closer look at our element shows that three physical effects contribute to Φ_{Out}: both glass–air surfaces and the glass medium (Fig. 2.2). If the surface function at the input side is $z_1(r)$ and at the output side is $z_2(r)$, if d is the axial glass thickness and n the refractive index of the glass, then

$$\mathrm{OP}(r) = \{z_1(r) + n(r) * \{d - [z_1(r) + z_2(r)]\} + z_2(r)\}.$$

Using $z_1(0) = z_2(0) = 0$ and $n(r) = n(0)$ we obtain

$$\Phi_{\mathrm{Out}}(r)/(2\pi/\lambda) = \mathrm{OPD}(r) = -[n(r) - 1] * [z_1(r) + z_2(r)].$$

Note, we ignore any z dependence of the refractive index.

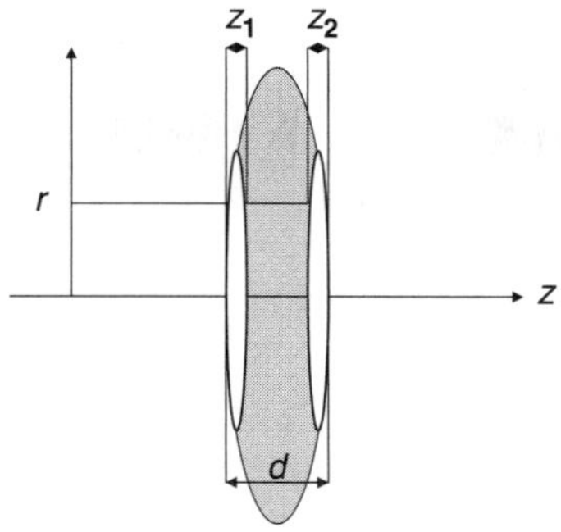

Figure 2.2 Phase contributions.

2.1.2 Optical design and tolerancing

The optical designer has to calculate the surface shape functions and glass thickness, and he has to select the glass material to obtain the "ideal" phase term Φ_{Design} to fulfill his specifications. The designer then has to specify the amount of wavefront degradation that he can accept. This allows him to tolerate quantitatively the fabrication errors.

By variation of the phase term above with respect to z_1, z_2, and n, we obtain

$$|\delta\Phi_{\text{Tol}}(r)|/(2\pi/\lambda) = |[n(r) - 1] * [\delta z_1(r) + \delta z_2(r)]| \\ + |\delta n(r) * [z_1(r) + z_2(r)]|,$$

which has a contribution from two terms:

- The "perfect" material parameter n multiplied by surface tolerance values δz_1, δz_2 and
- Material tolerance value δn multiplied by "perfect" surface parameters z_1, z_2.

Obviously, process *and* material tolerances contribute equivalently to $\delta\Phi_{\text{Tol}}$.

2.1.3 Production and metrology errors

The production process introduces shape errors on both sides of the component and the refractive index of the delivered material shows deviations from the catalog value $\delta' n$. These errors must be properly added to the measurement errors, resulting in a forecast of the expected production errors.

The design office calculates $\delta\Phi_{\text{Prod}}$ from the reported production errors and compares it with the allowable tolerance value $\delta\Phi_{\text{Tol}}$. If the production value is greater than the tolerance value, the component is normally rejected. If it is an expensive component, the following possibilities exist:

- The component is accepted, because other components perform better than specified.

- The component can be remachined by the factory.
- Special adjustment means during system assembly may be used to compensate for the wavefront error.

2.1.4 System performance criteria

The design and tolerance work for realistic systems has to be performed at the system level, including all optical elements collectively, for many wavelengths and for many object points. This obviously leads to more complicated quality measures than the simple phase error. However, we will outline in Sec. 3.2 that all quality criteria can be traced back to physical phase errors, which we presented above for illustration purposes.

2.2 Definition of Aspherical Optical Elements

Aspherical optical surfaces deviate more or less pronouncedly from the spherical shape of standard optical surfaces. They are used in optical systems to increase imaging quality, to reduce construction size or the number of elements, to save weight, to simplify the assembly process, or to reduce the overall manufacturing costs.

Aspherical optical elements can be produced in several configurations: as one aspherical surface on a substrate (e.g., a parabolic reflector), as a combination of aspherical surfaces with spherical surfaces (e.g., aspherical lenses) or as a combination of several aspherical surfaces (e.g., bi-aspheric lenses, free shaped prisms; see http://www.olympus.co.jp/en/news/2004a/nr040126fslue.cfm).

As will be shown in Sec. 2.2.2, aspherical surfaces can be described by continuous mathematical functions. They can be rotationally symmetric, axially symmetric, or completely asymmetric (free-form surfaces). Dependent on the production volume, on the degree of asphericity, and on the required tolerance values, aspherical elements can be manufactured by a variety of production methods (Chapter 5), for example, by casting and injection molding of plastics, by blank pressing of glass, or by precise machining (diamond turning of metals or polymers, grinding and polishing of metals, optical glasses, crystals, ceramics).

2.2.1 Basic characteristics of aspherical elements compared with spherical elements

2.2.1.1 Quality of the surface form

Spherical surfaces are characterized by a constant curvature value and thus can be manufactured using large-format tools. These tools are state of the art and operate, when properly driven, over a long period of time without significant quality degradations. Additionally, the tooling heads move in a rather stochastical way, which avoids the generation of "zonal" artifacts in the surface structure. Consequently, very high form accuracies can be achieved, even with relatively simple machines.

In the case of aspheres, the local curvature changes across the surface, requiring small tooling heads for grinding and polishing. These tools are more sensitive to the deteriorations that destabilize the process. Very accurate machine kinematics and complex correction procedures are required, and the risk of generating artifacts is rather large. Additionally, very precise measuring methods with accuracies in the range 500 nm to below 1 nm are indispensable. Because several correction loops must often be performed, artificial ripples in the surface structure cannot be avoided completely and must be carefully tolerated.

2.2.1.2 Quality of the surface texture

The small-area working tools mentioned, in combination with the deterministic tool path, with little room for stochastic movements, tend to decrease the quality of the surface texture. In order to achieve the same high degree of polishing as obtained with spherical surfaces, more technical efforts are necessary. For example, grinding must be performed with smaller grain sizes, of 10 μm down to 3 μm, and with small tool pressure, leading to long working times. The polishing times are also much longer than those needed for equivalent spherical surfaces. Recent progress in polishing technology, such as magnetorheological polishing techniques (and the appropriate polishing fluids), which are used for finishing all kinds of optical surfaces, yields both high-quality surface form and texture.

2.2.1.3 Quality of positioning in optical systems

A spherical surface is, due to its rotational symmetry, uniquely described by its center of curvature. This has the advantage that a lens with two spherical surfaces has a unique optical axis, which is the line connecting both centers of curvature. The centring of such a lens, that is, the alignment of its optical axis with respect to a mechanical axis, can be performed after the manufacturing of the optical surfaces with virtually no limitation of the centring precision.

Aspherical surfaces, in contrast, have only one datum axis given by the design. In the case of rotationally symmetric surfaces, it is the symmetry axis; but in the case of asymmetric surfaces, it is an axis that hits the surface at a certain point and with defined direction to the normal of the surface at this point. Therefore, it must be guaranteed during the manufacturing process that the centring of the second surface with respect to the first axis is in tolerance. Thus, when combining an aspherical surface with a spherical surface, the center of curvature of the sphere must ideally lie on the axis of the asphere, but for the case of bi-aspheric lenses, both axes must ideally be collinear. The permissible deviations from the ideal case have to be specified.

Another aspect that has to be considered is that the spatial position of the axis of an asphere depends on the method with which the form deviation is estimated [1].

Due to these constraints, special techniques and equipment are necessary for the manufacturing and testing of aspherical optical elements (see Chapters 5 and 6).

2.2.2 Mathematical representation of aspherical surfaces

2.2.2.1 *Basic equation according to ISO 10110—Part 12*

The standard ISO 10110—Part 12 describes surface functions of second order with axial symmetry as

$$z = f(r) = \frac{\frac{r^2}{R}}{1 + \sqrt{1 - (1+\kappa)\left(\frac{r}{R}\right)^2}} + \sum_{n=2}^{m} A_{2n} \cdot r^{2n},$$

where r is the lateral coordinate, z the sagitta error, and R the paraxial surface radius. The conic constant κ is 0 for spheres, -1 for parabolas, <-1 for hyperbolas, between -1 and 0 for oblate and >0 for prolate ellipses. Details and the complete mathematical description can be found in Chapter 10.

2.2.3 Specifying tolerances for aspherical optical elements

2.2.3.1 *Surface form*

Tolerancing specifies the maximum permissible deviation values of the manufactured actual form from the designed or theoretical form. Figures 2.3 and 2.4 show measured profiles. The global deviation, shown in Fig. 2.3, may be understood as the deviation of a best-fit radius from the theoretical value. This can be tolerated similarly to spherical surfaces, according to ISO 10110—Part 5, by specifying the permissible value of the sagitta error.

Rotationally symmetric deviations, as shown in Fig. 2.4, can be limited by indicating the permissible rotationally symmetric irregularity provided by this standard. Nonrotationally symmetric deviations can also be limited by specifying the permissible total irregularity.

As can be seen in the measured profiles, additional local deviations with strong gradients occur, which must be limited by an additional tolerance for the maximum allowable angular deviation of the local normal from the theoretical normal. This deviation is called "slope error" (or surface tangent error).

Parts 5 and 12 of ISO 10110 give the rules on how to indicate these form tolerances in the drawings of optical elements. The standards also specify the units of

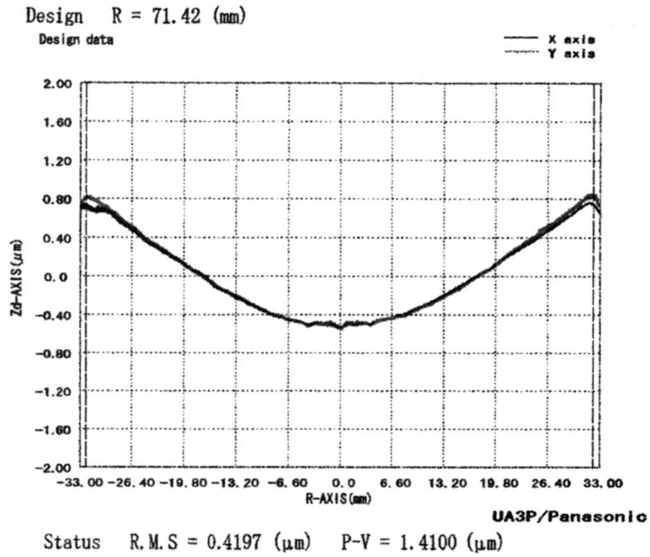

Figure 2.3 Global deviation from the specified surface shape function (rotationally symmetric asphere).

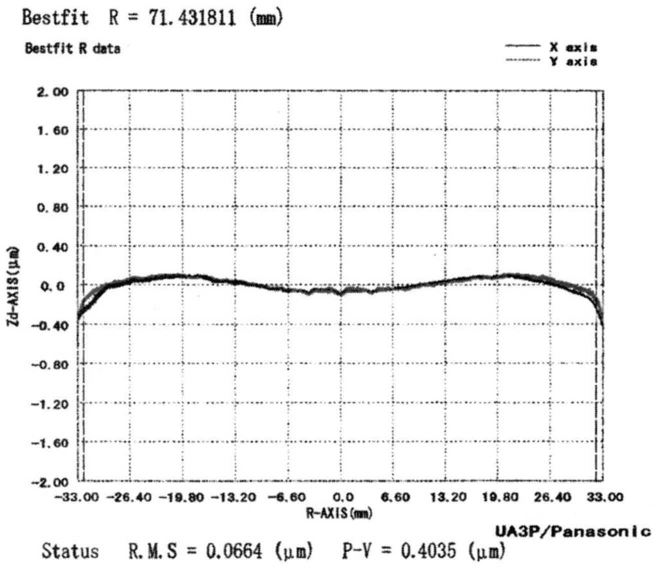

Figure 2.4 Local deviation from the best-fit radius, which varies by +0.012 mm from the specified value.

tolerance indications and give information on testing of optical elements, especially by interferometric methods.

Alternatively, the permissible form deviations can be specified according to ISO 1101 as tolerance zones, inside which the manufactured surfaces must be contained. The boundary surfaces of the tolerance zone are tangential surfaces to spheres, the diameters of which are the tolerance width, and the centers of which are located on the theoretical surface. As previously stated, a permissible slope error must also be indicated, to limit spatial oscillations of the surface within the boundary surfaces.

2.2.4 Surface texture

The tolerance for surface texture is indicated according to ISO 10110–Part 8. The required quality is specified by indicating one of four polishing grades. The polishing grades are related to certain maximum allowable numbers of pits in the surface, which can be detected by scanning a given distance on the surface, for example, by using a stylus with an appropriately small tip radius.

The tolerancing of the surface texture of aspherical surfaces is the same as for spherical surfaces.

2.3 Drawing Indications

Figure 2.5 shows a drawing of an aspherical lens element. The design equation, with its constants and coefficients, is given in the field of the drawing. The coordinate axes are indicated in the drawing. An abbreviated table with some function values is shown for information. It is especially useful to check for the correct signs of the constants and the coefficients. The indications are arranged in tabular form, according to ISO 10110-10. This prevents the drawing from being overloaded. The indications refer to the left and right surface and to the material data, given at the center of the table. The permissible form deviations are specified following the error code 3/, and tolerances for the position deviations of the surfaces follow the error code 4/.

The form tolerances of the asphere are given according to ISO 10110-5 as 3/4(0.8/0.4), which means a sagitta error of 4 fringes (@ $\lambda = 546\,\text{nm}$), a total irregularity of 0.8 fringes, and a rotational symmetric irregularity of 0.4 fringes are permissible. Because the axis of the asphere is the datum axis, no tolerance for the tilt angle is specified following error code 4/. The runout of the outer cylinder is limited to ≤ 0.005, according to ISO 1101.

For the slope tolerance, no error code exists. Therefore, the tolerance is indicated as a text note in the field of the drawing, according to ISO 10110-12.

2.4 Information Exchange over Aspherical Elements

Currently, information about how to describe and characterize aspherical optical elements is produced and distributed in many different formats. This extends from using different mathematical formulas in Optic Design programs and the deduced technical drawings, to different user interfaces of measurement instruments (e.g., interferometers, profilometers) and manufacturing machines (e.g., generators,

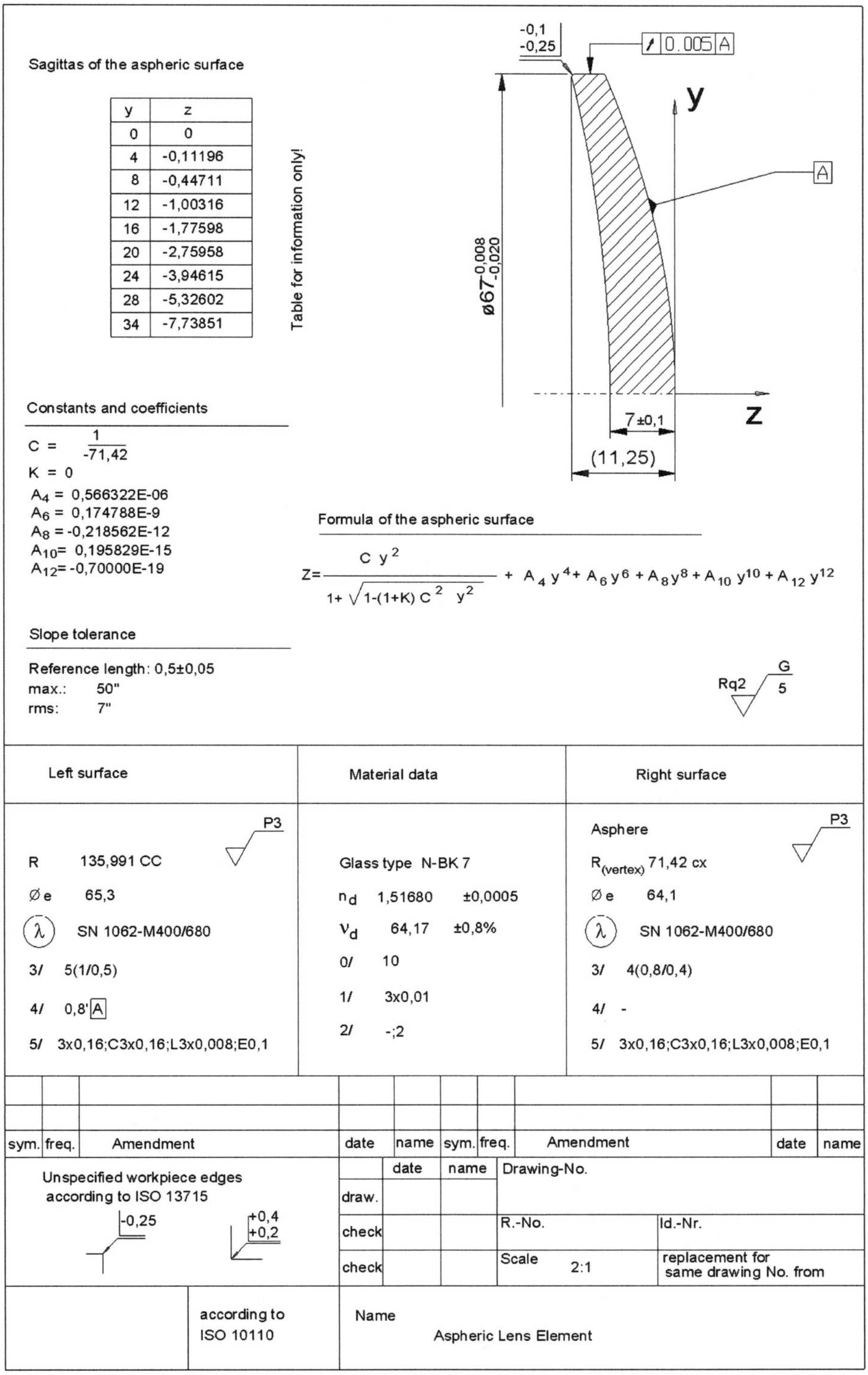

Figure 2.5 Drawing of an aspherical lens element according to ISO 10110.

polishing machines). It may be that the data in the drawing for an aspherical surface can be put directly into one device, while for another device the signs of some parameters have to be altered. Although the correct understanding and handling of the data is still under the control of companies, communication between different institutions is often difficult, time-consuming, and risky as a result of this inconsistent representation.

Unfortunately, the strict obedience of the relevant international standard ISO 10110–Part 12 in its present version does not give definitive safety, as it does not contain a sign convention. The recommendation of this standard to indicate the coordinate system and to add a numerical table of some surface function values in the technical drawing should therefore be considered.

2.5 Study about Surface Errors

In the following, we want to demonstrate the importance of the comments made above, mainly the need to communicate extensively between design and production. It will be shown later, in Chapter 5, that a large variety of cost-attractive manufacturing technologies exist today. However, each method has its performance limits, which lead unavoidably to residual surface deviations from the ideal form. The designer is well advised to know these limits in advance. This enables him to judge how reliably his specifications can be realized by the fabrication process.

2.5.1 Aspherical laser collimator

We consider an application with one plano-convex aspherical lens. The lens should image the emitting area of a laser diode to infinity (i.e., to collimate the laser beam in Fig. 2.6). The lens has a focal length F = 30 mm and an f-number of 1.8, equivalent to a free aperture diameter of about 16 mm.

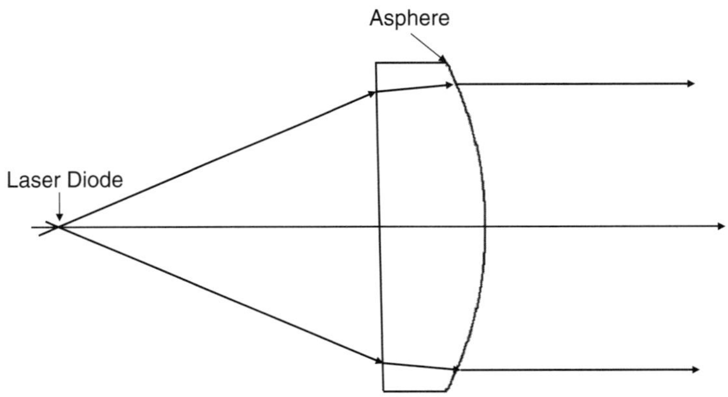

Figure 2.6 Aspherical lens for laser collimation.

2.5.2 Comparison of different surface-finishing technologies

What production technology should be chosen? In Fig. 2.7 we show characteristic surface deviations from the ideal shape taken from samples of about 1 in. diameter, which were produced by two "cold" surface-finishing processes (methods A, B) and by a "hot" precise glass pressing method C.

The ordinate in the figure expresses the amount of deviation from the predefined surface form in nanometers. All three variants show problems at the edge of the sample, which in our special case is made irrelevant by making the samples some millimeters larger than the required "effective" diameter.

More important are the clearly visible spatial periods. We identify in all three cases structures with a typical period length of about 4 mm, with an amplitude of ±350 nm in case A, ±250 nm in case B, and ±50 nm in case C. Even if the data of methods A and B can be significantly improved today by better production means, we will use them to illustrate how production errors can cause deterioration of system performance. (Note: a sinusoidal structure with an amplitude of ±150 nm and a spatial period of 4 mm corresponds to a maximum slope angle of 30 arcsec.)

2.5.3 Coherent beam propagation

Do the residual errors influence the collimation quality, and what fabrication process, A, B, or C, would be acceptable for this special application? For this purpose we model the measured errors as sinusoidal phase gratings of amplitude ±150 nm and ±300 nm, respectively, with a spatial period of 4 mm on the convex aspherical surface, and consider the intensity distribution at a distance of 25 m, 50 m, and

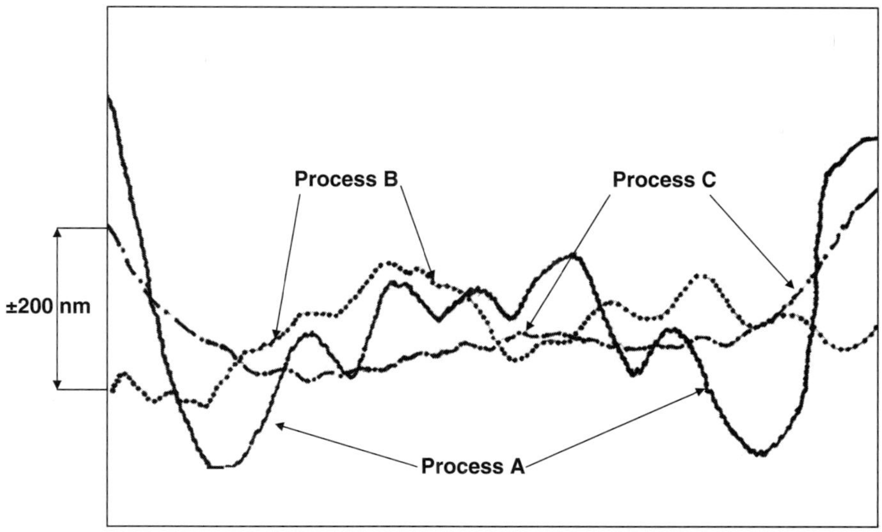

Figure 2.7 Residual surface deformations for different manufacturing processes. A: classical grinding and polishing; B: newer method; C: hot pressing.

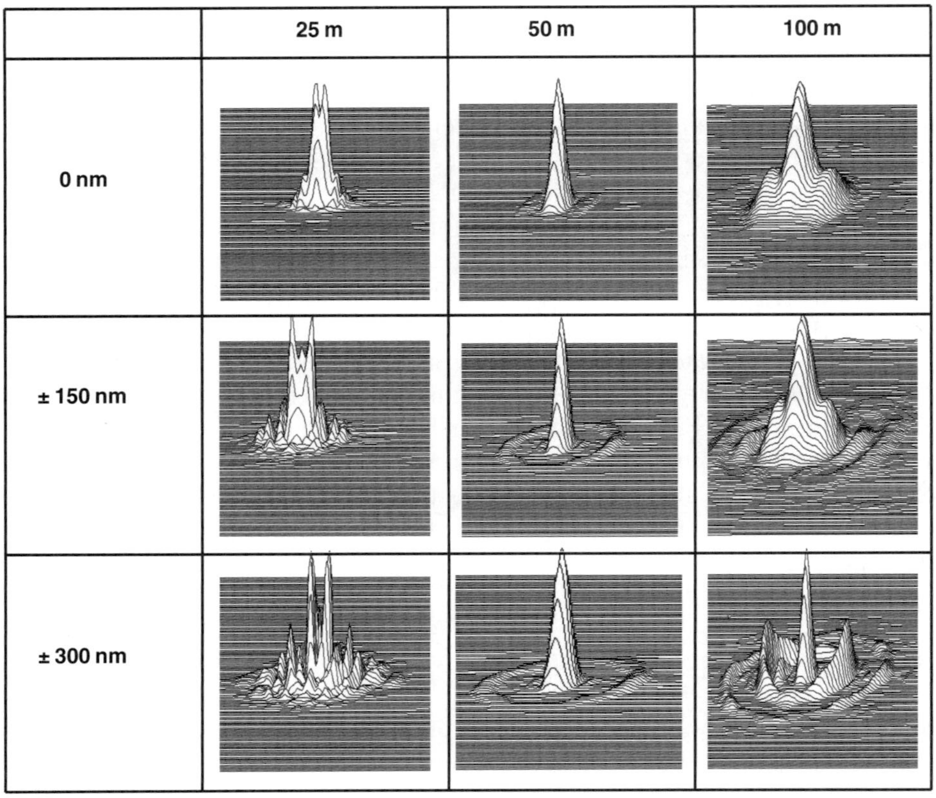

Figure 2.8 Intensity spot size at different distances from the laser source for sinusoidal gratings of 4 mm period and different amplitudes, present at the aspherical surface of the collimator.

100 m. The lens is best focused to 50 m, and the active spot area of the laser diode is 0.6 μm × 1.2 μm. In Fig. 2.8 we present the calculated spot intensity for cases A, B, and C at the three distances. We see from the simulation of the coherent beam propagation that larger amplitudes of the sinusoidal surface error lead to side lobes around the laser main spot, and more severely to double peaks at lower distances. It should be remembered that distance-dependent intensity fluctuations already occur for a plane wave, incident on a perfect lens, by diffraction at the aperture [2]. The illumination of the aperture by a Gaussian TEM00 Laser mode minimizes these fluctuations if the diameter is made large enough.

The question is "To what degree can the selected detection concept handle such intensity deformations?"

2.5.4 Application case: Line marking on sport fields

The collimated laser radiation is used as a guiding beam for a manually pushed line-marking device. The goal is to paint white border lines on sports fields, where the

Figure 2.9 Beam riding for line marking (*Courtesy*: Beamrider Ltd, Malvern; Fachhochschule Nordwestschweiz, Prof. Gottwald).

line length could be as long as 120 m. Independent of the erratic movements of the device, the straightness of the line position must be held within some millimeters tolerance over the full distance (Fig. 2.9). To this end, several sensor elements on a platform, movable normal to the line path, lock on the stationary laser beam. Because the platform also carries the painting nozzle, a straight line occurs independently of the actual movements of the mobile device. For robust device operation, the emitted "Gaussian" laser intensity profile should be without distortions over the full distance. The numerical simulations indicate that degradation amplitudes below ±150 nm would be acceptable.

In conclusion, aspheres produced by methods A and B would be insufficient, but those produced by method C would be suitable.

2.6 References

1. H.W. Randall, R.C. Brost, D.R. Strip, R.J. Sudol, R.N. Youngworth and P.O. McLaughlin, "Considerations for tolerancing aspheric optical components," *Applied Optics*, Vol. 43, No. 1, pp. 57–66 (2004).
2. M. Born and E. Wolf, *Principles of Optics*, Cambridge University Press, Cambridge, 1999.

Chapter 3

Applications

In this section, we present more details about imaging physics, what criteria describe best the image quality, and what makes aspheres so attractive to the community of designers.

3.1 Physical Considerations

The design of an optical system has to ensure that the specifications for image quality are fulfilled inside the 3D working volume at the object side. This volume is defined as the product of the field of view (FoV) and the usable depth of focus (DoF). The image quality for all object points, often expressed by spatial and spectral aberration values, must be controlled within the spectral bandwidth of the received light.

The aberrations can be physically understood as optical wavefront deformations of an object point source. Such a source emits a spherical wavefront, which is picked up by the entrance pupil of the optics. When traveling through the system, the wavefront is unavoidably deformed by diffraction but also by imperfect design or by manufacturing errors. Then a distorted spherical wavefront leaves the exit pupil, causing an aberrated or blurred intensity spot in the image plane. Because diffraction is a physical phenomenon of light as information carrier, we must accept, in any case, a degraded image. An optical system unavoidably acts as a low pass filter, which cuts off higher spatial frequencies, that is, higher spatial structures.

To minimize the aberrations, the designer needs a minimum number of optical parameters to vary, such as surface shape, lens thickness, and the glass values of all the lenses. But what are the quality criteria for which an optical system has to be optimized?

3.2 Image Quality

A good image must look sharp, brilliant, and must be stable. Sharpness means that the image has well-defined black and white edges, and that even fine details are resolved with good contrast. The image must look "fresh" and brilliant, and the colors should be undisturbed. We accept an image as "stable" if we observe no color shading when moving the object slightly out of the plane of best focus. Our eye would react very negatively if a slightly defocused image gets colored "comatic" errors. On the other hand, a good design can manipulate the image impression, especially when the object moves around in the object space.

First, we have to decide how the object FoV should look. In the case of geodetic instruments, the FoV is kept "equi-angular," which means that the magnification changes linear with object distance. These anallactic telescopes are simple, robust, compact, and traditionally used to obtain the object distance by analyzing the magnification.

When we try to track an object very precisely, we prefer to keep the magnification independent of the object distance for accuracy reasons. This results in an "equi-width" FoV, which needs, like a zoom lens, at least two movable lens groups. The same holds when we have to follow a fast-moving target. A distant independent magnification would facilitate the software treatment. Another similar example would be a wide FoV at near distances, but a small FoV at large distances (panfocal) to inspect rooms. What FoV should be chosen can be answered easily. How large must a sensor pixel be when projected into the object space at a certain distance?

Next we may ask how the image should look, that is, what are the important quality features? For pointing applications where the user preferably watches points near to the optical axis (like in geodesy), the "on-axis" image is made extremely brilliant, and its neighborhood is kept slightly less brilliant to facilitate intuitively the user's aim. On the other hand, when designing observation instruments for military or medical surgery applications, we design the image at the edge of the field of view to be extremely brilliant, because professional users always expect problems at the periphery, not in the center!

A good design also satisfies the user's esthetic expectations. When the user tries to focus onto an object, the image normally gets sharper and sharper, until the best focus is reached. But the user is much more impressed if the image remains unsharp during focusing, but suddenly "jumps" into maximum sharpness. This helps him to focus more efficiently. We see that quality is seldom expressable by static parameters; it is much more a dynamic process, often psychologically driven. The designer has to study the user's hidden preferences. Such an unusual optical behavior as described is possible by carefully designing higher-order aberrations, an art that companies develop over decades, and for which aspheres now offer new and extremely valuable possibilities.

To achieve the quality specifications, the designer must control simultaneously many system aberrations. Aberrations can be defined as geometrical ray intersections, and their standard classification in terms of astigmatism, coma, and spherical aberration can be found in any textbook. They are the "fine" structure of the image of an object point and depend on the field angle, the defocus, the wavelength, and also the point where they leave the pupil. A critical design could require control of up to 100 aberration values, especially their mutual dependence when a system parameter is varied. We have to handle highly nonlinear systems, which for optimization need powerful algorithms, and also a lot of the designer's intuition and expertise. Each ray aberration can be expressed by an equivalent wavefront deformation, given in λ.

Therefore, all requirements for the materials, the surface shape processes, and the measurement tools can be readdressed for physical wavefront considerations, from which we started our explanations.

The question arises, "What practical advantages do we expect from aspherical components?"

3.3 Case Study

Let us consider a single spherical lens. Its imaging properties, expressed by its focal length, depend on the glass material on both surface radii and, less importantly, on the lens thickness. Thus, three geometrical parameters can be varied. In the case of more complicated specifications (e.g., if high resolution within a wide angular FoV and a large spectral bandwidth range is required), 10 or more spherical lenses may

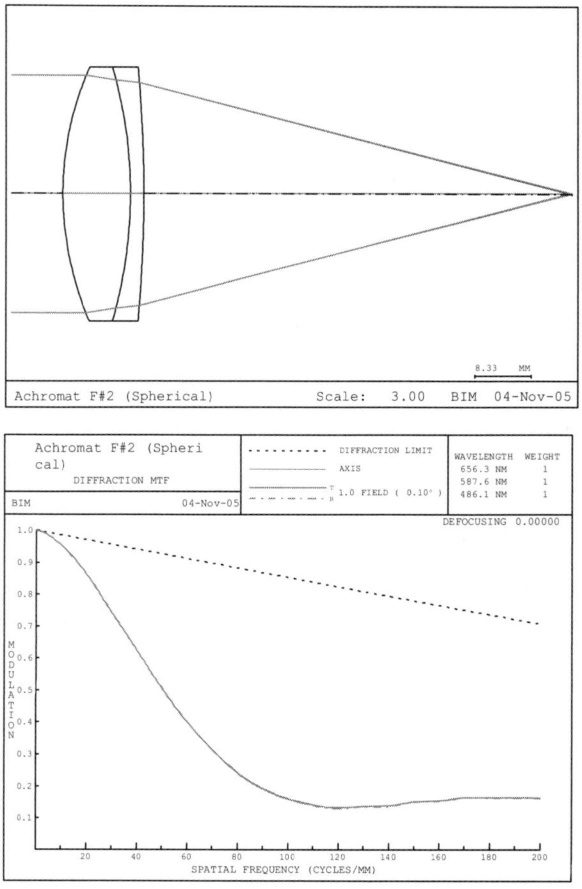

Figure 3.1 Doublett F/# = 2.

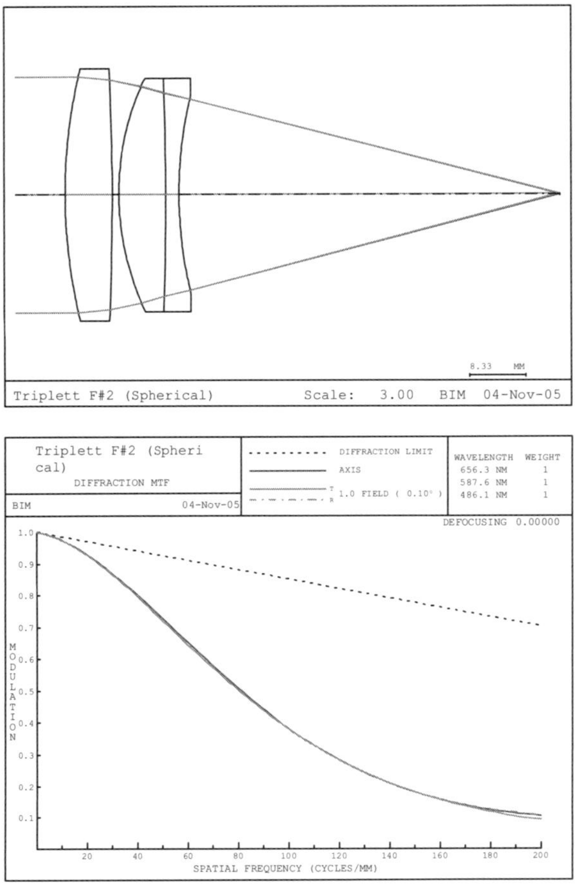

Figure 3.2 Triplett F/# = 2.

be needed. Such a large ensemble of lenses is, however, difficult and expensive to produce and, more seriously, may cause severe light transmission losses. Especially in the deep blue spectral region, where glass material becomes absorbing, this can lead to unacceptably low transmission values.

Back to our question: "Can fewer optical elements deliver the same number of parameters, which a spherical design needs to reach the specifications?" The answer is "Yes," if we put more independent geometrical parameters on one optical surface than just only the curvature value. But this would result in surface shapes described by nonspherical functions. Obviously, these are more difficult to produce than normal spherical lenses. It will be shown in this book that modern surface-finishing technologies allow the manufacture of aspherical surface shapes today in a reliable and also cost-effective way.

For illustration purposes we show in Fig. 3.1 a spherical achromat with f-number 2, corrected for the visual spectrum, but with a weak contrast of 0.2 at 100 lp/mm. The contrast at this frequency can be increased by a factor of 2, if an additional spherical lens is added (Fig. 3.2). The same contrast improvement is

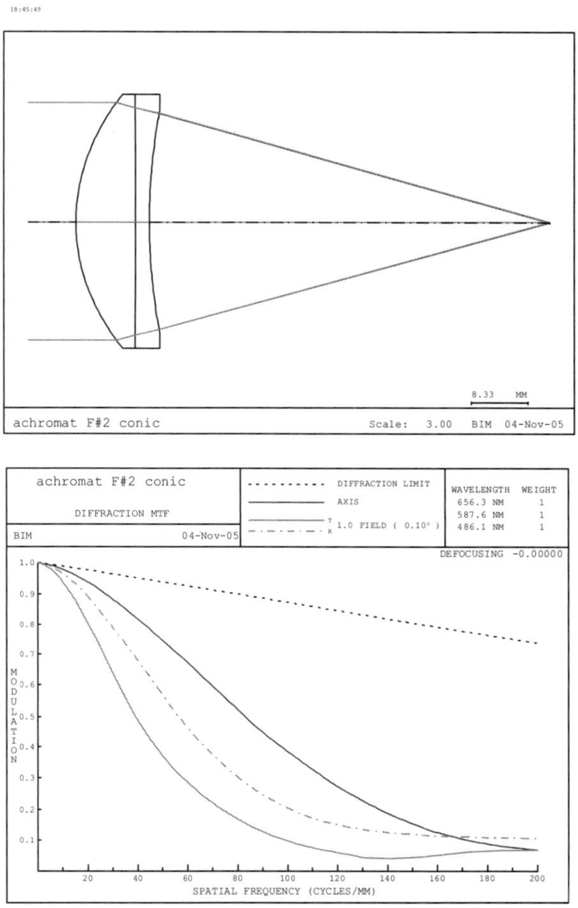

Figure 3.3 Doublett F/# = 2 (surface 1,3 conical).

also achieved on-axis by conically aspherizing both outer surfaces of the doublett (Fig. 3.3). To correct the sine condition, we have to add aspherical coefficients of the fourth order (Sec. 2.2.2) to obtain the required contrast value (Fig. 3.4).

3.4 Design Drivers

The reduction of the number of optical components is only one reason to insert aspheres into optical systems. Other important design drivers are

- To increase the imaging quality (resolution, distortion), which cannot be achieved by a pure spherical design (example: deep-UV-lithography);
- To reduce the construction size (example: photographic zoom lenses);
- To save weight, because one asphere is perhaps lighter than several spherical components yielding the same optical performance (example: IR-optics at 1–5 μm and 8–12 μm, made of "heavy" germanium or silicon material);

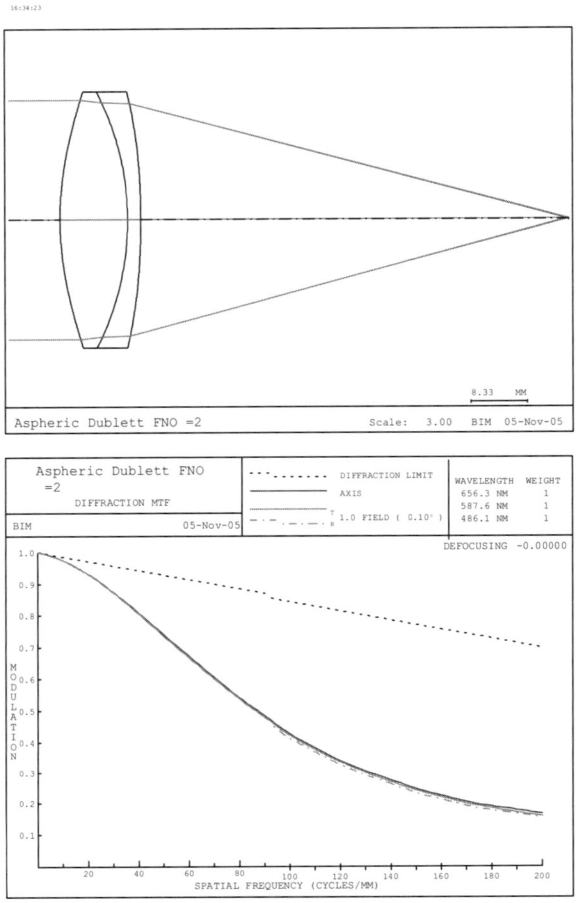

Figure 3.4 Doublett F/# = 2 (surface 1,3 aspherical of fourth order).

- To improve the total light transmission by reducing the number of optical elements (example: fluorescence microscopes with high transmission demands in the blue and UV spectral range);
- To simplify the assembly process (Chapter 8).

All these drivers can be combined to reduce the overall manufacturing costs.

***Example*:** IR-optics in the 1–5 μm and 8–12 μm spectral range can be easily made by numerical controlled machining (diamond turning), because diffraction effects at the residual mechanical grooves are ignorable at these larger wavelengths.

***Example*:** Laser collimators for CD- and DVD-players; and optics for "digital projection."

3.5 Classifications

In addition to the mathematical description (Chapter 10), aspherical components are often classified in a pragmatic physical/technical sense

- As refractive, reflective, or diffractive components, if the light deflection is caused by lenses, mirrors, or holographic elements;
- As "conical" (Fig. 3.3), "higher aspherical" (Fig. 3.4), or even "free-form" components, if the surface deformation is rotational invariant around the optical axis or may lack any symmetry; and
- As "on-axis" or "off-axis" components, if axially centered or decentered to the optical system axis.

Our application examples in Chapter 11 will present several variants and combinations of these classifications.

3.6 Technical Challenges

The production and assembly of aspheres requires, in general, processes with significantly narrower tolerances.

3.6.1 Centering

To demonstrate the different situations for spherical lenses and aspheres see Figs. 8.3 and 8.4 in Sec. 8.5. A spherical lens has two centers of curvature, M1 for the first surface and M2 for the second. The line between M1 and M2 is the optical element axis. To align the lens with respect to a preset system axis is rather easy: M1 by mechanical adjustment and M2 by simply "rolling" the second surface around M1. This is much more difficult with an aspherical lens (Fig. 8.3). If the first surface is rotational symmetric, then its normal vector at $r = 0$ must coincide with the system reference axis achievable by mechanical adjustment. However, the orientation of the second surface is then fixed. In practical situations, one has to find a compromise between the tolerated misalignment of both surfaces 1 and 2. Thus, different to spherical lenses, the assembly tolerance is determined by the preprocess of the component manufacturing.

3.6.2 Stability criteria

Any surface deformation by thermal or mechanical influence (e.g., vibrations) affects simultaneously all aspherical parameters and thus degrades the optical quality much more strongly than a spherical lens. Mechanical efforts to stabilize the position and alignment of aspheres over the life-cycle have to be undertaken. The same holds for the aspherical shape–any mechanical stress of the mounting on the component has to be strictly avoided, so as not to introduce unwanted aspherical coefficients.

3.6.3 More complex metrology

To verify the correct shape and mounting of the aspheres, one needs special measurement tools like interferometers and wavefront sensors (see Chapter 6), which are expensive instruments and need special operator skills.

Aspheres need a much higher level of controlled technologies for production and metrology and to ensure stronger stability requirements over the whole lifetime of the optical system.

3.7 Application Spectrum

In Chapter 11, "Applications" in Part II, authors from different industrial companies will illustrate why they use or plan to use aspheres for their special business tasks. Applications range from mass production of consumer optics to special objectives for lithography, space communication, and airborne sensing instruments. In Table 3.1 we summarize the application fields, the main drivers, and also the production status.

Table 3.1 Application fields, main drivers, and production status.

Application fields	Advantages and drivers	Status
Large quantities		
Illuminations	Better imaging quality with one element; cost reduction	Production
Laser collimator	Better imaging quality with one element; cost reduction; beam stability	Production
Photo-optics	Necessary for zoom systems; cost reduction; better imaging quality; smaller construction length	Production
Large-format film lenses	Necessary for zoom systems; cost reduction; better imaging quality; smaller construction length	Production
Small quantities		
UV-lithography	Better imaging quality; higher transmission at UV wavelengths	Production
Aerial survey	Better correction of distortion and telecentricity; cost, weight and size reduction	In preparation
Space communication	Lightweight; compact layout; radiation resistance	Production
Correction plate for mirror telescope	Lightweight; compact layout; radiation resistance; better imaging quality	In study

Chapter 4

Materials of Aspheres

The introduction of aspherical lenses into optical designs has generated a significant impact on the size, weight, and performance of optical systems. However, because the same key optical properties are as important as for spherical optics, material selection and the development of materials for aspheres is not driven by new material requirements but rather by cost issues. As differences between spherical and aspherical lenses are mainly caused by production, new development is driven by the production technology used, such as precision molding, diamond turning, and computerized numerical control (CNC) machining.

Although costs are the driver, materials for aspheres are still specified by their main properties according to their applications and production or processing requirements. The most important physical parameters of any optical material are the refractive index n, the Abbe number v_{d}, and different partial dispersion values P. Note that the Abbe number $v_{\mathrm{d}} = (n_{\mathrm{d}} - 1)/D$, where D is the dispersion term $D = n_{\mathrm{f}} - n_{\mathrm{c}}$. Thus, small Abbe numbers describe high-dispersive materials.

The values of these parameters are material dependent and thus vary in a wide range. This allows the designer to properly select and combine different materials to optimize optical systems. Other important material parameters to be considered are the transmission values and the scattering characteristics in the ultraviolet (UV), visual (VIS), and infrared (IR) parts of the spectrum. Stress optical coefficients K and birefringence are of secondary relevance but clearly must be considered by the designer. In some high-end applications like UV wafer steppers for microlithography, the intrinsic birefringence can play a significant role.

Chemical resistance of materials against water, acids, and bases are of relevance, not only for a specific application, but also with respect to processing steps like grinding, polishing, and cleaning.

Mechanical properties, such as hardness (Knoop HK) and Young's modulus (E), are important for grinding and polishing, but they also determine the scratch and stress resistance of optical devices.

Thermal parameters, such as the thermal expansion coefficient α, thermal capacity C_{p}, thermal conductivity λ, heat resistance, and the thermal shift of optical properties, must be known when analyzing a lens system with respect to temperature changes. The thermal dependence of the material viscosity of glasses and polymers must be taken into account in production, for example, in precision molding for glasses or injection molding for polymers.

For low-T_g glasses and the precision molding process, the so-called index drop is particularly important. It defines the change of the refractive index and Abbe number during the pressing process, which have to be known to reach precise optical positions for the final lenses.

Some physical quantities describe how the optical system behaves in practical situations when environmental conditions change. With the use of Youngs modulus E, the Poisson ratio μ, and the density ρ, we obtain the following terms:

- Specific thermal stress

  $$\varphi_w = \frac{\alpha \cdot E}{1 - \mu}$$

  for the maximal expected stress in glass for a spatial local temperature difference of 1 K;
- The specific heat conductivity

  $$\kappa = \frac{\lambda}{c_p \cdot \rho}$$

  describes heat diffusion in materials. The actual heat flow also depends (besides on κ) on the mechanical boundary conditions, for example, whether the lens mount is kept at a constant temperature or, in contrast, is actively heated. It allows calculation of the temperature gradient inside lenses as a function of time. Consequently, the local refractive index of glass varies with time, which could have a serious impact on the imaging properties of a lens system. A profound explanation of the basic physical modeling, together with software routines for many constraints, can be found at http://www.leicageosystems.com/corporate/de/ndef/lgs_4045.htm.

This short overview indicates how relevant material properties are for system performance. Very often, trade-off decisions must be made if a system performs optically perfectly, but not thermally. Also, compromises between cost and performance must be found. The perfect material does and will not exist in the future, but a large variety of material types is available to create the ideal system. The most often used optical material classes are glasses, crystals, polycrystalline ceramics, and polymers.

To date, the use of glass ceramics has been restricted to reflective components. However, the use of mirror systems, particularly in combination with aspherical surfaces, is growing tremendously in many application fields, where optical systems with a large aperture, but a small field of view and polychromatic illumination, are required. Besides astronomy, which is the traditional domain of reflective systems, instruments for space, lithography, and military applications are prominent examples.

Polymer-nano-composites and fluids offer interesting features and may attract more attention in the future, but they are not really in use today due to their lack of maturity as a technology. However, fluids in particular could play a role in adaptive optics of the next generation.

The different material types can be classified primarily by refractive index and Abbe number, which describes the dispersion. In Fig. 4.1 typical areas of the different material classes are shown. Fluids and polymers are located mainly in the lower right field of high dispersion and low refractive index values. Glasses are typically above this field, and only crystalline materials have the potential to exceed the glass field limit to higher refractive indices. The two lines, the so-called magic lines, indicate borderlines. Beyond these lines, normal glasses ($*$ dotted line) or crystals ($**$ hatched line) are not stable or do not exist at all.

As mentioned above, optical materials are characterized by many physical parameters. Very important are the values of the partial dispersion values PgF ($ng - nF/nF - nC$) and PCs ($nC - ns/nF - nC$). In Fig. 4.2, we show the "anomal partial dispersions," that is, the distance to the normal dispersion line in the $\upsilon_{\mathrm{d}} - P$ diagram (Fig. 9.2). Each glass positioned at (n_{d}, υ_{d}) is characterized by a vector of two dispersion values P, with PgF pointing in the n_{d} direction.

The primary glass characteristics given by the figures (n_{d}, υ_{d}) provide two degrees of freedom, allowing color correction at two spectral lines, which is known as achromatic correction. Enhanced color correction of the remaining secondary spectrum can be achieved by additionally taking the individual glass anomalities into account.

In terms of color correction, aspheres will not directly improve the optical performance compared to spheres, as they are only improving the spherical

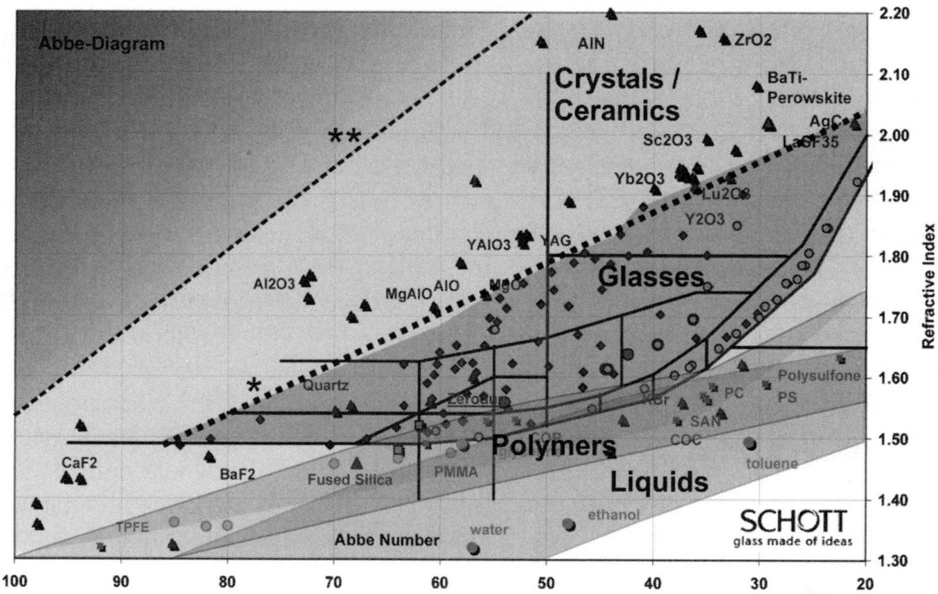

Figure 4.1 Abbe diagram of optical materials (fluids, polymers, glasses, crystals).

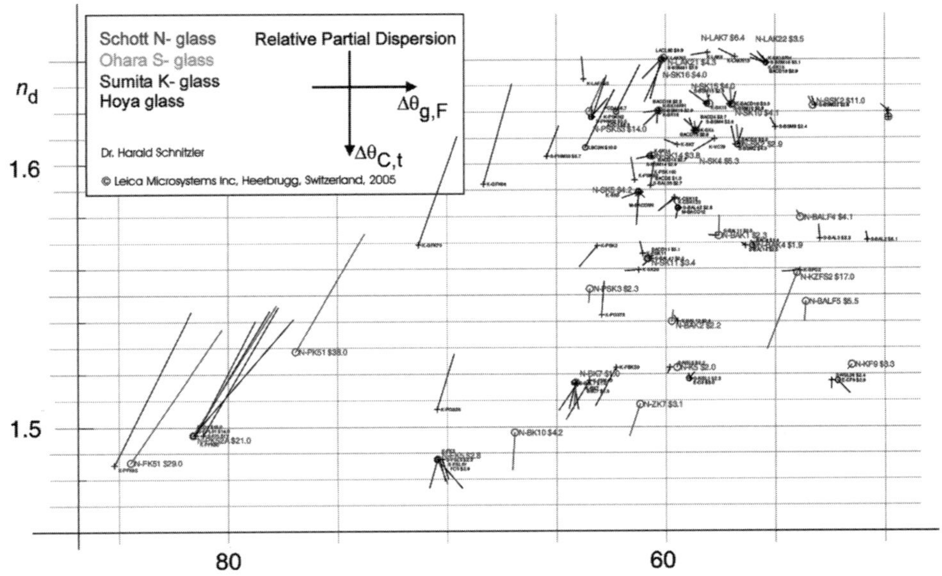

Figure 4.2 $n_d - \upsilon_d$-PgF PCt diagram (courtesy H. Schnitzler, Leica Microsystems AG).

aberration. However, they can have an indirect influence on color correction as they are introducing a further degree of freedom to complex optical design.

Apochromatic color correction requires a skilled implementation of available anomal glasses by the optical designer to even out the spectral behavior over a broad spectral range. Quite often, anomal glass types with ΔPgF > 0 and ΔPCt < 0 (bar directed to the top-right) used with lenses of positive refractive power help to flatten the secondary spectrum, and vice versa with lenses of negative refractive power.

Another important optical property, the optical transmission, is wavelength dependent. Special high-transmittive materials exist for UV light (microlithography), for visual light (consumer optics), and for IR light for night-vision systems. In digital optics with CCD sensors, high transmission of blue light is particularly important due to the low detection sensitivity of the sensor at short wavelengths.

Figure 4.3 representatively shows the transmission of different materials at 300 nm. For glasses (diamonds) the transmission at higher refractive indices decreases. Polymers in general have low transmission in the UV. Additionally, they show solarization effects at lower wavelengths, causing further transmission losses due to degradation effects during light exposure. For crystals (circles) with both a low (e.g., CaF) or relatively high refractive index (e.g., YAG, Yttria), high transmission values are obtained. In the UV region below 250 nm, fused silica, phosphate glasses, CaF or LuAG crystals, as well as Spinel optoceramics can be used.

In the three IR bands (1–14 μm), for example, germanate glasses, chalcogenide glasses, ZnS, ZnSe crystals, or optoceramics such as AlON, YAG, yttria, or Spinel show advantageous properties (Fig. 4.4).

One of the critical mechanical parameters is hardness (e.g., measured as Knoop hardness, HK), which helps to estimate the processing costs for cutting and grinding.

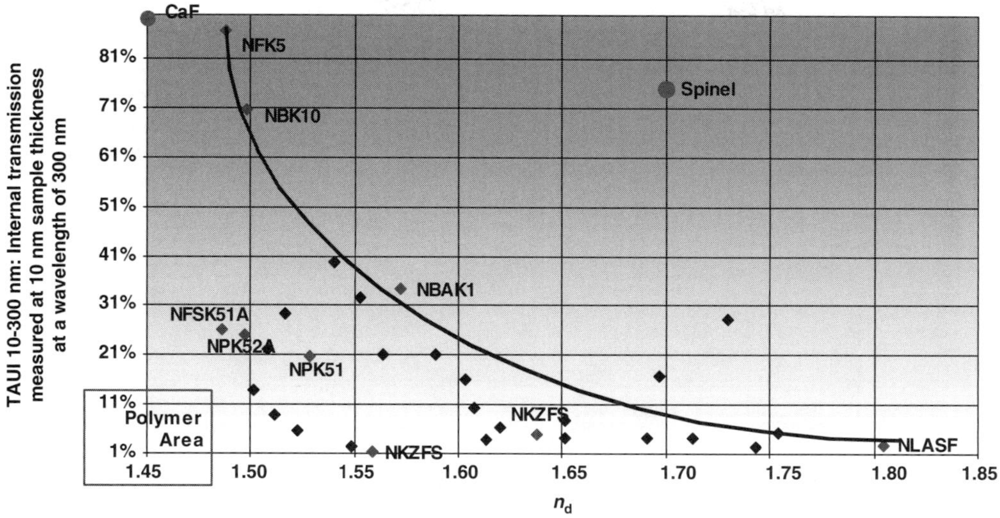

Figure 4.3 n_d vs. transmission at UV 300 nm (10 mm thickness).

Generally, crystals and ceramics show much higher Knoop hardness values than glasses (Fig. 4.5). For the polishing process, which is a combination of mechanical abrasion and chemical reactions, no simple relationship between the fundamental material properties exists. Therefore, the polishing behavior is determined empirically in most cases.

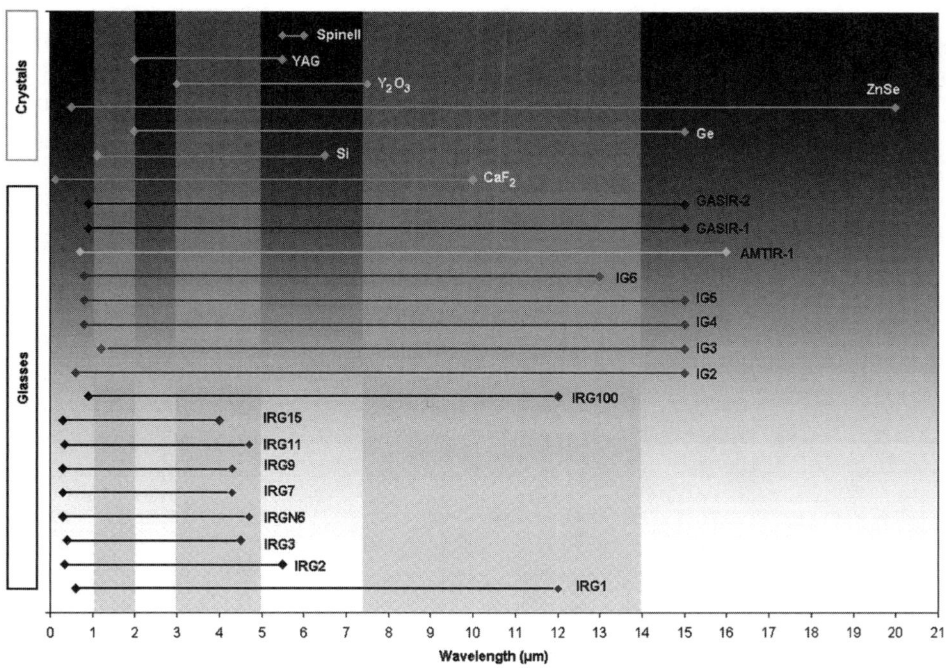

Figure 4.4 IR transmission of materials.

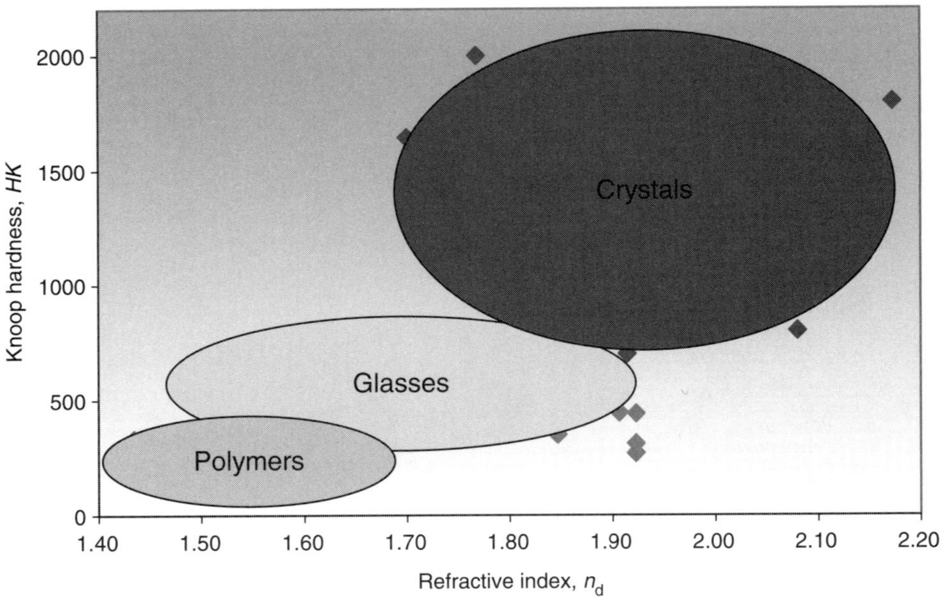

Figure 4.5 Graph of n_d vs. hardness, *HK*.

Typically, polymers have very low hardness values, which is a clear disadvantage. This can only be partially overcome by special hard coatings. Another problem is the strong thermal dependence of optical parameters such as the large dn/dT. Asymmetric heating of an instrument (for example, by strong sunlight exposure) could introduce irreversible lens deformations, not acceptable in metrology instruments like theodolites. On the other hand, the low weight and low

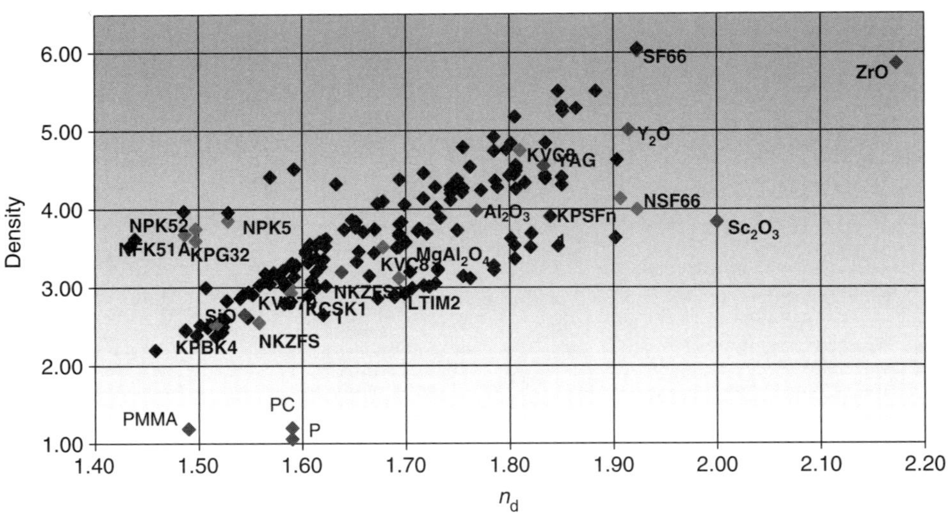

Figure 4.6 Correlation between refractive index and density.

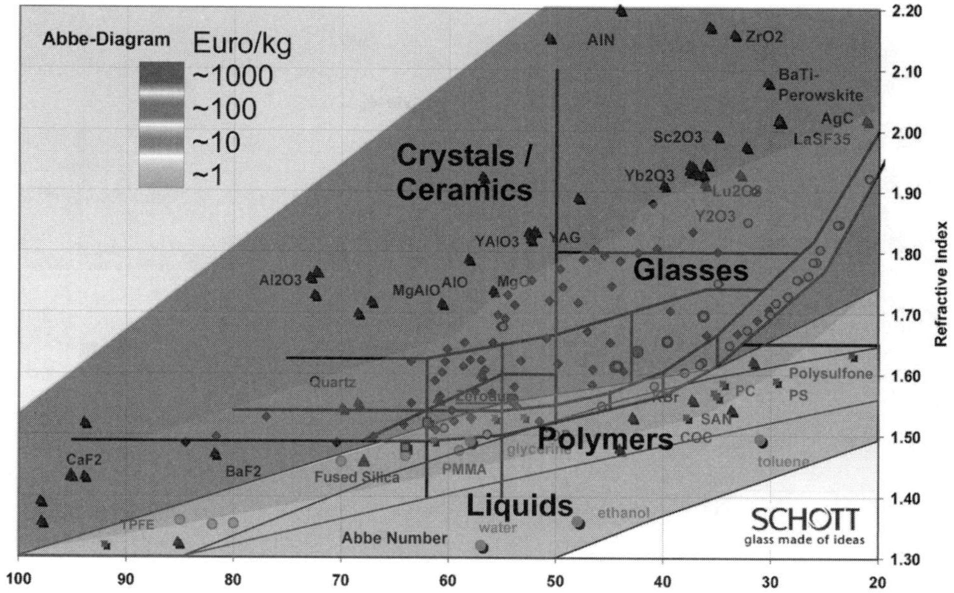

Figure 4.7 Principal price sectors of materials.

costs are attractive for hand-held viewing instruments. Nevertheless, medium- to high-quality consumer instruments are based on glass lenses.

The density of optical materials varies widely from $<1\,\text{g/cm}^3$ for fluids and polymers to $>5\,\text{g/cm}^3$ for special inorganic materials. There is a principal trend that higher density materials have higher refractive indices and dispersion values (Fig. 4.6).

There is also a large price variation for optical materials. Polymers, at 5.20 Euro/kg, are at the low-cost side. Together with injection-molding techniques, one obtains lens costs more in the cent than in the Euro region. Glass aspheres for consumer applications produced by precision molding cost several Euros. Aspheres of larger diameter for industrial optics cost more than 100 Euro per lens. CNC machined lenses in low volumes can cost several thousand Euros. Figure 4.7 gives a rough overview of the general price range (per kg material) of the material classes. In specific cases, significant deviations from these price regimes exist; for example, special IR-glass can cost more than 10,000 Euro/kg. High prices typically scale with high optical performance.

4.1 Glasses

In general, all glasses can be used for aspherical lenses. For standard glasses, the manufacturing is performed by CNC grinding and magnetorheological finishing (MRF) polishing. The contour is controlled by tactile and holographical measurements and finished in several optimization loops per lens. This process can only deliver small lot sizes at high costs, typical for industrial optics.

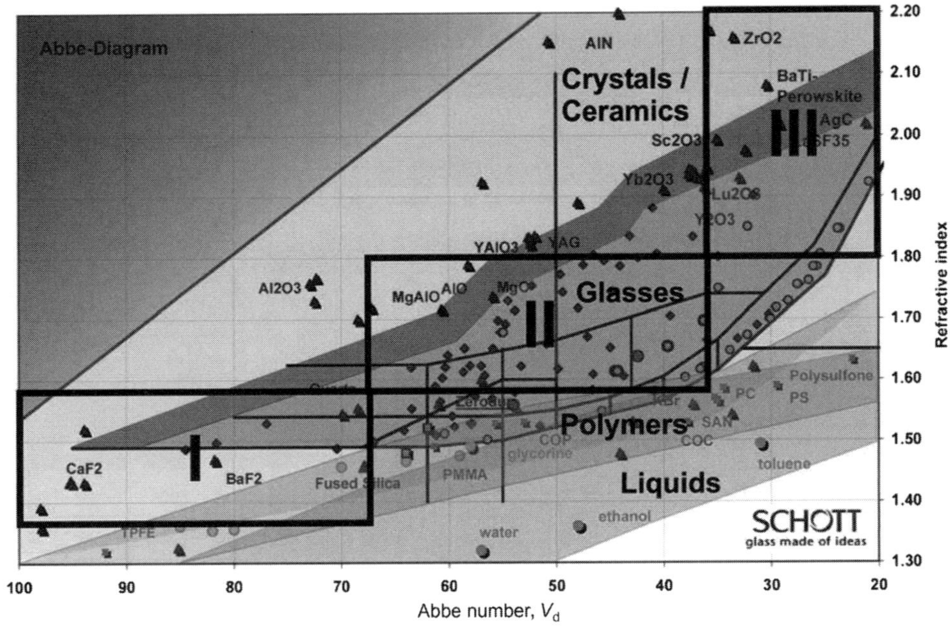

Figure 4.8 Preferred low-T_g glass areas.

The so-called "low-T_g glasses" are of special interest, because they can be combined with a low-cost, high-precision molding process. The focus areas (I, II, III) for low-T_g glasses are enclosed in Fig. 4.8 by black boxes and can also be found in the Tables of Part II (12.1.6, 12.2.6 and 12.3.6) for low-T_g glasses. In all these areas, glasses with reduced transformation temperatures are available, so that aspheres can be directly pressed out of suitable preforms at moderate temperatures (typically lower than 650°C). Main drivers for such developments are consumer optics like digital still cameras and camera phones.

Glasses with high IR transmission (like chalcogenides) have relative low T_g values and can also be used in precision molding processes for aspherical lenses. They are mainly used in military and security systems.

4.2 Polymers

Two types of optical polymers can be defined, the thermoplastic and the duroplastic polymers. Duroplastics are cast and cured in molds. The viscosity of thermoplastics is thermally changeable, allowing hot forming processes.

Aspherical polymer lenses can be produced very cost-effectively via injection-molding processes. In specific cases one can use diamond turning, but mainly for fast prototyping. The weak points of polymers are the strong temperature dependence of their optical properties, insufficient long-term stability, sensitivity to radiation impact and humidity, outgassing, and transmission loss in the blue and UV part of the spectrum. Also, the variation of the refractive index from production lot to lot has to be controlled. Although the refractive index of glass can be fine-tuned and controlled by an annealing process, this is not possible for polymers. Typically used polymers are PMMA, polystyrene (PS), polycarbonates (PC), cyclo-olefin polymers (COC, e.g., Zeonex®), CR39, and the resin MR-8. Because the optical position of these polymers is not controllable with high accuracy during production, lens systems often have to be corrected during assembly (melt-dependent production).

New materials are being developed to overcome some of the disadvantages mentioned. Composite materials of polymers with inorganic nanoparticles are particularly promising. Refractive index values significantly higher than 1.65 have been recently demonstrated for thermoplastics. The major problem, not yet solved, is the high amount of scattering, mainly caused by the insufficient dispersion of the nanoparticles.

4.3 Glass Ceramics

Zero-expansion glass ceramics such as Zerodur® from SCHOTT are typical materials to be used in reflective optics (mirrors) with asperical shapes. Normally, the dimensions of such mirrors are outstanding (1–4 m in diameter). Typical application fields are astronomy and LCD projection systems. Key properties are large achievable sizes, zero expansion and expansion homogeneity, as well as very low bubble and inclusion numbers in large volumes. The combination of aspherical geometries and lightweight structures is one important future trend.

4.4 Single Crystals and Polycrystalline Ceramics

Single crystals have unique optical properties but are expensive to produce and difficult to process, which limits them to high-price applications. In most cases, cubic crystals are preferred due to their lack of birefringence. A material with very low dispersion and high UV-transmission is calcium fluoride, used for high-end lens systems in photography and microlithography at low wavelengths (193 nm, 254 nm). In the IR area, sapphire, Spinel, ZnSe, and ZnS are typical candidates.

The production of aspherical lenses from crystals requires CNC machining or diamond turning. Therefore, high material costs, together with high machining costs, lead to expensive lenses, affordable only in industrial or military instruments.

Chapter 5

Processing Technologies

The advantages of aspheres in optical designs, such as better optical performance and a reduced number of elements, as well as the difficulties in fabricating them, have been known for several years [1, 2]. New design approaches [3] based on aspherical shapes and varying production lots from 10^2 to 10^6 lens elements require rapid, cost-effective fabrication processes. Optic designers will never make use of aspheres in their designs if not convinced that the fabrication processes will deliver the required number of lenses in time. The economic fabrication process is one of the key elements for making aspheres attractive for the optical designer.

5.1 Processing of Aspheres: The Historical Approach

5.1.1 Overview

In Fig. 5.1, past progress in generating and polishing is shown. Major milestones are the realization of the copy principle, the application of the computer, and the invention of fine correction methods without a polishing pad. The different kinds of polishing and generating are explained in the following sections.

5.1.2 Generating

As an introduction, we start with the principle of manufacturing spherical optics, which is based on the identity of the surface functions of complementary bodies, here tool and workpiece, as shown in Fig. 5.2. The lens (above) is moved over the rotating, full-sized stiff tool (below). The tool, which is in contact with the overall surface of the lens, generates a negative copy of its shape on the lens.

Using the same kind of relative movement between lens and tool, the 2D symmetry of the surface is lost for aspheres. In this case the area of contact between a stiff tool and the workpiece is reduced to a line or even to a point contact; for example, the meridional line, as an axis of symmetry, can be used as a line of contact.

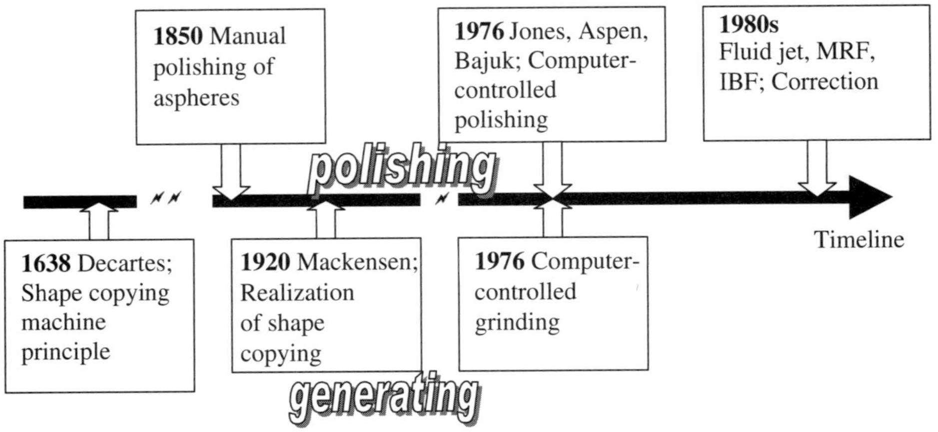

Figure 5.1 Historical progress in generating and polishing processes.

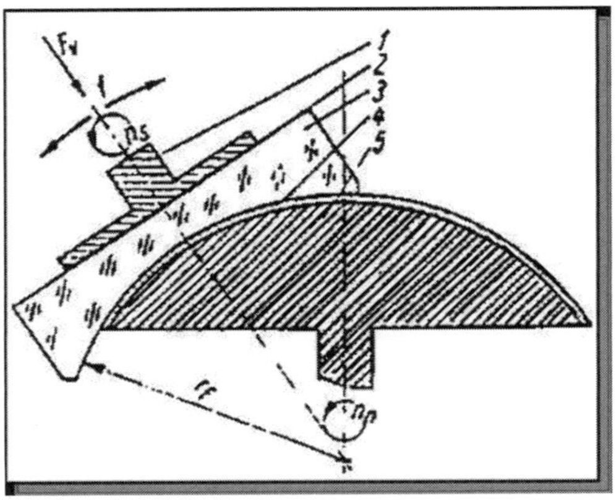

Figure 5.2 Principle of generating or polishing spheres using the symmetry between tool and workpiece (lens).

Descartes [4] was the first to take these considerations into account, and he designed the first shape-copying machine where a grinding stone was used as a master. The tool for generating the master was moved along the desired shape, guided by a push rod, but the shape accuracy was limited by insufficient guiding and bearings. Building a template of the contour line of the designed asphere, Fig. 5.3, was another approach for generating aspheres. The wheel, which is moving along the shape, is transferring, via a guiding system, the contour line to a grinding wheel of the same size. The wheel itself serves as a low-pass filter for manufacturing errors of the template.

Similar to the Descartes method, Mackenson designed and built (c. 1920) at Zeiss a machine to manufacture aspheres [5]. The principle is shown in Fig. 5.4. The asphere (1) rotates and has a line contact to the grinding wheel (2). The shaping tool allows permanent refiguring and is moved in an adapted polar coordinate system, which provides higher accuracy when compared to copying a simple template.

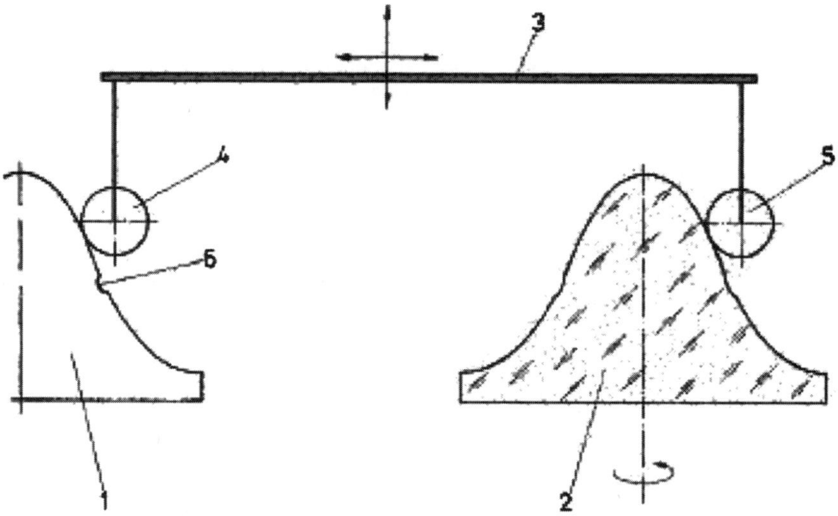

Figure 5.3 Principle of a copy machine for aspheres.

The introduction of the computer around 1976 revolutionized optical fabrication. The technical possibility of continuous path control by computer initiated modern generating and polishing methods. With regard to generating, two approaches were chosen by the scientists and engineers at that time. The first approach was to keep most of the fundamental ideas from the time before the computer era. One example of such a machine is shown in Fig. 5.5. The aspherical element is located below on a rotary table. The large vertical axis is rotating around the origin of the best-fitting sphere. A position control system is moving the vertical axis forward and backward depending on the actual angle φ. This generates an additional shape

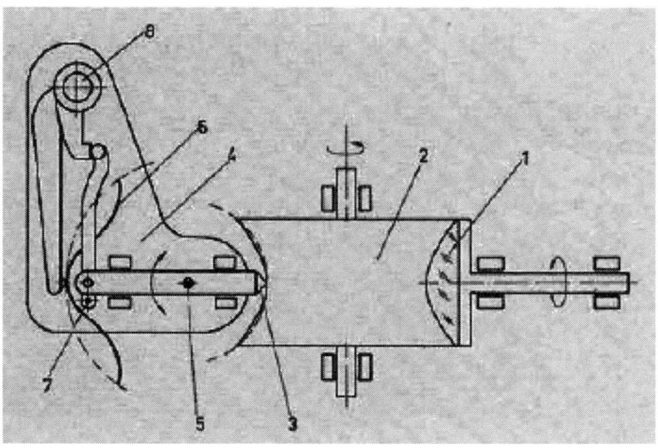

Figure 5.4 Principle of shape-copying machine as designed by Mackenson in the 1920s [5].

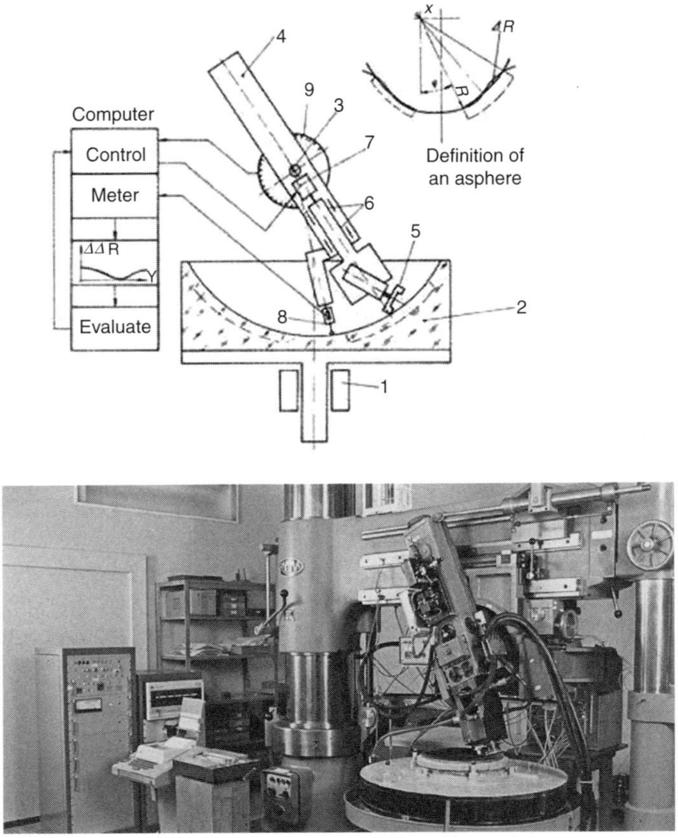

Figure 5.5 Principle and setup of one of the first computer controlled grinding machines built by Zeiss.

correction $\Delta R = \Delta R(\varphi)$. As computers were rather limited in their performance in those days, only the control of ΔR was possible, not that of R, but nevertheless it represented a huge evolutionary step toward better accuracy.

The second approach was to achieve the required accuracy with computing power and resolution of the numerically controlled drives. This approach [6] became more and more the industrial standard, with increasing clock rate, larger memories, and better resolution of the gages.

5.1.3 Polishing

As formulated by Preston [7] in 1927, the removal of material by polishing is a function of tool pressure, relative velocity between the tool and optical element, and the polishing time,

$$\frac{dz}{dt} = C_{\mathrm{p}} \cdot \frac{L}{A} \cdot \frac{ds}{dt},$$

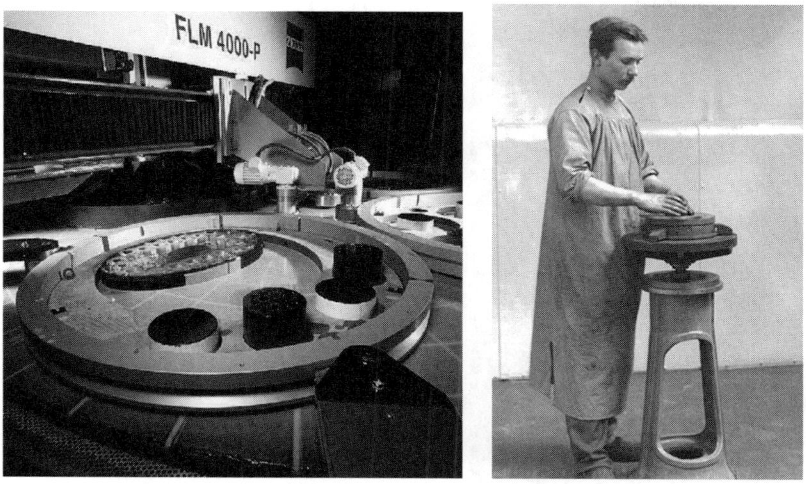

Figure 5.6 Left, plano polishing machine as an example of full aperture polishing; right, skilled craftsman polishing aspheres.

where $\frac{dz}{dt}$ is thickness change over time or removal rate (m/s), C_p is the Preston coefficient, L is the load [= total normal force (N)], A is surface area where the removal takes place (m^2), and $\frac{ds}{dt}$ is the relative velocity of the work piece to the tool (m/s).

Whether it is a microscope lens element or an astronomical mirror that has to be polished, full aperture polishing (Fig. 5.2) is the best-known process in lens manufacturing, developed over centuries for flat and spherical surfaces. The complete optical effective area is processed at the same time; that is, the polishing tool is permanently everywhere on the work piece. Figure 5.6 illustrates full aperture polishing of an elliptical lightweight mirror on a plano polishing machine. The mirror is polished face down on a large rotary table with polishing pitch. The relative velocity (ds/dt) depends on the rpm of the rotary table and the load force is the net weight of the mirror.

Due to the change in the local curvature, full aperture polishing with stiff tools is not possible for aspheres. Thus, polishing aspheres, before the computer era, was an art performed by highly skilled craftsmen (Fig. 5.6, right). In the 18th century the optical quality of refractive telescopes was significantly improved by manually polishing aspherical corrections on plane or spherical elements, but by trial and error.

The introduction of the computer allowed the calculation of the removal rate based on the parameters of the Preston equation, but also the numerical control of the movement of small subaperture tools. Local computer-controlled polishing (CCP) was developed between 1968 and 1976 (Jones et al. 1968–1982 [8]; Aspen et al. 1972; Bajuk 1976), in particular to polish conical aspheres for astronomy (Fig. 5.7). The tool, made of flexible material, was moved across the optical surface, and the tool itself performed a special kind of motion to get the required relative velocity. Based on the required local removal of material, a mathematical algorithm calculated a dwell time or a pressure map of the surface. The tool was numerically

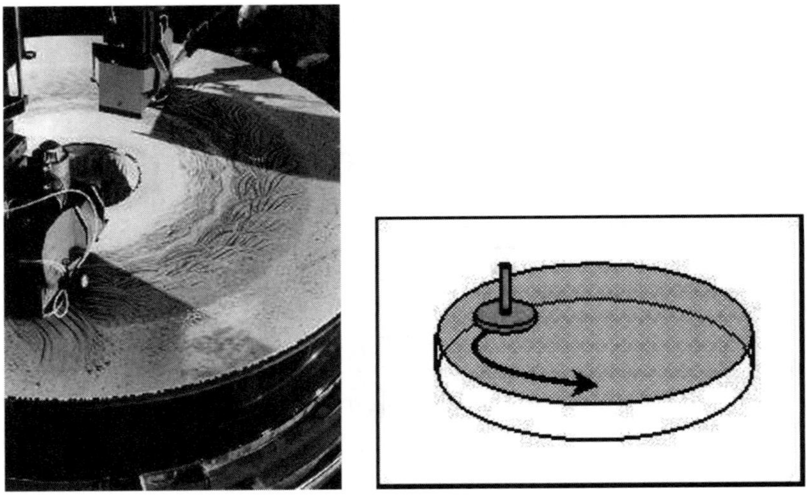

Figure 5.7 Left, computer-controlled polishing (CCP) process of the NTT ESO 3.5 m mirror; right, principle of the CCP process.

controlled when moving over the optical surface according to the calculated dwell time or pressure map. Exactness, how the locally polished surface matched the required shape, depended on the algorithm, the starting parameters, and the tool size.

Several parameters define the quality of the subaperture polishing process: mechanical stability, and the constantness of the polishing slurry and tool wear. To overcome the intrinsic limitations of such a chemo-mechanical removal process, alternative technologies for controlled material removing have been studied, including fluid jet polishing, ion beam figuring, and magnetorheological finishing. Common to these new technologies, which are based on fluids or ion beams, is the minimization of tool wear. For ultra-precise optics, such as for lithography, the complex computer-controlled subaperture processes are still indispensable.

5.1.4 Forming

Parallel to the progress in generating aspheres in the 1970s, molds for forming aspheres out of glass or plastic were available with increasing quality. These molds for the glass blank pressing process as well as the molds for plastic injection molding were manufactured by grinding, milling, or turning, and subsequent polishing. These reforming processes were used for small, low-level optics. Higher precision optics was produced at that time by a hybrid technology, where a thin plastic layer is reformed on a polished glass sphere.

5.2 Overview Processing

Figures 5.8 and 5.9 give an overview of the different types of processing available. The classical fabrication is listed in more detail in Fig. 5.9. The characteristic features of each process step are discussed in Table 5.1.

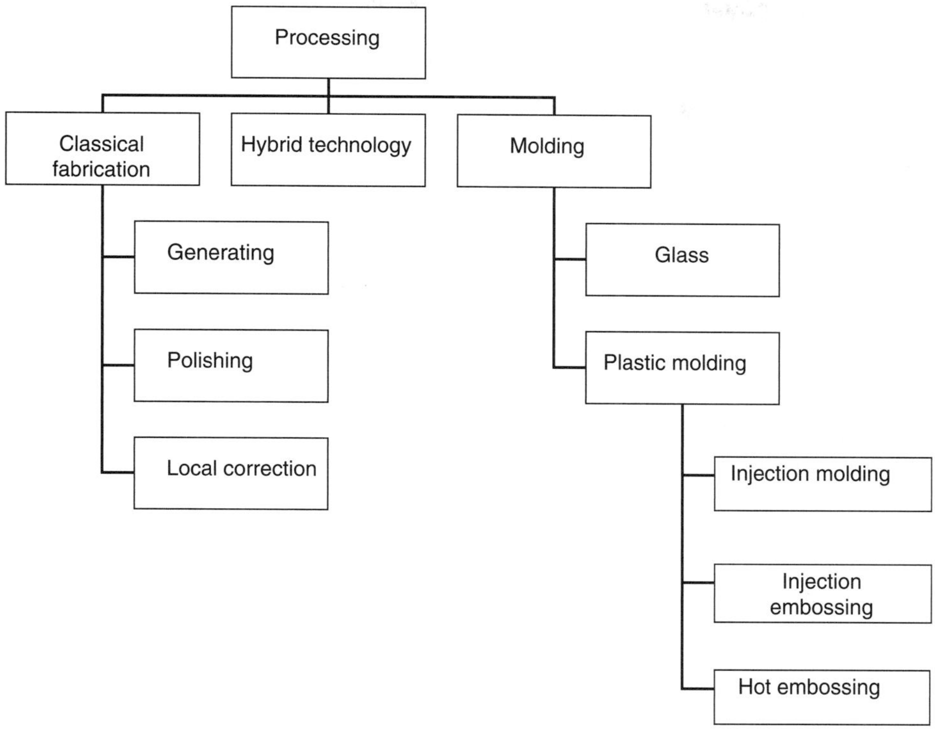

Figure 5.8 Overview of different kinds of process technologies.

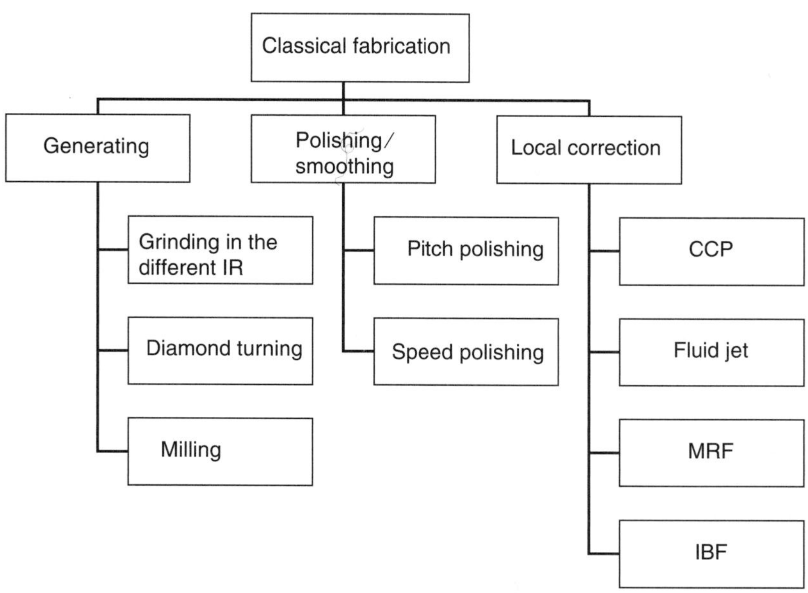

Figure 5.9 Detailed structure of classical optics fabrication.

Table 5.1 Overview specifications and characteristics of the different processes (typical values in production).

Process	Batch size	Size (mm)	Shape deviation (nm) PV	Surface roughness (nm) rms	Advantages	Limits	Main cost driver
Grinding	$<10^4$	2–400	1000	50–1000	Fast generating process	Subsurface damage	Size, accuracy
Diamond turning	$<10^3$	2–400	100	5–20	No subsurface damage, for IR sufficient roughness	Surface roughness	Size, accuracy
Speed/pitch polishing	$<10^4$	10–300	300	0.2–0.5	Very low surface roughness, fast polishing process	Correction of local surface deviations	Size
CCP	$<10^3$	5–8000	30	0.5	30 years experience	Tool wear, edge roll-off	Size
MRF	$<10^3$	5–500	10	0.3	No edge roll-off, no tool wear, low damaged surface layer	Center artefact for r–φ tool path	Size, fluid
Fluid jet	$<10^3$	5–240	30	0.5	No edge roll-off	Stability of foot print	Size
IBF	$<10^3$		5	0.2	No edge roll-off	Low removal rate	Vacuum
Precise glass pressing	10^4 to $>10^6$	0.5–35	1–5	2	High volume	Size, accuracy	Cycle time, accuracy
Injection molding	10^4 to $>10^6$	0.5–200	1–10	5	High volume, complicated shapes possible	Birefringence, micro structures	Accuracy
Injection/hot embossing	10^4 to $>10^6$	0.5–200	1–10	5	Low birefringence, optics including mounting	Thickness	Cycle time accuracy

CCP: computer controlled polishing; MRF: magnetorheological finishing; IBF: ion beam figuring.

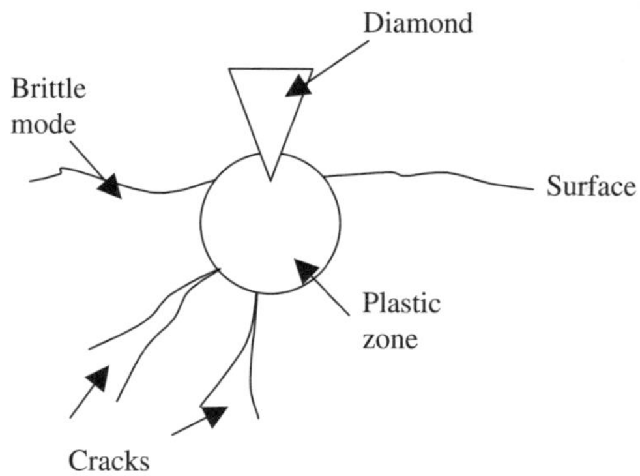

Figure 5.10 Principle of material removal within the plastic (ductile) and the brittle modes.

5.2.1 Generating

The principle of the generating process is shown in Fig. 5.10. The material removal can be understood by analysing the crack generation beneath a sharp indenter [9], for example, a single diamond. Small loads result in a plastically deformed zone. Increasing the load creates radial and lateral cracks. If a lateral crack reaches the surface, a chip of material is removed (brittle mode). The remaining radial cracks determine the depth of the subsurface damage.

Working only in the regime of the plastic or ductile removal mode requires highly stiff and accurately controlled machines. Plastic removal can be carried out by fixed abrasive grinding or single-point diamond turning (SPDT), depending on the material properties. Brittle materials like glass are processed by ductile grinding; moulds based on nickel are processed with SPDT. Details of SPDT are discussed in Sec. 13.9. Details of the grinding process are described in Sec. 13.1 "Zonal Grinding Process."

Related metrology is described in Sec. 6.5.1, "Surface form measurement of non polished optical surfaces."

5.2.2 Polishing

Smoothing the surface after the generating step is the task of polishing. Polishing has to remove subsurface damage and to improve the surface roughness to the required level. For parts that are generated by ductile grinding or SPDT, the subsurface damage is nearly zero, and the amount of material to be removed is very small. Only smoothing of the surface is required. For parts generated by conventional grinding, like most optical elements, the subsurface damage layer determines the amount of material that has to be removed by polishing. The polishing process can be described by the Preston equation [7]. Research work on chemical mechanical planarization (CMP) indicates a polishing gap of several micrometers [10]. To keep the gap constant, as shown in Fig. 5.11, the polishing pad has to be flexible enough to adapt itself to the local curvature of the asphere within a tolerance of a few micrometers.

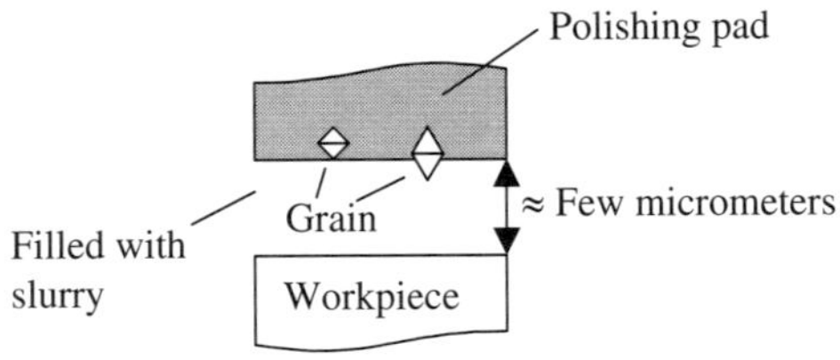

Figure 5.11 Schematic of the gap during the polishing process.

To fulfill the condition of a nearly perfect match of the pad tool to the aspheric workpiece, subaperture tools for polishing are used in most cases. Details of the polishing are described in Sec. 13.2, "Zonal polishing process."

5.2.3 Local correction

Due to the nonperfect polishing step or based on the required final surface specification, the aspheric element has to pass a third process step, local correction. Residual surface deviations from the nominal shape have to be removed by this process step. As shown in Fig. 5.7, a small subaperture tool is moving across the surface in a meander or spiral path. The tool itself is characterized by a tool function, which is the removal rate at a fixed position (Fig. 5.12). The desired removal $R(x)$ is the difference between the nominal and real shapes.

For the local correction, the tool described by function $c(x)$ has to be moved in a computer-controlled fashion across the surface, such that the desired mass removal $R(x)$ will be achieved. One parameter to optimize the tool path is the dwell time. If $s(x)$ is defined as the inverse velocity (slowness; to be determined), then the equation for local correction can be written as

$$R(x) = \int c(x - x')s(x')dx' \tag{5.1}$$

Tool function measured

$s(x)$ to be determined

Desired removal e.g. surface

Map from interferometry

where $R(x)$ is the desired removal (e.g., surface map from interferometry), and $c(x - x')$ is the tool function (to be measured). This equation, which convolves the tool function with the inverse velocity function, has to be solved for the slowness. To get a stable convergence of the local correction process, the following conditions have to be fulfilled:

- Stability of the process, tool wear and slurry (if applied);
- Tool function (process, symmetry);
- Accuracy and stability of the machine;

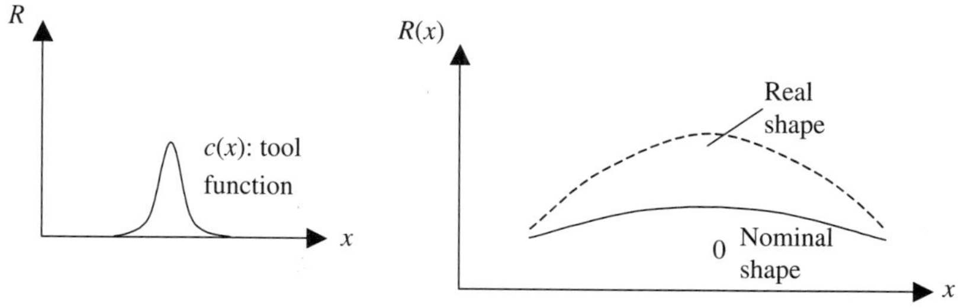

Figure 5.12 Left: tool function; right: real and nominal shape.

- Performance of the algorithm; and
- Accuracy of the related metrology.

The goal of all research and development work in this field is to determine the optimum parameters mentioned above. This holds for all methods involved: CCP, MRF, fluid jet, or IBF. A short overview of these different process types is presented in the following. Details are discussed in Fluid jet polishing (Fähnle, Sec. 13.6), Magnetorheological finishing (MRF) (Haag-Pichl, Sec. 13.3), Robot polishing (Schwarzhans, Sec. 13.5), Computer controlled polishing (CCP), and Ion Beam figuring (Börret, Secs. 13.4 and 13.7).

5.2.4 Computer-controlled polishing (CCP)

The principle of CCP is shown in Fig. 5.7. As already mentioned, CCP was the first local correction method and is characterized by the following features:

- The tool is a flexible subaperture pad.
- Removal is generated by a chemo-mechanical process, as shown in Fig. 5.11.
- The parameters are the relative velocity between tool and workpiece, tool size, and the applied pressure.
- The local removal rate at the surface (spot, footprint, or influence function) depends on the relative velocity and pressure distribution.
- The working with a flexible tool across the edge of the part has a strong impact on the pressure distribution of the pad and the related removal rate. The varying removal rate results in an imperfect shape at the rim for CCP polished parts.
- Tool wear is critical for the stability of the process.
- The CCP process is dwell-time controlled.

The CCP method is applied in customized setups by several optical companies and is also commercially available from Zeeko/Satisloh.

5.2.5 Fluid jet polishing (FJP)

If the flexible subaperture pad of the CCP process is replaced by a nozzle that applies a premixed abrasive slurry jet (Fig. 5.13), one gets the fluid jet polishing

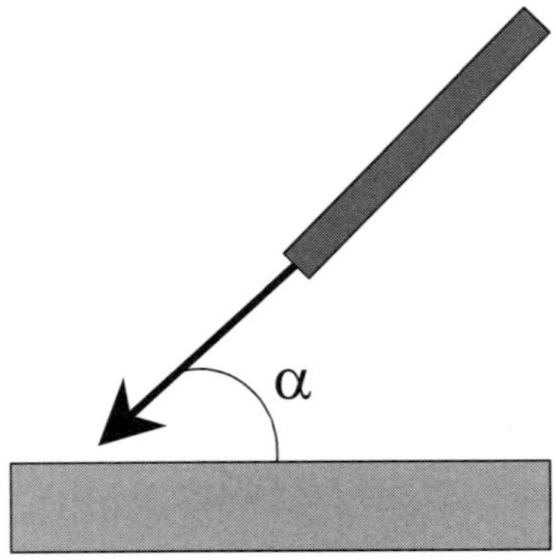

Figure 5.13 Principle of FJP, side view (O. Fähnle, Fisba Optik).

(FJP) process. The method is commercially sold as an option for the Zeeko CCP polishing machine. FJP is characterized by

- The pressure of the fluid (3–20 bar);
- The shape of the nozzle; and
- The orientation of the slurry beam.

Compared to CCP there is no wear of the tool itself and a nearly constant removal rate by working over the rim of the part.

5.2.6 Magnetorheological finishing (MRF)

If the flexible subaperture pad of the CCP process is replaced by a wheel with a magnetorheological liquid on it, one gets the MRF process. The principle is shown in Fig. 5.14.

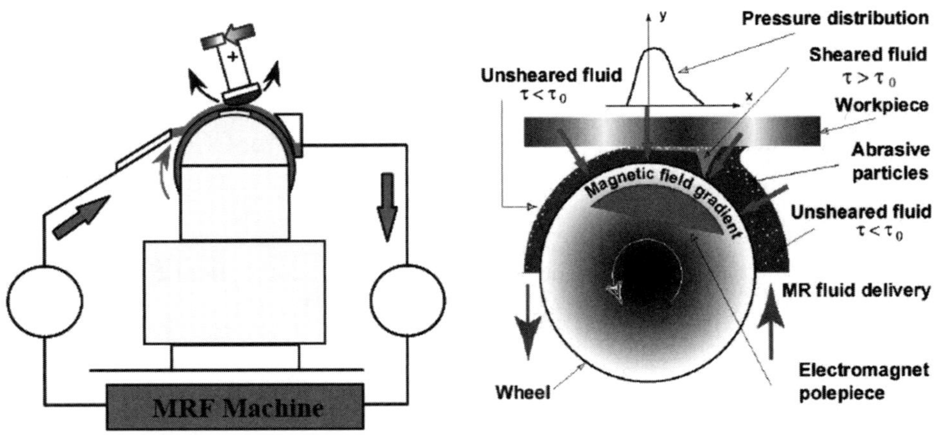

Figure 5.14 Principle of the MRF process. The material is removed by the sheared MR fluid (W. Kordonsky, QED Technologies).

The MR liquid, which is transported through a converging gap, is used to remove material by creating a tailored shear–stress distribution. The typical removal area has a size of about 1 cm^2. MRF is characterized by

- Varying stiffness of the fluid by applying different magnetic fields;
- The removal rate correlated to fluid viscosity; and
- The adaptation of the tool size by varying the gap.

Compared to CCP, there is no wear of the tool itself (like FJP), perfect adaptation of the liquid to the workpiece shape, and a nearly constant removal rate by working over the rim of the part [only small (mm) surface deviations at the rim].

5.2.7 Ion beam figuring (IBF)

The IBF method is characterized by replacing the flexible subaperture pad of the CCP process by an ion gun (Fig. 5.15). Clearly, IBF must be operated in a vacuum chamber. IBF is some kind of "sandblasting" at the atomic level. Atoms from the workpiece surface are removed by a sputtering process. The ions are generated in an ion source from a plasma of rare gas (e.g., argon) and are extracted by a negative voltage from the ion source and accelerated. The ions hit the workpiece surface with kinetic energy of several keV. Inside the material the ions are stopped by absorption. One to two atoms of the workpiece material per incoming ion are quarried out due to transferred momentum. Based on the physical laws of elastic and inelastic scattering, the sputter rate of the ion beam can be calculated very accurately. The absence of chemical interaction is one of the reasons for the stability and predictability of the process.

The typical removal areas vary between some mm^2 to several cm^2. IBF is characterized by

- The necessity of working inside a vacuum chamber and
- The ion beam energy and ion beam current.

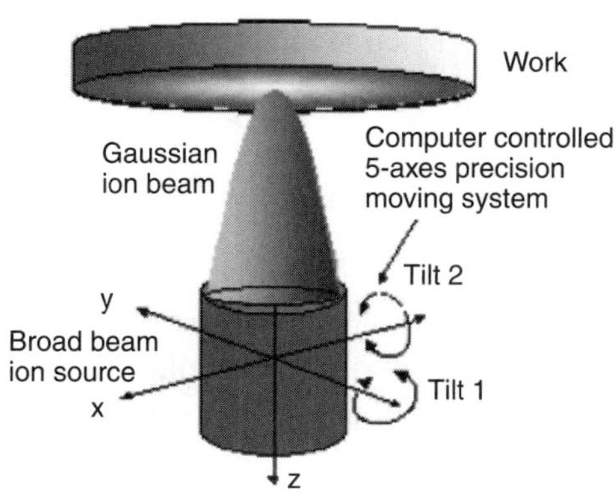

Figure 5.15 Principle of IBF. The ion beam as a tool is moved in a computer-controlled fashion across the surface.

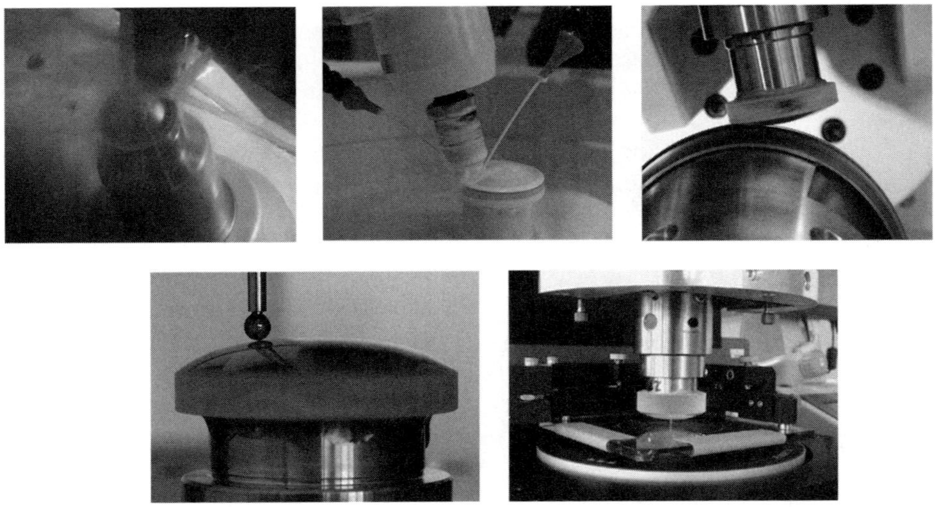

Figure 5.16 Example of a process chain for processing aspheres, including related metrology.

Compared to CCP there is no wear of the tool itself (like FJP, MRF), a perfect adaptation to the workpiece shape, a nearly constant removal rate by working over the rim of the part, and no print through (quilting) for lightweight structures.

5.3 Process Chain for Processing Aspheres

An example of a process chain for classical fabrication of aspheres is given in Fig. 5.16. Here, the generating process is performed with a Schneider ALG 200, polishing with a robot polisher, and local correction with Q22 MRF. The surface is measured using a tactile method after grinding, and interferometrically after polishing and local correction. Details of the tactile measurement are discussed in the Sec. 14.1, and details on interferometry are found in Sec. 14.2.

5.4 Hybrid Technology

Replication techniques become more and more cost-effective with larger quantities. For one of these technologies, the hybrid-press method, a highly transparent optical polymer is pressed with an aspherical mold on a polished glass sphere. The replica technology [11] is a cold molding process for creating a variety of hybrid optical components. The thickness of the transparent organic layer is about 1–30 μm. Because more than 99% of the lens thickness is glass, the element has similar optical and thermal properties to generated and polished aspheres. The disadvantage of these replicated elements is the increased sensitivity to humidity and scratching. Diameters are possible up to 50 mm.

5.5 Molding

Molding aspheres out of plastic or glass is the preferred method for the mass consumer market. Typical applications are mobile phones and digital cameras.

5.5.1 Precision glass molding

In precision glass molding (PGM), the glass is heated up to temperatures greater than T_g, the glass transition point, where the material becomes soft and can be reformed. Compared to other hot forming processes of glass, the viscosity for PGM is quite high. The large viscosity reduces effects like shrinkage, but results in a longer cycle time. Specially developed low-T_g glass types for process temperatures between 350°C and 650°C are used. Depending on the final application preforms, such as precision gobs or balls or near shape, preformed elements are the base material for the molding process. The process steps are as follows:

- The preform is put into the aspherical master.
- The preform and master are heated up to the process temperature.
- The preform is pressed into the final shape.
- The new element and the tool are cooled down by keeping the pressure at a high level.
- After cooling, the new element is removed from the master and the cycle is started again.

Master forms are typically made from ceramics, such as tungsten carbide, by classical fabrication processes. Typical cycle time for PGM is between 15 and 20 min. Details are discussed in "Precision Glass Molding" (Sec. 13.8).

5.5.2 Plastic molding

For plastic optics, three kinds of processes are possible (Fig. 5.17):

- Injection molding,
- Injection embossing, and
- Hot embossing.

For injection molding, the optics shape is completely filled with liquid plastic under pressure. For injection embossing, a plastic drop is injected into an extended shape and pressed onto the master shape. For hot embossing, a plastic disc is heated up and pressed onto the master shape. Details of plastic molding are discussed in the

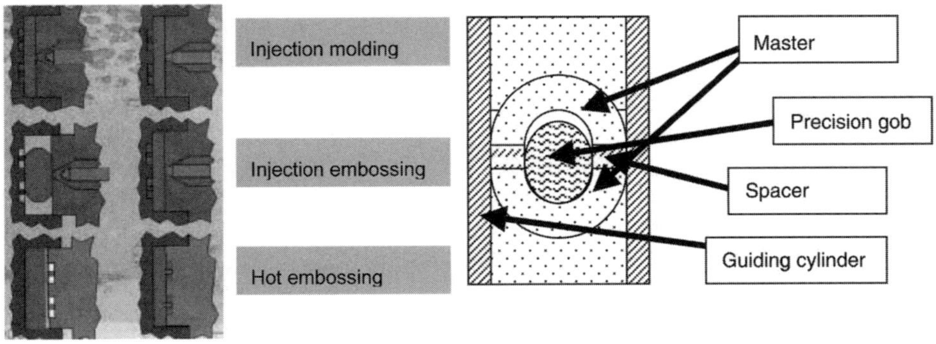

Figure 5.17 Left: different kinds of plastic molding for glass; right: the blank press method.

contribution of Viaoptic, (Sec. 13.10 Injection Molding of High-Precision Polymer Optics). Masters are typically made of Ni-coated stainless steel by ultra-precision machining. Details are discussed in the contribution of Aixtooling or Fraunhofer IPT (Sec. 13.9) about tools for precision glass molding.

5.5.3 Correlation—final surface quality—surface processing

The final quality of the optical element is related to the characteristics of the manufacturing process. Regardless of which process is used, replication, grinding, or polishing, the special properties of the tool and the process determine the final surface quality.

To improve processing of the optical elements, the correlation between surface structure and tool, process characteristics have to be determined. One possibility for describing shape and roughness measured by different metrology devices is the relation of height amplitude to surface structure (spatial wavelength), as described by Franks [12], or the power spectral density (PSD), as described in the contribution of H.J. Tiziani. By applying Fourier analysis to the height profile, the amplitude spectrum of the surface is determined and plotted against the related surface wavelength or wavenumber. As an example, the topography and related PSD of a diamond turned workpiece are shown in Fig. 5.18.

The characteristic surface structure of tool and process is clearly seen as a peak in the PSD plot. The optics manufacturer must know if the characteristic surface structure can be smoothed out by the subsequent polishing process. A new approach consists of describing the polishing tool by an amplitude reduction function as shown in Fig. 5.19. An amplitude reduction of 1 is equal to a reduction of the related surface wave to 0, an amplitude reduction of 0 keeps the related surface wavelength unchanged. The full aperture tool in Fig. 5.19 reduces the short waviness and the surface roughness quite well, but it degrades the overall shape of the optical

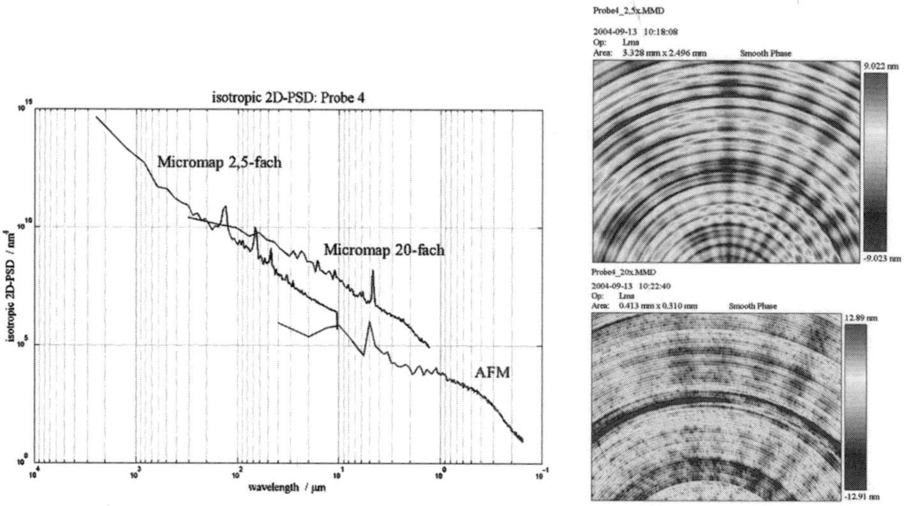

Figure 5.18 Power spectral density plot and related surface topography for a diamond turned master. The characteristic structures from diamond turning are clearly visible in the PSD plot.

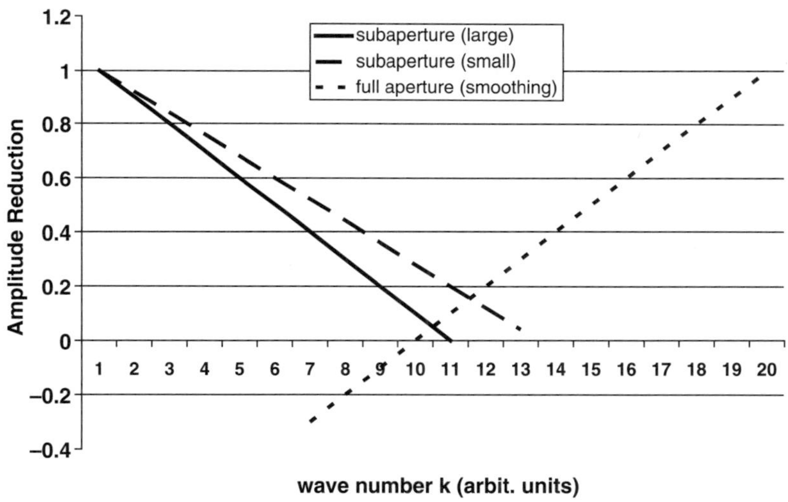

Figure 5.19 Reduction of the amplitudes depending on the surface wavelength for different tool sizes. An amplitude reduction of 1 is equal to a reduction of the related surface wave to 0; an amplitude reduction of 0 keeps the related surface wavelength unchanged.

element. In contrast, the subaperture tool is perfect in correcting the overall shape, but it cannot improve the surface roughness at high wavenumbers. The potential of describing optical elements and tools by amplitude wavelength plots will also be discussed in the Sec. 9.5.

5.6 References

1. R.E. Wagner and R.R. Shannon, "Fabrication of aspherics using a mathematical model for material removal," *Applied Optics*, Vol. 13, No. 7, pp. 1683–1689 (1974).
2. E. Heynacher, "Asphärische Optik–Warum sie gefordert und wie sie gefertigt wird," *Zeiss-Informationen*, Vol. 24, pp. 19–29, 1978/79, Heft 88, Carl Zeiss, Oberkochen.
3. C. Beder, "Aspheres for high speed cine lenses," *Proc. SPIE*, **5962**, pp. 59620V (2005).
4. R. Descartes, *La Dioptrique*, Paris, 1638.
5. F. Twyman, "Non-spherical surfaces," Chap. 10 in *Prism and Lens Making*, Hilger-Watts Ltd., (1952).
6. A. Cox and M.F. Royston, *Proc. Conf. on Optical Instruments 1961*. Chapman & Hall Ltd, London, pp. 330 (1952).
7. F.W. Preston, "The Theory and Design of Plate Glass Polishing Machines," *J. Glass Tech.*, Vol. 11, pp. 124 (1927).
8. R.A. Jones and P.L. Kadakia, "An automated interferogram analysis technique," *Applied Optics*, Vol. 7, pp. 1477 (1968).
9. O.W. Faehnle and H. van Brug, "Novel approaches to generate aspherical optical surfaces," *SPIE*, Vol. 3782, pp. 170–180 (1999).
10. S. Anjur, D. Apone, C. Barns, C. Gray, V. Manno, M. Monsour and C. Rogers, "In-situ friction and pad topography measurements during CMP," *MRS Symposium Paper*, Spring (2004).
11. J.J. Braat, A. Smid and M.M.B. Winjakker, "Design and production technology of replicated aspherics objective lenses for optical disk systems," *Applied Optics*, Vol. 24, pp. 1853 (1985).
12. A. Franks, "Nanometric surface metrology at the National Physical Laboratory," *Nanotechnology 2*, pp. 11–18 (1991).
13. J. Lehmann, "Fertigungstechnologie für diffraktive Elemente," *Diploma Thesis*. University of Applied Science, Aalen (2005).

Chapter 6

Metrology

High-precision fabrication technologies for spherical, but for aspherical surfaces in particular, have been significantly improved within the last few years, mainly driven by developments in the semiconductor market. Aspheric surfaces would be even more attractive if production costs could be further reduced. Their main advantages in optical systems are a better image quality or a reduced number of optical surfaces while maintaining image quality. An improved metrology is necessary, and an in-process metrology for production and quality assurance would be desirable.

6.1 Measurement of Optical System Performance

For the measurement of optical system performance, classical methods are available. The most frequently used will be mentioned briefly.

The optical transfer function (OTF) is frequently used for image quality analysis and is preferable for sensor systems. Both parts of the complex OTF, the amplitude term (called the modulation transfer function, MTF) and the phase term (called the phase transfer function, PTF), should be measured. The OTF is mostly obtained as a Fourier transform of the measured point or line spread functions at different field angles. The MTF gives the contrast ratio of a line pattern in the image versus object space for different spatial frequencies and therefore indicates whether the optical system fulfills the resolution specifications. The MTF, as a function of the spatial frequency, gives more useful information about the performance of the optical system, compared to the classical resolution test. The PTF itself is extremely important for the designer to be able to judge comatic errors in his layout. It is also very important for the assembly process. In both cases of a perfect optical system, the PTF should be zero. Any nonzero PTF values result from asymmetries (for example, asymmetric point spread functions) and are serious hints for decentered lens elements.

Another test method used in astronomy and microscopy is the "star test," where the image of a pinhole is analysed in shape and position [3] (Sec. 16.2.4).

The quality of polished surfaces and optical systems in workshops is mainly tested by interferometric methods. Twyman–Green, Fizeau, or Mach–Zehnder arrangements can be used. The result is an interference pattern. From the automated analysis of the fringe pattern, the type of the errors can be more easily identified and quantified than with OTF measurements. However, coherent and monochromatic light is needed [3].

Another alternative is the Shack–Hartmann wavefront sensor, which yields the derivatives of the local wavefront by measuring the transverse ray aberration. The wavefront itself is obtained by integrating the measured wavefront gradients.

Transmission and stray light need to be measured too.

6.2 Measurement of Individual Surfaces

For the measurement of individual spherical or aspherical surfaces and elements, different techniques are known to be used. An important requirement is the measurement of surface quality, for example, surface form (shape), waviness (including surface defects), roughness, as well as the coating properties. Furthermore, the central thickness and lens position, as well as the centering state must be measured with high precision. For the thickness or lens position measurements, tactile as well as optical techniques can be used. Optical techniques can be based on confocal or interferometric techniques such as white light concepts [4–7]. For measuring the centering error, a collimator is frequently used. When rotating the lens or lens elements, the reflected image of the surfaces or element should be stationary; the mechanical axis and the optical axis, defined by the line joining the two centers of a lens, coincide. The measuring principle for spherical and aspherical lenses is basically the same; the centering techniques may be different (Sec. 16.3.4), and the tolerances for aspherical surfaces may be more severe.

It was found that depolarization effects due to material stress or mechanically induced stress by cementing or mounting of lenses need to be analyzed, especially for crystalline optical materials at very short wavelengths (e.g., 193 nm and 157 nm), where crystalline optical materials are used.

For high-quality lenses, material properties such as the refractive index and its homogeneity in all three directions need to be determined with high accuracy, $\Delta n/n \leq 10^{-5}$.

The most expensive and time-consuming task is to determine the surface form (shape) of general aspherical surfaces. Therefore, form measurement of polished surfaces is discussed in more detail in the templates on Metrology (Sec. 14.2 and 14.3). Roughness measurements are essential in the short wavelength region, <300 nm, where the microroughness needs to be in the subnanometer region. Light scattering is caused by surface roughness and needs to be measured (Duparre in Surface/Microstructure inspection; Sec. 14.4). We will concentrate in the following on the measurement of the form and microroughness. It will be shown that calibration of the tools and the traceability of the results are important considerations.

6.3 Surface Metrology

For aspherical surfaces, the characterization of the height profile and its deviations from the ideal surface form are discussed in Sec. 2.2. We will concentrate on the measurement of aspherical optical surfaces, especially form, roughness, and waviness. Profilers are mainly used to measure the surface finish as well as the surface roughness, but they can also be used to measure the form (geometry), especially for single aspherical surfaces (ground or polished). The inspection area of profilers is limited to tens of millimeters, but the minimum field of view can be as small as a few tens of nanometers. It should be noted that we can split the profiling methods used into contact and noncontact types. Stylus probes represent the contact measurements, and have a mechanical tip, looking at the height variation of the probe when scanning across the surface under test. For the noncontact measurements, the tip is replaced by a focus sensor based on interferometry, the confocal principle, or by capacity measurements. The height measurement can be based on the inductive principle or on interferometry.

For the form (shape) measurement of polished surfaces, interferometry is frequently chosen. For nonpolished as well as for polished surfaces, the surface could be scanned with a pointwise working profiler.

Form-testing instruments mostly based on interferometry are characterized by a larger measuring area but a lower spatial resolution. Slope measurements, where the height is not directly measured, but the slope is measured, could be an alternative. The height is obtained by the integration of the slope. In the last few years, various surface slope-measuring principles for topography measurements have been discussed and used. The slope measurement is independent of height variation. Furthermore, fast computers have drastically reduced the time needed for the calculation of the topography from local slopes. Slope measurement is performed by wavefront sensors and deflectometers (including scanning autocollimators), as well as by differential and shearing interferometers.

6.3.1 Characterization of optical surfaces

The characterization of optical surfaces can be done either in the spatial or spatial frequency domain. The most intuitive description is in the spatial domain, because we experience a 3D space by considering two spatial dimensions and the height. By applying a Fourier analysis to the height profile, the amplitude spectrum of the surface is determined. Then the stochastical surface is characterized by the power spectrum, the Fourier transform of the squared-height variation. In this case the instrument response function of the metrology equipment can be asserted more easily.

The most common classification of optical surface topography considers three spatial regions, namely form, waviness, and roughness [1, 2] (Table 6.1). Form, also called shape, geometry, or figure, refers to a general macroscopic shape, characterized by spatial wavelengths >1 mm. The form of optical components is usually obtained by grinding or lapping, which ensures that errors of the surface shape are

Table 6.1 Classification of surface topography.

Roughness (high spatial frequencies)	Waviness (mid spatial frequencies)	Form (low spatial frequencies)
$\Lambda \leq 20\,\mu\text{m}$	$20\,\mu\text{m} \leq \Lambda \leq 1\,\text{mm}$	$\Lambda \geq 1\,\text{mm}$

kept ≤500 nm. The final surface generation steps usually involve polishing, where random and quasi-periodical height variations (called roughness and waviness) occur. They are spread over the whole surface and are collectively termed surface texture and localized imperfections. Those remaining microscopic defects of texture and localized imperfections are collectively referred to as the surface finish. There is, in general, no unique definition of the microstructure of optical surfaces, at what spatial period "roughness" becomes "waviness" or "waviness" turns into "form." This depends on the applications. Furthermore, roughness, waviness, and form cannot be separated easily, and most polished surfaces have a combination of all three. The spatial wavelength regions can be separated from each other by spatial filtering.

6.4 Measurement of Surface Roughness and Waviness

There are different methods to measure roughness and waviness. One method measures light scattering, either by measuring the total integrated scattering (TIS) or the angle-resolved scattering (ARS). The surface roughness should be smaller than the wavelength of the light used. Details will be discussed in the contribution by Duparre in Sec. 14.4, "Surface/Microstructure Inspection."

For the geometrical measurement of the form, roughness, and waviness, a pointwise working stylus instrument can be used for scanning the surface. Care needs to be taken, because the measured data are the convolution of the surface function under test and the tip surface function, that is, by the stylus tip radius, which smoothes the results and acts like a low-pass spatial filter, as indicated in Fig. 6.1. The influence of the low-pass spatial filtering is shown for a focused light spot used in a contactless optical measuring technique. The most precise method to measure the roughness of nonconducting materials is the atomic force microscope (AFM), where lateral dimensions ≤5 nm can be resolved. The depth resolution can be ≤0.05 nm. The drawback is the small field, typically 10 μm × 10 μm or 100 μm × 100 μm. The fields can be extended by different techniques such as stitching (combining overlapping areas); however, a high precision stage is needed.

In addition to AFM, optical techniques based on microscopic fringe projection, the confocal principle, or white light interferometry can be chosen to measure waviness and the low-frequency part of roughness. The results of the different methods must be carefully compared. One very promising way is by considering the power spectral density (PSD), that is, the Fourier transform of the autocorrelation of the microroughness function. The measurements data must be deconvolved by the stylus tip function or, in the case of an optical remote sensing method, by the

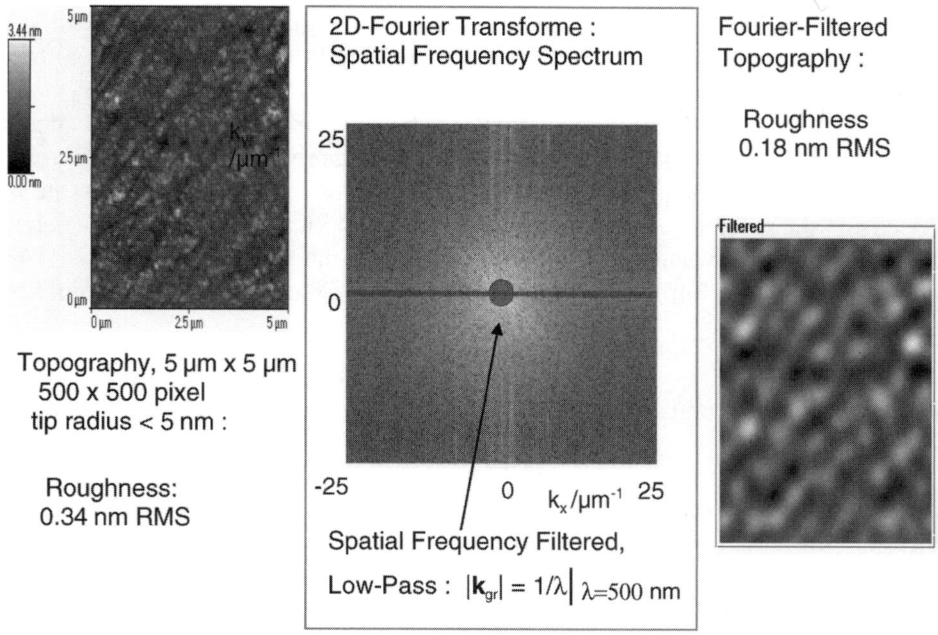

Figure 6.1 Influence of the low-pass filter on roughness measurement.

intensity distribution of the light spot. This function depends on the light wavelength and the numerical aperture (NA) of the focusing lens. Attention is required when using different filtering processes or when cutting higher frequencies. For optical measuring techniques, we have to consider the low-pass filtering proportional to 1/wavelength, shown in Fig. 6.1. To demonstrate the result of the optical low-pass filtering, a surface is measured with an AFM with a tip radius ≤ 5 nm. An rms roughness of 34 nm was measured. Considering NA close to 1, a perfect optical system would measure only 18 nm (rms) because the high frequencies are lost. Figure 6.1, on the right, shows the deconvolved roughness function.

Light scattering methods are very useful for studying microroughness. The results obtained can be related to the results obtained by profilometry by using the power spectral density [7–9].

The relation between the bidirectional scattering distribution function (BSDF) and the power spectral density $S_z(f)$ can be written as [8]

$$\mathrm{BSDF}(\theta_s) = \frac{16\pi^2}{\lambda^4} Q \cos(\theta_i)\cos(\theta_s) S_z(f),$$

where f is the spatial frequency, Q is an optical factor, λ is the wavelength, and θ_i and θ_s are the incidence and scattering angles, respectively. It should be noted that

$$\mathrm{BSDF}(\theta_s) = \mathrm{ARS}(\theta_s)\frac{1}{\cos(\theta_s)},$$

where $\mathrm{ARS}(\theta_s)$ is the angle resolved scattering function.

It is important that environmental influences like vibrations are reduced and that the surface, especially for optical measuring methods, needs to be free from any contaminations such as particles or films.

Figure 6.2 presents some limitations of the measuring methods. The spatial resolution and some typical ranges are shown, together with the rms height resolution. It should be noted that the spatial resolution of optical measuring systems depends on the resolution (smallest spot size) of the optical systems used, which in turn depends on the wavelength (λ) of the light and the numerical aperture (NA). A typical diffraction limited spot diameter (D) is deduced from the Airy function (circular aperture) as

$$D = 1.22\lambda/\text{NA},$$

and the two-point resolution, again by circular aperture, is given by

$$e = 0.61\lambda/\text{NA},$$

where λ is the wavelength, and NA is the numerical aperture. It should be mentioned that the width of the point spread function (image of a point) could be considered at half the maximum height, for instance depending on the response of the detector.

Furthermore, the resolution depends on the pixel size of the detector projected into the object space. The numbers given are therefore typical for $\lambda = 633$ nm and a NA of 0.5. The limits in Table 6.2 and Fig. 6.2 depend very much on the measuring conditions and result from some practical experience. In fringe projection, the depth limit is given by the depth of focus; but in white light interferometry, it is given by the depth of the fringe contrast, and the height range is typically given by the

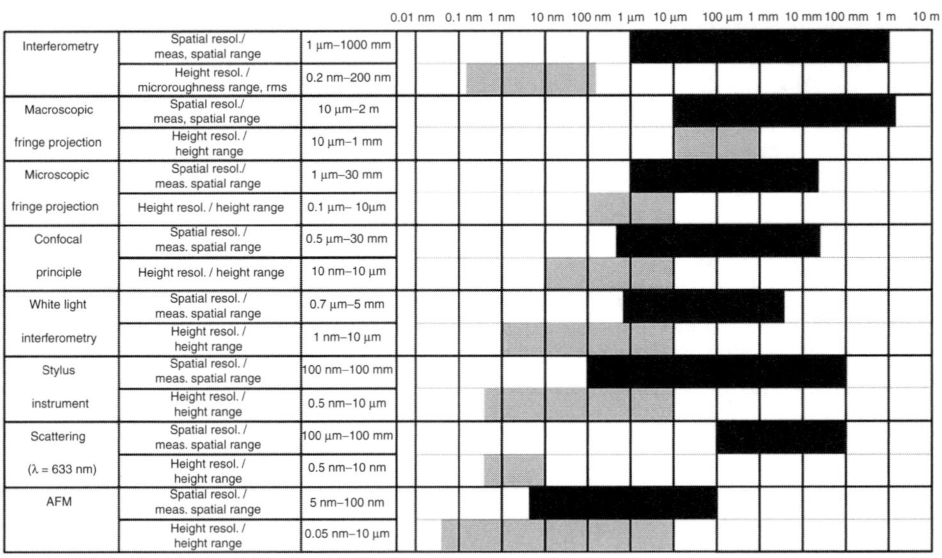

Figure 6.2 Spatial resolutions, the practical range of the field, and the rms height resolution, together with the practical height range of potential techniques used for the measurement of optical surfaces ($\lambda = 633$ nm).

Table 6.2 Comparison of 3D surface measuring methods.

Measuring principle	Height resolution	Camera images (pixels)	Diameter of measuring field
Confocal principle (microscope)	10 nm–10 μm	50–200	1 μm–30 mm, parallel process
White light interferometry	1 nm–10 nm	100–1000	50 μm–1 mm
Microscopic fringe projection	0.1 μm–10 μm	4–8	1 mm–30 mm
Stylus instrument	0.5 nm–400 nm		1 mm–100 mm, scan speed 0.1 mm/s
Atomic force microscope	0.05 nm–0.1 nm	Pointwise	5 μm–100 μm, scanning
Scattering	0.5 nm–10 nm		0.1 mm–100 mm
Macroscopic fringe projection	0.01 mm–1 mm	4–8	50 mm–2 m

range of the piezo or the vertical moving stage. It should be noted that the resolution can be higher than indicated in Fig. 6.2 for special applications in high-precision measurements. In Table 6.2 and Fig. 6.2, some limits of some measuring instruments and methods are summarized and will be discussed briefly. The stylus instruments represent the most common group of contact surface profilers. A height profile is determined by moving a small–tipped probe across the surface and sensing the height variations of the tip. The surface roughness can be determined with rms height resolutions down to 0.5 nm, but typically 5 nm, and lateral resolution of 0.1–0.5 μm depending on the shape and dimensions of the stylus. The measuring range can be 100 mm or more by scanning. It is not designed to measure very rough surfaces; therefore, the measuring height range is given as 10 μm (a rule of thumb as for the lateral range) as shown in Fig. 6.2, where a comparison and some limits of the mainly used different methods are given. As discussed earlier, the output of the measurement is the convolution of the surface and the tip. When the tip is smaller and sharper, the surface can be followed more precisely, but the local force on the surface can lead to surface deformations or even damage of the surface.

To increase the accuracy of the measurement, a vibration isolation system must be used. Stylus instruments are very popular surface profile instruments for measuring microroughness and surface form of polished and nonpolished surfaces with high accuracy and high lateral resolution. It is, however, a contact measurement and the measuring time is long.

The AFM is designed to measure surface structures with atomic resolution. It is part of the group of scanning probe microscopes to which the scanning tunneling microscopes belong. In a tunneling microscope, the current is measured when a tip is moved towards the surface. The tunneling microscope can be applied for conducting surfaces only. By contrast, the AFM works for any kind of surface. It works in two modes, repulsive or attractive. In the repulsive mode, the scanning tip is in a cantilevered position, making contact with the surface, and its deflection is measured. In the attractive mode, no contact is made with the surface under test, and the tip must be kept 2–20 nm above the surface. The high-frequency vibration of the tip caused by the attractive atomic force is measured. The resolution in this mode is somewhat lower than in repulsive mode. The height resolution obtained with an AFM is of the order of 0.05 nm for a field of $100 \times 100\ \mu m^2$, and the lateral resolution is 5 nm for small microroughness of the object, $\leq 1\ \mu m$.

Optical focus sensors resemble the actuator of a CD player, where an optical beam with high NA is adjusted such that it is being focused on the surface. The height of the scanned profile is measured by the actuator displacement. The 3D topography is measured by scanning the surface. The two representative methods to be mentioned are the confocal and the white light interference principles. The lateral resolution is given by the smallest spot size, which can be smaller than 1 μm for white light (or short coherence) interferometry and even slightly smaller for the confocal principle. The height range is limited by the scanning range in the vertical direction. For shape measurements with white light interferometry, care needs to be taken with respect to the coherence length of the light source. The rms depth resolution is 1 nm for white light interferometry, or 10 nm for confocal, respectively. The field for the confocal principle can be extended by using arrays of pinholes or microlenses to improve light efficiency or fibers. In Fig. 6.2, the lateral range is given as 30 mm without stitching but using parallel processing and microlens arrays. The optical focus sensor principle can also be used for form measurement of aspherical surfaces, but lateral scanning is needed.

6.5 Surface Form Measurement

There are different methods to measure the form (shape) of optical surfaces, either polished or nonpolished. It should be noted that interferometric measurements are applicable only for microroughness $\leq 0.25\ \mu m$ for an interferometer wavelength of 633 nm. The range can be extended by using IR light or oblique incidence.

6.5.1 Surface form measurement of nonpolished optical surfaces

For nonpolished optical surfaces being grinded or lapped, a stylus instrument is often used to achieve a measuring accuracy of about 1 μm or better. The technique is also applied for single polished spherical and aspherical surfaces where the mechanical tip is moved across the surface and the position coordinates are taken by mechanical, inductive, or interferometric sensors. Accuracies in the subnanometer region can be obtained. The stylus instrument can be integrated into a precision machine with high-precision spindles. However, the measuring time is long. Figure 6.3 shows a schematic of a test sample. Detailed information will be given in the contribution by B. Dörband in "Tactile Profile Measurement" (Sec. 14.1). We will concentrate next on the interferometric testing of polished optical surfaces.

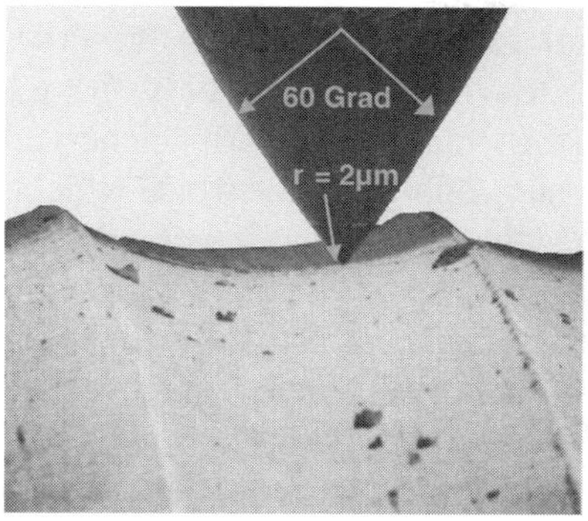

- The roughness measurement starts at the tip of the diamond
- The shape of the diamond is a conus
- The angle is 90 resp. 60 degree
- The tip has a radius of curvature of 2 µm, 5 µm, or 10 µm
- The measuring force of the test-needle is between 0.6 and 2 mN

Figure 6.3 Diamond tip in tactile measurements.

6.5.2 Surface form measurements of polished optical surfaces

There are different techniques to measure the surface form. Interferometry is the best known and most precise optical measuring technique, and the Shack–Hartmann wavefront sensor is an equivalent alternative for many applications. A stylus instrument is an another technique for measuring individual surface forms. More flexible measuring techniques will be discussed in the contribution on "Interferometry" by Tiziani (Sec. 14.2 and 14.3). They are mainly based on interferometry. Deflectometry is an interesting method, where the local slope of the test surface is obtained by measuring the local reflection angle of the beams or rays falling on the object under test. These can be derived from the displacement of the scanning laser spot on a position-sensitive detector in the focal plane of a lens. Another approach is to observe a grating-like structure reflected off the test surface. The image of the grating-like structure will be deformed according to the test object shape. The surface form is obtained by integration of the surface gradients, the slope angles. There is some similarity to a wavefront sensor where the local slope of the wavefront arriving in the pupil of the optical system is measured in the focus of the microlenses [11–13].

6.6 Interferometric Testing

Interferometry is the preferred metrology for testing not only polished plane, spherical, or aspherical single surfaces, but also the performance of whole systems. Classical interference arrangements frequently used are those based on Twyman–Green, Fizeau, or Mach–Zehnder principles [3].

When measuring a spherical surface, the illuminating "perfect" wavefront of the interferometer is focused into the center of curvature of the test surface. After autoreflection, the wavefront deformation, which propagates back to the interferometer, is twice the unknown surface deformation. By interference, that is, by superposition with the ideal reference wavefront, a fringe pattern is generated; fringe deformations visualize the wavefront distortion we are looking for. If the deformation is not too strong, the automated fringe analysis software determines the surface deformation.

From these data, it is easy to derive all the relevant data needed for the surface characterization according to ISO 10110 (e.g., the deviation from the best-fitting sphere).

In the case of aspherical test surfaces, the illumination with a spherical wavefront may lead to phase deviations of the autoreflected wavefront, which are too large, leading to a fringe pattern with fringes that are too dense to be detected. The fringe pattern cannot, at all or partially, be retrieved by the analyzing software. In this case, a Null corrector element is needed. Its purpose is to deform the incoming ideal wavefront from the interferometer in such a way that it corresponds to the ideal surface shape of the test surface. Then the autoreflected wavefront carries again only the actual deviation of the aspherical surface from its ideal function, the same as described by testing spherical surfaces. The difference to testing spherical surfaces is that the reference wavefront is no longer a perfect spherical wave but a complicated wavefront, which in turn is also more difficult to calibrate. The Null corrector can be a specially designed lens system or a computer-generated hologram (CGH).

For weak aspherical surfaces, that is, where the departure from the best-fitting sphere is only a few microns, one might omit the Null corrector. In such a non-null test configuration, retrace errors must be carefully considered.

For parabolic, or hyperbolic, and to some extent also for elliptical surfaces, some shapes' special interferometric test arrangements can be used.

When including all the features just mentioned, the interferometric techniques can reach form accuracies, that is, deviations from the best-fitting sphere, of a few nanometers.

Figure 6.4 shows a Twyman–Green arrangement for testing spherical and aspherical surfaces. For testing spherical surfaces, the CGH will be omitted.

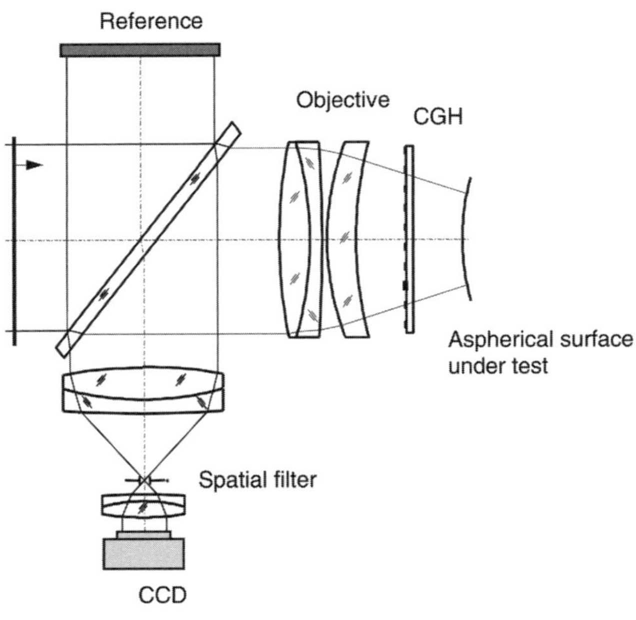

Figure 6.4 Twyman–Green arrangement for testing spherical and aspherical surfaces.

The Null lens in Fig. 6.4, a CGH for instance, is used for adaptation of the spherical wavefront generated by the collimating objective into an aspherical one such that the rays hit a perfect test surface perpendicularly.

6.6.1 Interferometric testing of aspherical surfaces with CGHs

A Twyman–Green setup for testing spherical and aspherical surfaces in reflection is shown in Fig. 6.4. For testing spherical surfaces, the center of curvature of the sphere coincides with the focus point of the spherical wave. With the CGH as a Null lens, the arrangement is used for testing aspherical surfaces as shown and described. Adjustment is more difficult for aspherical surfaces. For rotationally symmetric systems, as shown in Fig. 6.5, there are ten degrees of freedom. The CGH, the surface under test, as well as the spherical illuminating wave, need to be centered and need to be put into the correct position and corrected with respect to tilt. Furthermore, adjustment errors of the surface under test and the CGH lead to similar fringe patterns; the errors are therefore difficult to separate. Different structures for different additional functions, such as adjustment and calibration, can be written on the same CGH. Hence, the adjustment is much improved. To write a reflection hologram at the periphery helps to simplify the hologram alignment (Fig. 6.6).

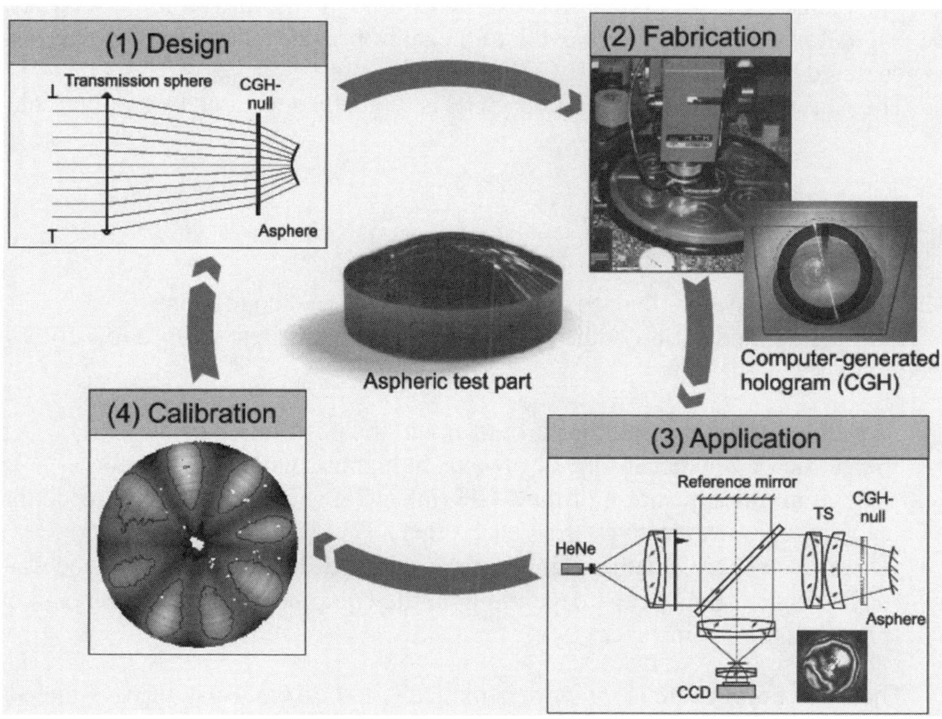

Figure 6.5 Testing procedures with CGH taking into account production and calibration.

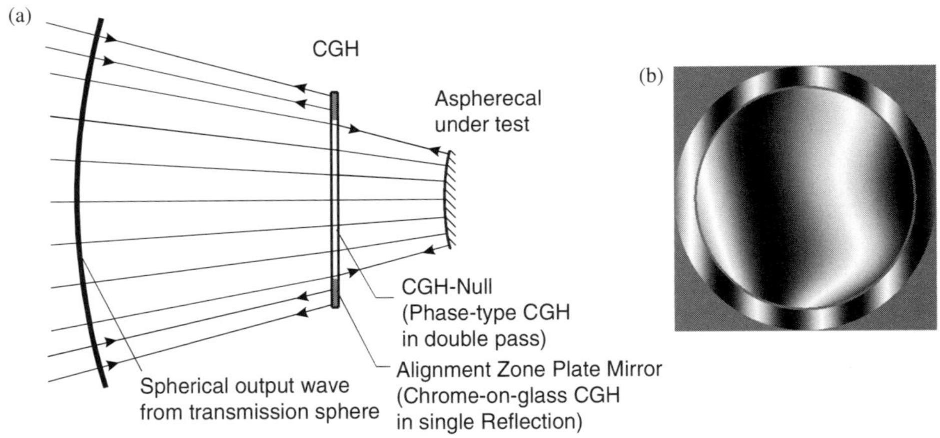

Figure 6.6 Test arm with CGH including a hologram at the periphery for adjustment of the setup.

6.6.2 Design and production of CGHs

CGHs are very appropriate for application as Null lenses and are frequently used for testing aspherical surfaces and systems. The basic principle is the generation of the wavefront such that the appropriate rays hit the perfect surface under test perpendicularly. The generation of the corresponding phase function that transfers the original plane or spherical wave into the appropriate aspherical wave needs to be generated by diffraction in the CGH.

The phase function $\Phi(x, y)$ of the CGH is typically described by a polynomial

$$\Phi(x, y) = \frac{2\pi}{\lambda_0} \sum a_{nm} x^m y^n,$$

where λ_0 is the wavelength used, and a_{nm} are polynomial coefficients.

For the implementation of the CGH in an experimental test setup, a few criteria need to be satisfied:

- Selection of the type of interferometer and the positioning of the CGH;
- Selection of whether in-line or off-axis holograms will be chosen;
- Selection of the position of the CGH (the CGH should not be put inside the caustic region where rays cross each other);
- Decision as to whether amplitude or phase holograms are needed—technological limits exist depending on the equipment available (E-beam or photo plotter for instance).

The testing procedure is shown schematically in Fig. 6.5. First, the CGH needs to be designed, taking into account the fringe density, diffraction orders, and the fabrication limits.

To avoid very dense structures on the CGH, a hybrid design is frequently chosen, as shown in Fig. 6.5, where a lens system is chosen to generate a spherical wave as used for testing spherical lenses. Therefore, the power does not need to be compensated by diffraction. One drawback of using CGH as part of a Null optic is the additional generation of coherent stray light due to unwanted diffraction orders. Frequently, off-axis holograms are chosen to spatially separate the diffraction orders at some point in the setup. The desired order is then selected with a mechanical diaphragm. Allowing an off-axis arrangement adds a powerful means to reduce the influence of unwanted diffraction orders, but it increases the fringe density in the CGH and adds more complexity in centering the test setup. Therefore, in-line CGH's are frequently used (Fig. 6.5) [14, 15].

It should also be noted that for maximum contrast the amplitude of the reference and measuring wavefronts should be equal. It should be remembered that for binary amplitude gratings the diffraction efficiencies in the plus and minus first order is 10% only, and 25% of the incident light goes in the zero order; the even diffraction orders are zero if the grating structure width is half the period. For a binary phase grating, 40.5% are diffracted into the plus and minus first orders. The zero order can be eliminated for an optical depth difference of $\lambda/2$ (top and bottom) between the structures.

The theoretical diffraction efficiency can be written by normal incidence of the illumination for the different orders of diffraction;

$$\eta = (A_0^2 + A_1^2 - 2A_0A_1 \cos\varphi)q^2{}_{\mathrm{D}}\mathrm{sinc}^2(mq_{\mathrm{D}}),$$

where A_0 and A_1 are the reflection amplitudes of the ridges and bar of the binary structure, respectively, φ is the phase shift, m is the diffraction order, and q_{D} is the relation bar to period, also called the duty cycle (frequently $q_{\mathrm{D}} = 0.5$). It is therefore evident that amplitude holograms, on chromium for instance, are used for a single transmission, whereas for double transmission, binary phase holograms are used.

Care needs also to be taken with respect to distortion occurring by imaging the aspherical surface onto the detector; the CGH should be positioned, if possible, close to the surface under test.

For the requirement with respect to precision for generating CGHs, it should be noted that position errors of the structures can be derived from the following relation, where the wavefront error is $W_{\mathrm{PD}}(x, y)$:

$$W_{\mathrm{PD}}(x, y) = -m_{\mathrm{R}}\lambda_0\xi(x, y) \cdot \nu(x, y),$$

which is proportional to the spatial frequency $\nu(x, y)$ and the distortion vector $\xi(x, y)$, and the diffraction order m_{R} is frequently chosen to be plus or minus one. For a wavefront error of $\lambda/100$, the position error of the structure should not exceed $p/100$, where p is the fringe period.

For a spatial frequency of 500 Lp/mm (line paits/mm, dimension mm^{-1}), $\Delta\xi$ should be $\leq$20 nm. Figure 6.6 shows an interferogram of a Null test setup that uses a combined hologram: an adjustment ring at the periphery together with the Null hologram in the central part. No fringes can be seen at the outmost ring, which indicates that the centering and alignment were nearly perfect.

It is therefore evident that the fringe density in the CGH should not be pushed to the extreme in order to be able to fabricate the CGH, reduce the influence of residual distortion errors, and to avoid vectorial effects that occur when the structures are of the same order of magnitude, as the wavelength of the incident light. Calibration of the complete setup is important and will be discussed together with more details on the testing procedure in "Metrology" by H.J. Tiziani (Sec. 14.2).

Today, the deviations from the best-fitting sphere are stronger; in addition, aspheric functions may include more higher order polynomial terms, the Null optics become complex, and one needs even more accurate means to produce them. The idea of using CGHs is to synthesize the wavefront adaptation as "programmable" diffractive elements. The first CGHs were plotted by large-scale plotters and then photographically reduced to diffracting structures close to or below 1 μm (period 2 μm) [14]. Today, they are directly written on coated glass or quartz plates with very high optical quality (homogeneity, parallelism, and flatness), using e-beam technology or a high-precision photoplotter. The diameter of the CGHs produced with e-beam technology is typically $\leq$150 mm, while those produced with a photoplotter are $\leq$300 mm. Due to programmability, a wide range of wavefronts can be generated; however, the limited density of the structures to be written limits the slope of the deviation to be compensated. The smallest width of the structure is of the order of 1 μm. Amplitude or phase holograms with higher diffraction efficiency can be produced using lithographic techniques. There are very few providers of CGHs for industrial applications, but delivery time has been improved recently to be typically one week.

Unfortunately, the design and fabrication of Null optics is expensive and therefore not always the most economical way for testing prototypes or small quantities of aspherical lenses. Because a Null element can be optimised for one specific aspherical surface function only, one looks for alternatives to generate a Null correcting function to be used for different aspherical surfaces.

Figure 6.6(a) shows a schematic of the test arm of an interferometric test setup with CGH. An additional hologram for adjustment was used at the periphery. In addition, a further hologram to be used for calibration can be added (Sec. 14.2). Figure 6.6(b) shows the resulting interferogram with a perfectly adjusted setup.

Several alternative technologies are under investigation to test the aspherical shape without Null optics. The methods are based on stiching (stepping) interferometry, using partly overlapping fields. The method is time-consuming and needs high-precision mechanics for the lateral or longitudinal displacements. An optical method described in "Interferometry" by Tiziani (Sec. 14.2) uses tilting the wavefronts to reduce most of the problems. Multiple-wavelength interferometry and the use of a deformable membrane mirror compensating for parts of the wavefront are other alternatives to be described in the contribution "Interferometry." Other methods use deflectometry as well as the Shack–Hartmann principle to be described next.

6.7 Surface Form Measurement with a Shack–Hartmann Wavefront Sensor

The Shack–Hartmann sensor is an attractive alternative to interferometry. The surface under test is imaged by relay optics onto an array of microlenses with a 2D-CCD sensor in the common focal plane. The spot produced by each lenslet is shifted on the CCD depending on the local wavefront gradient.

Any aspherical surface can be imaged onto the microlens plate if an appropriate optical relay system is available. Good zoom systems are recommended to achieve the flexibility required. The measured wavefront is a product of the unknown surface wavefront and the coherent transfer function of the relay optics. The latter must be known to retrieve the desired surface information. It should be noted that the Shack–Hartmann sensor is a slope-measuring device. From the slope data, the wavefront is obtained by numerical integration. For strong aspherical surfaces, a preshaping of the illumination wavefront by an element equivalent to a Null lens is required. More information can be found in the contribution "Shack–Hartmann wavefront sensor" by Tiziani (Sec. 14.3).

The method is robust, compact, and cost attractive, and its accuracy competes well with interferometric techniques. A drawback is the limited lateral resolution, which is typically more than an order of magnitude lower as compared with interferometric techniques.

6.8 Comparison of Methods

For testing form and waviness as well as the roughness of ground or lapped surfaces, the advantages of the tactile profile measuring methods using 3D measuring machines had been shown (Sec. 14.1). Shape accuracy is in the order of 0.4 μm for standard high-quality machines. However, with special machines, nanometer resolution can be obtained. For spherical or aspherical surfaces, the stylus instrument can be integrated into a precision machine rotating around the center of curvature of the best-fitting sphere of the test surface.

For polished surfaces, high-precision stylus instruments can be used for surface form, waviness, and roughness measurements with nanometer resolution. Scattering methods are very useful for the measurement of roughness of smooth surfaces.

For testing the form of polished optical surfaces, interferometry is the state-of-the-art method. When testing aspherical surfaces, CGHs might be required as Null optics. Their design and manufacturing is still a complicated and expensive process and limited to individual aspherical surfaces. Therefore, alternatives that are more flexible are being avidly investigated, as discussed in the contribution on "Interferometry" by Tiziani (Sec. 14.2). The stitching method as well as the Shack–Hartmann wavefront sensor are other alternatives; but again, for the Shack–Hartmann wavefront sensor, a shape and pupil adaptation by an optical system is needed. The dynamic range can, however, be extended. Deflectometry is another alternative to be investigated [11–13]. To reduce the cost and to improve the efficiency of the production of aspherical surfaces, "in-process" metrology is desirable. Different methods are already under investigation and will be available in the future.

6.9 References

1. ISO 11562, 2.6, 3.2 (1996).
2. ISO 3274, 4.4 (1996).
3. D. Malacara, *Optical Shop Testing*. 2nd ed., J. Wiley & Sons, New York (1992).
4. H.J. Tiziani, "Optical Metrology of Engineering Surfaces-Scope and Trends," Chap. 2 in *Optical Measuring Techniques and Applications*, Rastogi, P.K. (ed.), John Wiley & Sons, New York (1997).
5. H.J. Tiziani, A. Rothe and N. Maier, "Dual-wavelength heterodyne interferometer for high precision measurements of reflective aspheric surfaces and step heights," *Appl. Optics*, Vol. 35, pp. 3525–3533 (1996).
6. T. Wilson, *Confocal Microscopy*. Academic Press, London (1990).
7. J.M. Bennett, "Surface Roughness Measurement," Chap. 2 in *Optical Measuring Techniques and Applications*, Rastogi, P.K. (ed.), John Wiley & Sons, New York (1997).
8. J. Stover, *Optical Scattering: Measurement and Analysis*. 2nd ed., SPIE Optical Engineering Press, Bellingham, WA (1995).
9. M. Bass, Chap. 7 in *Handbook of Optics*. 2nd ed., Vol. 1, McGraw Hill, New York (1995).
10. J.M. Bennett and L. Mattson, *Introduction to Surface Roughness and Scattering*, Optical Society of America (1989).
11. R.D. Geckeler and I. Weingärtner, "Sub-nm topography measurement by deflectometry: flatness standard and waver nanotopography," *Proc. SPIE* **4779**, pp. 1–12 (2002).
12. G. Häusler, C. Horneber and M. Knauer, "Phase measuring deflectometry applied to measure reflective surfaces," *Annual Report*, University of Erlangen (2000).
13. A.S. van Amstel, S.M. Bäumer and J.L. Horijon, "Optical figure testing by scanning deflectometry," *Proc. SPIE* **3739**, pp. 283–290 (1999).
14. B. Dörband and H.J. Tiziani, "Testing aspheric surfaces with computer generated holograms: analysis of adjustment and shape errors," *Appl. Optics* Vol. 24, pp. 2604–2611 (1985).
15. S. Reichelt, C. Priss and H.J. Tiziani, "Absolute interferometric test of aspheres by use of twin-generated holograms," *Applied Optics*, Vol. 42, pp. 4468–4479 (2003).

Chapter 7

Coating Technologies

7.1 Introduction

Optical coatings are usually multilayer film structures used to obtain desired transmission, reflection, and absorption from surfaces. The characteristics may be due to the intrinsic property of the material (e.g., metal reflectors) or due to interference effects. Today, optical thin films are involved in numerous optical systems, where they are key to ultimate system performance. For most applications, high accuracy is required for both the optical and nonoptical properties, including low-loss energy balance, mechanical and thermal behavior, damage threshold, and nonlinear properties. Great progress has been made in almost all aspects. However, there are still problems and barriers to progress that need to be overcome. Identification and possible solutions to these problems represent an important aspect of research and development.

7.2 Market and Business

7.2.1 Global market for optical coatings

In the "Future for Optical Coatings," BCC, published February 2003 [1], the market for optical coatings has been investigated. In telecommunications, multilayer thin-film stacks have made increased bandwidth available for fiber-optic communications. The telecommunications segment in 2003 represents the fastest-growing segment of the coatings market and is an active area with many new entrants.

In the transportation segment of the market, coatings are used on windows, panels, and mirrors. Margins in all auto-supplier markets are limited; therefore, vehicle makers strive to minimize part costs. The electronics and instrumentation segment of the market includes many of the standard coatings applications, including camera lenses, computer and television screens, flat panel displays, and laser optics. Segment revenues represent the largest percentage of the total. The medical category includes coatings deposited on eyeglass lenses. Such coatings are a value-added feature of the lens suppliers and as such are captive revenues. Uses for optical coatings in defence applications include the optical systems for electro-optic and night vision systems. Military spending was in decline throughout the 1990s, but after September 11, 2001, it is now increasing.

7.2.2 Coating types

Coatings can be divided into application categories: antireflective, filters, mirrors, electrochromic, and transparent conductor coatings. Antireflective (AR) coatings are the most common type of coating. Most AR coatings are simple single-layer MgF_2 or four-layer broadband coatings. Simple AR coatings are amongst the lowest priced of coatings but have the highest volumes.

7.2.3 Coating costs

Coatings are labor-intensive to make and are often processed in batch modes. Yields for complex coatings are low. The total amount of material required is small. In the majority of cases, material costs are not a significant amount of the coating cost.

Overhead includes the cost of maintenance and operation of the capital equipment required for coating. This is the most significant factor in coating cost. A rule of thumb, according to one industry source, is that a coating machine will cost roughly 15% of its purchase price to operate annually.

7.2.4 Global markets

The coating market is divided into regional market shares. Asia and the North American market (principally the United States) represent the largest share of the market. Most of the optical shops in the United States and Europe are small and derive a significant portion of their revenues from custom design. In Asia, large high-volume operations, which minimize labor costs, produce large volumes of simply coated optics that can be sold at a minimal price.

7.3 Deposition Technologies, Coating Design, and Monitoring

7.3.1 Deposition technologies

Some of the basic technologies in optics, which directly lead to a significant increase in the performance of optical systems, comprise advanced coating technologies. The most important technologies are shown in Fig. 7.1.

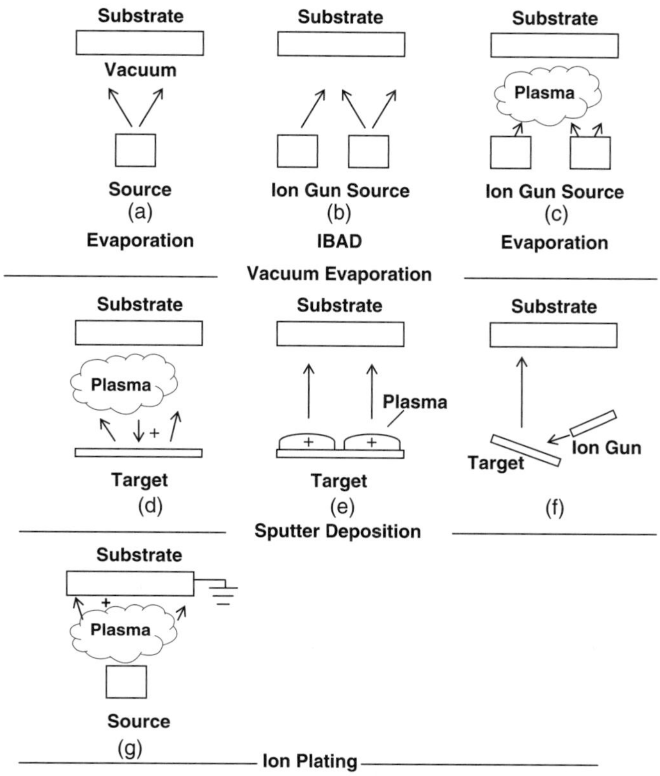

Figure 7.1 Physical vapor deposition (PVD) technologies (after Donald M. Mattox).

7.3.1.1 Boat/electron-beam evaporation (Fig. 7.1a)

Classical evaporation is a mature and well-studied technique that is common in commercial processes. Most optical coatings for optics and ophthalmics are deposited in this manner. During deposition, substrate heating is required to densify the coating. The performance of the evaporation sources is computer-controlled. Shutters start and complete the deposition process. Normally, the deposition rates as well as the film thickness are measured by the quartz crystal method. In order to improve the adhesion and the mechanical properties of the deposited films, the parts are plasma-treated by a glowing process prior to the actual coating step, as well as preheated.

7.3.1.2 Ion-assisted deposition (Fig. 7.1b)

In all versions, the film is bombarded directly by reactive gas ions and noble gas ions. These ions transfer momentum to the surface atoms and improve the mobility of these atoms. Dense coatings can be achieved without substrate heating. Plastics can be coated.

7.3.1.3 Plasma ion-assisted deposition (Fig. 7.1c)

The Leybold APS is a powerfully energetic deposition technology and is mature worldwide in optics. Beside momentum transfer, the evaporated material is partly ionized in the vapor phase and can therefore direct kinetic energy towards the substrates via a bias voltage. Moreover, APS can be used for reactive ion etching of plastics to produce high-performance antireflective structures [2].

7.3.1.4 Ion plating deposition (Fig. 7.1g)

Plating yields extremely dense and hard coatings and is mature in the area of hard and wear-resistant coatings. In optics, ion plating is used by a limited number of experts.

7.3.1.5 Sputtering (Figs. 7.1d and e)

Sputtering is a mature technology and is widely used in the optical coating industry. Most metal thin films are sputtered. Sputtering is a cost-effective way to deposit a film with a thickness determined by controlling time. Today, sputtering is the workhorse for flat and large areas. Sputtering is also the deposition method of choice for ultra-precise mirrors for extremely short wavelengths, for example, for EUV lithography at 13.5 nm.

7.3.1.6 Ion beam sputtering (Fig. 7.1f)

Initially, ion beam sputtering (IBS) was developed for the coating of high-quality mirrors for laser gyroscopes, which are dependent on extremely low backscatter values. IBS is a very stable and reproducible process, which can be well controlled and guarantees low defect rates in the growing layers. Today, IBS is also used for EUV mask mirrors, telecommunications filters, and other high-performance laser coatings.

7.3.1.7 Plasma impulse chemical vapor deposition (PICVD)

Originally, PICVD was developed for coatings inside tubes. The technology makes it possible to produce multilayer systems on 3D substrates. It is one of the most reliable CVD-based techniques for industrialized optical coatings and is currently used to produce millions of cold light reflectors for general lighting, digital projection, and other applications.

7.3.1.8 *Atomic layer deposition (ALD)*

The CVD technique ALD is used to fabricate ultra-thin and conformal thin-film structures for many semiconductor and thin-film device applications. A unique attribute of ALD is that it uses sequential self-limiting surface reactions to achieve control of film growth in the monolayer or submonolayer thickness regime. Planar ALD batch tools are used today for high-volume display coating. Low deposition rates are compensated by large areas that can be coated in one run. ALD is receiving attention for its potential applications in precision optics.

7.3.1.9 *Sol–gel techniques*

Sol–gel deposition of coatings involves covering the surface of a substrate with a liquid layer that is then dried and heated to form the final film. Coating is typically achieved by spinning or spraying the solution onto the substrate or by dipping the substrate into a reservoir of the coating solution. For precision optics and ophthalmics, sol–gel coatings do not have the quality and layer numbers necessary for optical applications. The allure of this process is the low capital investment required to establish the coating facility for large areas. However, there are no commercial supplies of sol–gel technologies. Vacuum-coating processes require the purchase of expensive vacuum systems that must then be maintained. However, the incremental cost for additional coatings fabricated by vacuum techniques declines precipitously once the capital cost is recovered. Also, many coatings can be made at a time. This is not the case with sol–gel coatings, which have significant material costs as well as high overhead costs.

7.3.2 Coating design [3]

Coating design is part of the manufacturing process of optical coatings. It generates the layer sequence for the production run, assists the monitoring of the deposition process during the run, and enables a reverse engineering after the run. Computer-assisted coating design, however, does not reduce the need for comprehensive knowledge, skill, and experience in thin-film optics, both in theory and in practice. Coating design needs thin-film design software consisting of packages for input, analysis, refinement, synthesis, and manufacturing assistance. Most frequently, the following packages are used: FilmStar (http://www.ftgsoftware.com), OptiLayer (http://www.optilayer.com), TFCalc (http://www.sspectra.com), The Essential Macleod (http://www.thinfilmcenter.com), and FilmWizard (http://www.sci-soft.com).

Besides design for performance, design for cost is very important. Recently, Alexander Tikhonravov [4] discussed a virtual deposition plant (computational manufacturing). This is a general structure of the software for computational manufacturing experiments. It is shown that computational experiments can be useful for checking feasibility properties of theoretical designs and for finding the most practical theoretical design for a given production environment.

The most widely used approach for optical coating design is to formulate the design problem as an optimization problem. However, in many cases the numerical solution of this problem is extremely difficult because of the large number of local minima in the merit function. Virtually all optimization methods are subject to convergence at local minima, and for this reason the search for the global minimum or even a good local minimum is a formidable problem. The solution of the design problem is greatly simplified when a good starting design is known.

For basic understanding, various useful graphical tools and methods for optical coating design, including the reflectance diagram, admittance diagram, and triangle diagram, are important. These tools give insight into how optical coatings function and how they might be designed to meet given requirements [5].

7.3.3 Monitoring

Process control in industrial deposition systems for precise optics is commonly realized by quartz crystal monitoring or optical monitoring using a single (variable) wavelength. Increased accuracy in optical thickness can be achieved using methods such as most-sensitive-wavelength monitoring. However, variations in optical constants can only be compensated for a particular wavelength. Because optical measurements are often preferable with regard to quartz monitoring for process controlling, a broadband monitoring system for *in situ* measurement of the optical performance can be used. Such systems are particularly well suited to monitoring highly dynamic processes, such as surface treatment of plastics by plasma etching and fast growth of metal island films. In many cases, such a high time resolution is not required, so transmittance measurements may be performed directly on samples at the rotating substrate holder.

For most coating designs, the advantages of broadband optical monitoring can only be shown for stable processes. Even in the case of a single-layer coating, a small absorption and a small inhomogeneity can lead to a similar transmittance. For this reason, the complexity of the assumed physical model of the coating should always be as simple as possible.

Inhomogeneous layers, such as so-called gradient index layers and rugate filters, represent prospective thin-film designs. Inhomogeneous layers are superior to classical coatings for antireflection over large angles of incidence. Manufacturing such systems in practice requires calculation, deposition, monitoring, and characterization of optical coatings with a well-defined continuous refractive index profile, along an axis that is perpendicular to the film surface. Today, new design and reengineering tools that take dispersion and absorption into account are available.

7.4 Multifunctional Coatings on Plastic Optics [6]

Injection-molded or hot-embossed polymer optics can replace glass optics as long as improved properties or lower costs can be achieved with the plastic parts. The problems with handling polymers in coating processes stimulate new coating or treatment techniques. One example is the AR-hard® design concept from Fraunhofer IOF Jena for antireflection purposes. AR-hard® coating is scratch-resistant because of its high overall thickness. Antireflection can be understood as an arrangement of symmetrical layer periods.

An alternative possibility for decreasing reflection on polymer surfaces is the use of appropriate layers with a decreasing effective index from substrate site to air. Investigations show that the application of special ion-bombardment conditions leads to stochastic antireflective structures on acrylic surfaces: so-called "NANO-moth eyes." The plasma source of a Leybold APS has been used to perform the etching step. The performance of the antireflective structure is much less sensitive to the angle of light incidence compared to interference coatings. In summary, this procedure should be favorably applied on curved and microstructured surfaces. Cost-effective mass production may be possible by direct ion etching for small optical parts as well as by replication of the structure onto larger parts.

7.5 Actual Topics [7–9]

Focus is currently on optical films that combine optical design with microstructural features tailored on the nanometer and micrometer scales, that deal with film properties ranging from optical nonlinearity and engineered bandgap to functional characteristics such as mechanical and chemical protection, electrical conductivity, gas and vapor permeability, and others. Evaluation of film stability and integrity in harsh physical and chemical environments, their compatibility with novel substrate materials (including organic polymers), business aspects, experimental designs, and industrial scale-up are important as well. Some actual trends include the following:

- Integration of coating technologies into production lines;
- Process control and monitoring;
- Optical coatings for IT (electronic–photonic convergence);
- Films for optical MEMS (CMOS compatibility);
- Coatings for optical sensors;
- AR films for micro-optical devices;
- Inhomogeneous and micro- and nanostructured films;
- Optical and protective coatings on plastics;
- Ophthalmic and light assembly applications;
- Coatings for short wavelengths (193 nm, 13.5 nm);
- Films for biomedical, pharmaceutical, and surface analytical applications (plasmon coatings);
- Coatings for displays (luminescent and emissive coatings, functional films);
- Transparent conductive films for VIS and NIR applications;

- Aerospace thin-film devices;
- Coatings for laser applications, including laser damage, fiber lasers, ultrafast coatings; and
- Metrology of optical films (new instrumentation and approaches).

7.6 Nanocoatings

Billions of years ago, molecules began organizing themselves into the unbelievably complex structures that capture light and convert it into chemical energy. Today, we use "nano" as a synonym for such processes. Nanophotonics deals with interactions between light and matter at a scale shorter than the wavelength of light. Nanoscale confinement of light can be performed by periodic dielectric structures with a repeat unit of the order of the wavelength of light. Such photonic crystals in one dimension are optical coatings. Nanoscale confinement of matter to make nanomaterials for photonics involves various ways of confining the dimensions of matter until optical properties change.

The formation of coatings on optics is a self-organized, atom-by-atom process, following the bottom-up principle in nature. In nature, modular building blocks can be assembled into target compounds with precise structural control at the picometer level through programmed sequences of synthetic steps. The possibility of mimicking the strategies successfully employed by nature to fabricate coatings will lead to nanocoatings with new multifunctional performances. Early examples are plasmon-boosted interference coatings and nanolaminates for scratch-resistant antireflective coatings on plastics. Nano is not new! New is the multidisciplinary cooperation between chemists, physicists, and biologists.

7.7 Summary

Optical coatings provide the means to engineer the properties of optical surfaces according to the various demands of an extremely broad range of applications in modern and future optical technologies. Besides the direct adjustment of the spectral transfer function, optical coatings are employed to optimize a variety of other surface characteristics, including for example environmental stability, abrasion resistance, or self-cleaning effects. The next generation of optical coatings will go even further and combine optical properties with other sophisticated features, such as a sensory functionality or an active control of selected transfer parameters. Today, optical coatings can be found in nearly every technical device, from ophthalmic glasses to cameras, binoculars, barcode scanners, or disc players, up to high-end products including complex optical systems for fundamental research, information, and laser technology. In many high-technology areas, the quality of the available optical coatings defines the technical limits of the optical systems and the efficiency of the related applications. Therefore, optical coatings are considered to be one of the critical enabling technologies governing further progress in many future developments and applications. As a direct key to the development of competitive innovations in the optical technologies, leadership in optical thin-film technology is an indispensable prerequisite for an economic area of highest technology status.

7.8 References

1. http://www.bccresearch.com/editors/RGB-187R.html
2. N. Kaiser, A. Kaless, P. Munzert and U. Schulz, "Nano-motheye antireflection pattern by plasma treatment of polymers," *Surface and Coatings Technology 20*, Vol. 1–4, pp. 58–61 (2004).
3. O. Stenzel, "New Challenges in Optical Coating Design," in *Advances in Solid State Physics*, Vol. 43, pp. 875–888, Springer-Verlag, Berlin, Heidelberg (2003).
4. A. Tikhonravov and M. Trubetskov, "Computational manufacturing as a bridge between design and production," *Appl. Opt.*, Vol. 44, pp. 6877–6884 (2005).
5. R.R. Willey, "Field guide to optical thin films," *Field Guide Series FG*07, SPIE (2006).
6. U. Schulz, "Coating on Plastics," in *Handbook of Plastic Optics*, Bäumer, S. (ed.) Wiley-VCH (2005).
7. C. Amra, N. Kaiser and A. Macleod (eds), "Advances in optical thin films II," Jena, *SPIE*, Vol. 5963, (2005).
8. N. Kaiser, "Optical coatings—trends and challenges for the present and the future," *Glass Coatings*, Vol. 3, pp. 44–50 (2005).
9. N. Kaiser and H.K. Pulker (eds), "Optical interference coatings," *Springer Series in Optical Sciences*, Vol. 88 (2003).

7.9 Further Reading

1. A. Macleod, *Thin-Film Optical Filters*, 3rd ed., Institute of Physics Publishing, Bristol and Philadelphia (2001).
2. R. Wiley, *Practical Design and Productions of Optical Thin Films*, Marcel Dekker Inc. (2002) [Is recommended for engineers].
3. A. Thelen, *Design of Optical Interference Coatings.* McGraw-Hill Book Company (1989).
4. H. Pulker, *Coatings on Glass.* Elsevier (1999).
5. H.J. Gläser, *Dünnschichttechnologie auf Flachglas*, Verlag Karl Hofmann (1999).
6. M. Ohring, *The Material Science of Thin Films.* Academic Press, Inc., Harcourt Brace Jovanovich Publishers (1992).

Chapter 8

Assembly Technologies

8.1 Relation between Design and Assembly

The complexity of optical systems depends on the specifications. Each design therefore requires a minimum number of lenses. The optical designer must exceed the specification values for two reasons: first to allow reasonable manufacturing and material tolerances and, second, to ensure that the imaging quality remains constant under environmental changes like temperature and air pressure fluctuations during the whole lifetime of the instrument.

On the other side, the design should have as few lenses as possible to minimize costs but also not be forced to spread the available tolerance budget on too many lenses.

We see that designing and tolerancing are trade-off processes, which require a lot of experience for the designer, good communication with the production engineers, and many iteration loops. Nevertheless, new parameters are welcome to relax the situation. This is part of the assembly concept, where by moving or tilting selected lenses in all three directions, one tries to get the system into the final specifications. In Secs. 8.2 and 8.4, we will call these adjustment parameters "compensators" to express their main function, that is, to compensate the residual errors of component production and of material uncertainties.

8.2 Review of Different Assembly Strategies

We describe several assembly strategies of increasing complexity.

8.2.1 Assembly of consumer optics with spherical lenses

Low production costs are the prime requirement in the business field of consumer optics. This demand is facilitated by specifications, which are easy to achieve by state-of-the-art technology. Lens components can then be directly mounted without additional adjustment actions. This leads to the following workflow:

Design office / Optional

- Reoptimization of the design using reported melt data from Schott
- Send to factory corrected values for the lens thickness

Lens factory

- Production of the optical elements, the mechanical mounts, and the tubes (housing) according to the tolerance specifications

Assembly group

- Fixation and glueing of the optical components into their mechanicals mount
- Centering of the mechanical cylindric axis with respect to the optical axis according to the specifications
- Assembly of mounted components into the tubes without additional adjustment

Quality assurance group

- Measure the image quality and the distortion

8.2.2 Assembly of high-end objectives with spherical lenses

When we ask for higher image quality, then more sophisticated specifications must be fulfilled, causing compensators to be considered. Sometimes it might be sufficient to adjust only focus, that is, to move the sensor into the plane of best focus. However, in most cases one needs at least two adjustments, such as shifting two lenses or two lens groups along the optical axis to correct focus and magnification. In the case of high-end systems with many lenses, it might be necessary to eliminate remaining errors of spherical aberration, field coma, field astigmatism, and image distortion, the so-called Seidel aberrations, which would require repositioning of more than two lens groups.

In the case of the large size cameras, the reposition of lenses is performed by changing the thickness of the mechanical spacers between the lens glass body and the mechanical seats. Typical values are some tenths of micrometers.

Image quality degradations also result from material parameter variations (see Sec. 2.1.1). In the case of airborne cameras, the tolerance calculations point out that the actual glass melt data, the true refractive indices of all glasses, must be known with an accuracy of 10^{-6}. This, however, would require expensive spectral measurements by the glass supplier. To avoid this, one reoptimizes the design with reported melt data of only 10^{-5} accuracy and readjusts in the finally assembled objective the position of two lens groups to correct focus and distortion.

In a similar way, we introduce compensators to eliminate lateral aberrations caused by residual decentering and tilt errors of lens groups. In most cases, it is sufficient to again move only two selected lens groups, but now laterally, to minimize comatic image errors and axial distortions.

All these fine-tuning operations require special measurement devices such as interferometers or wavefront sensors (Chapter 6). The monochromatic mode of these devices allows the analysis of only the monochromatic Seidel aberrations, but not the elimination chromatic errors, which is rather difficult. The best way to correct "chromacy" is to use glass material from highly reputable manufacturers like Schott or Ohara. Their glass-melting experience guarantees that the dispersion values, expressed by Abbe numbers, are kept in tolerance, as they depend mainly on the exact chemistry of the glass material. The small deviations of the absolute value of the refractive index can be explained by unavoidable small temperature variations during the cooling process of the glass melt and are easy to correct. Again, we summarize the workflow:

Design office
- Reoptimization of the design using reported melt data from Schott
- Send the factory corrected values for the lens thickness

Lens factory
- Manufacture the lenses and report the measured surface shape and thickness values

Design office
- Reoptimization using reported lens data

Start of optimization loop
- Send new values of the air gaps to the Mechanics factory and to Assembly

Assembly
- Mount lenses in modified mechanical structure and send to Quality assurance

Quality assurance
- Measure the image quality and the distortion
- Report "out of spec" values to Design office

Design office
- Calculate compensator motions along and across the optical axis

End of optimization loop

8.2.3 Assembly of high-end objectives with aspherical lenses

When assembling systems with aspherical elements, we have to be aware of some major differences when compared to systems with only spherical components. It is a fact that the decentering of aspheres, caused by lateral deposition or tilt errors, degrades the image quality more strongly than equivalent spherical elements. This should make the assembly of aspheres more difficult. But on the other hand, we have also mentioned that all technologies shaping aspherical surfaces must simultaneously ensure correct centering. Centering is the exact alignment of the aspherical coordinate system with respect to other optical axes (Sec. 2.2.1.3). Thus, all aspheres are delivered to the Assembly group with easily accessible auxiliary areas like plane or cylindrical facettes, which allow mounting into the housing or the mechanical structure without serious problems. Thus, the inherent sensitivity to decentering errors is transferred to the preprocess. The mechanical structure has to maintain the asphere's position and centering stability over the whole lifetime. The assembly workflow for high-end lenses will be more or less the same as described in Sec. 8.2.2.

8.2.4 Automated assembly of micro-optics

The trend to miniaturize optical systems is apparent in communications, metrology, remote tracking and sensing, life sciences (endoscopy), illuminations, to mention only a few. Aspherical elements are predestined for these application fields, where one needs to concentrate all the imaging parameters on few elements. To make effective use of the available space, the classical cylindrical optics layout is replaced by an arrangement in two dimensions on a plane plate (Sec. 16.4, "Micro-Assembly TRIMO").

This means that robot-controlled assembling methods with new strategies are required. The example of Fig. 8.1 is taken from a new assembly technique, called TRIMO (Three Dimensional Miniaturized Optical Surface Mounted Devices), which Leica Geosystems uses in production. A robot brings all elements into the correct position on a mounting plate, correctly orientated and soldered by laser heating. Because this process is not correctable after laser fixation, a closed loop assembly is applied to avoid the pile up of positioning errors (Fig. 8.2).

If Lens 1 is fixed, its deviation from the required position and orientation is measured to correct the precalculated target position and orientation of Lens 2. The

Figure 8.1 Planar arrangement of lenses and mirrors.

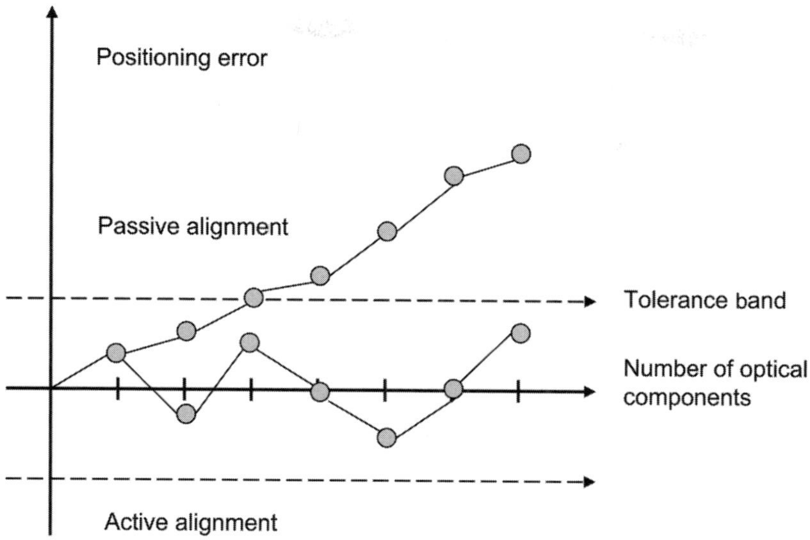

Figure 8.2 Active compensating avoids the pile up of residual positioning tolerances.

same procedure is repeated from Lens (n) to Lens ($n + 1$). Thus, all lenses act principally as compensators. In a practical situation, a combined action of soldering some elements at preset and some others at their corrected positions is recommended.

8.3 Errors and Tolerances

In the following, we estimate typical tolerance values for components and assembly, considering the four application cases of Sec. 8.2.

Table 8.1 Component tolerances.

Component tolerance	Consumer optics (spherical lenses)	HQ-optics (spherical lenses)	HQ-optics (aspherical lenses)	Micro-optics for photonics
Refractive index	$<10^{-4}$	$<10^{-5}$	$<10^{-5}$	$<10^{-4}$
Abbe number	<0.8%	<0.1%	<0.1%	<0.8%
Melt data	No	Yes	Yes	No
Radius deviation	$<2\lambda$	$<0.2\lambda$	$0.2\lambda - 2\lambda$	$<2\lambda$
Surface form error	$<\lambda/5$ rms	$<\lambda/20$ rms	$<\lambda/20$ rms	$<\lambda/5$ rms
Center thickness accuracy	±40 μm	±20 μm	±20 μm	±60 μm
Surface roughness	<10 Å rms	<5 Å rms	<5 Å rms	<5 Å rms
Lens wedge error	<±5 arcmin	<±20 arcsec	<±10 arcsec*	<±1 arcmin

HQ: high quality.

*Just one typical value; value depends on application and possibly might be different in some cases.

Table 8.2 Assembly tolerances.

Assembly tolerances	Consumer optics (spherical lenses)	HQ-optics (spherical lenses)	HQ-optics (aspherical lenses)	Micro-optics for photonics
Single lens				
Tilt error $\Delta\alpha$	$<\pm 5$ arcmin	$<\pm 30$ arcsec	$<\pm 10$ arcsec	$<\pm 30$ arcsec
Lateral decenter Δr	$<\pm 50$ μm	$<\pm 10$ μm	± 10 μm	$<\pm 1$ μm
Axial displacement Δz	$<\pm 50$ μm	$<\pm 10$ μm	± 10 μm	$<\pm 1$ μm
Lens group				
Tilt error $\Delta\alpha$	$<\pm 30$ arcsec	$<\pm 30$ arcsec	$<\pm 10$ arcsec	$<\pm 30$ arcsec
Lateral decenter Δr	$<\pm 50$ μm	$<\pm 2$ μm	± 10 μm	$<\pm 1$ μm
Axial displacement Δz	$<\pm 50$ μm	$<\pm 1$ μm	± 10 μm	$<\pm 1$ μm
Compensators				
Lateral position accuracy	NA	± 0.2 μm	± 0.2 μm	NA
Axial positioning accuracy	NA	± 0.2 μm	± 1 μm	NA
Operating temperature	0°C/+40°C	23°C ± 1°C or −10°C/50°C		−10°C/+40°C

HQ: high quality.

8.3.1 Component tolerances

The wavefront error introduced by the surface form errors is very severe when resulting from reflective than from refractive surfaces. It scales by a factor $2/(n-1) = 4$, using $n = 1.5$ for standard glass material. Mirrors must therefore be "better" manufactured by about a factor 2–3 than lenses with two surfaces. This holds for spherical and aspherical shapes.

8.3.2 Assembly tolerances

The data in Table 8.2 outline rather coarsely typical tolerance values.

8.4 Compensators

We mentioned the important role of compensators in reaching the final specifications. We add here some more comments.

For simple systems, all elements could be produced today with the necessary precision to allow assembling without further adjustments. Very often, an automatic and forced centering of all elements is achieved by the last flange, which presses all lenses in their mechanical seats to ensure the correct (and centered) position.

With increasing complexity, more adjustment parameters are needed, along and lateral to the optical axis, to eliminate residual static errors. We have shown that, in most situations, two lens groups are sufficient to act as compensators. The same holds for dynamic situations, when lenses must be moved in practise, for example, for autofocusing or in the case of zoom systems to vary magnification. Mechanical decentering effects due to the motion must then be tolerated and should be minimized during assembly. Zoom systems, where for the most part two lens groups are moved under computer control, are readjusted at both extreme magnification configurations: one drives the zoom into the position of highest magnification, adjusts best focus using compensator 1, then drives back into the position of low magnification and adjusts magnification using compensator 2. This is repeated until the specifications are reached.

We see that the use of compensators is generally expensive. What then are the advantages of methods like TRIMO (Sec. 16.4, "Micro-Assembly TRIMO"), which we described as an "all- or multicompensator" concept? At first glance we see negative facts. The placement of a lens depends on the actual positioning of the lenses already fixed and must be controlled by wavefront sensors. Therefore, one needs many "relay optics" to guide the measured wavefront from each lens to the stationary sensor. But the advantage is that we achieve extremely low total system errors. This is illustrated in Fig. 8.2, where in the case of a passive alignment strategy, all residual positioning errors could pile up and would require an expensive narrowing of all element tolerances. With an "all- or multicompensator" strategy, one would stay inside the tolerance band, even with lower component tolerances. The problem of using many relay optics is not a serious one—it can be avoided by a careful workflow analysis.

8.5 Alignment of the Optical Axis of the Aspherical Components

In Sec. 8.4, we mentioned the main difficulty of aligning or centering aspherical lenses inside systems. To better illustrate this, we consider an optical system consisting of many lenses, represented by a common optical axis (indicated by the dotted line in Figs. 8.3 and 8.4). One of the lenses, a single lens, should have an aspherical surface 1 of rotational symmetry, where the symmetry axis hits the center of curvature M1. The second surface should be spherical with M2 as center of curvature (Fig. 8.3). First, we see that M2 must lie on the symmetry axis of surface 1 to avoid a systematic decentering problem, a task performed by the lens

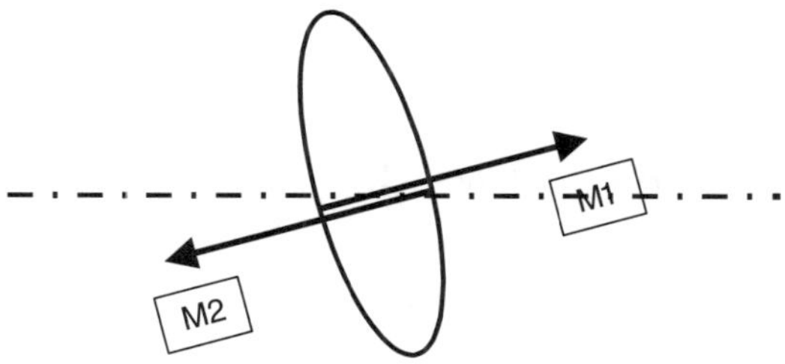

Figure 8.3 Centering aspherical lenses.

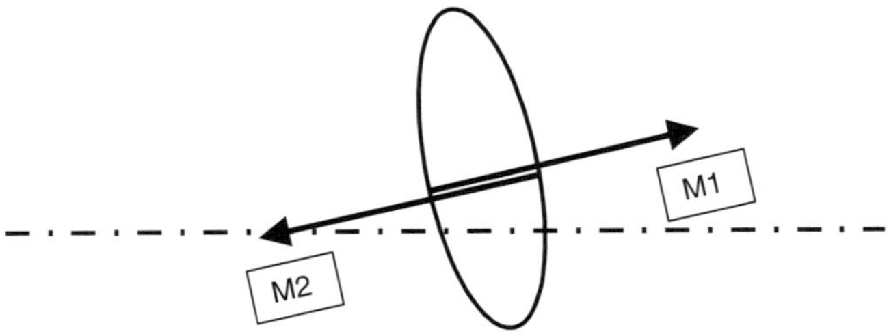

Figure 8.4 Centering spherical lenses.

supplier. Second, the lens has to be positioned and orientated during assembly so that the common system axis coincides with the symmetry axis of surface 1 within the tolerances.

The situation is more relaxed if the lens is spherical on both sides (Fig. 8.4). Both centers of curvature M1 and M2 must be as close as required to the common system axis. One way to achieve this is to move the lens until M2 lies on the common axis and then rotate around M2 until the deviation of M1 from the axis is within tolerance.

8.6 Monolithic Optics

Finally, we mention some interesting new technology that might be understood as a *unification* of surface shaping and assembly. The idea of merging both processes is driven by the tremendous progress made in computer-controlled machining. It allows the creation of complete optical systems including aspheres with one machine. The advantages are inherent high reliability, the concept of optical multifunctionality, and the integration of mechanical interfaces.

The basic idea of monolithical optics is to replace the traditional arrangement of refractive lenses in air by a sequence of reflective elements directly machined on a glass body. The glass body carries all elements, and thus replaces the mechanical mounts, but also guides the light propagation, normally done in air between lenses. In other words, a piece of glass is shaped in such a way that light is reflected from several surface locations to perform an imaging task. Such a properly shaped piece of glass we call a monolithic optical system.

Within the last ten years much progress has been made in producing glass monoliths for multifunctional applications (Assembly template: Sec. 16.5, "CNC-Machined Monolithic Optics," B. Reiss) [1]. These glass plates contain on both sides a reflecting plane, spherical or aspherical surfaces, directly machined and polished into the glass slab. The highly integrated glass component can be a complete optical $4f$-system with access to the pupil in the slab center from outside. In Fig. 8.5 we see as an example an angular encoder, where a rotating disc with a code pattern is placed at $2f$ in the pupil. Rotating the disc leads to a pattern shift on the CMOS sensor, which allows the determination of the unknown rotation angle. Large varieties of functionality, including a tilt [2] or tracking sensor, beam shaper, or even beam

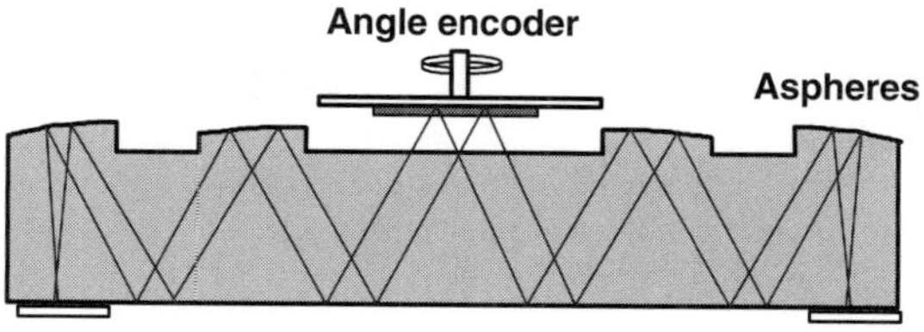

Figure 8.5 Angular encoder.

steerer have been realized. The main advantages are low costs, mechanical and thermal stability, and the possibility of machining even mechanical mounts by the same process.

When starting the project in 1998 at SwissOptic, it was not clear if glass could be treated like metal by CNC machines. Does it introduce local stress? Does it violate the glass matrix and thus facilitate the penetration of the polishing liquid into the bulk material, leading to uncontrolled oxidation of the coating after years? All these questions forced us to study the physics of surface treatment and to carefully optimize processing speed, pressure, tool shape, tool guidance, cooling, and so on; in short, to answer all the questions raised by the alternative methods discussed in Chapter 5, "Processing Technologies."

8.7 Technical Details

Details of the assembly can be found

- In Part II, Chapter 11 "Applications," Chapter 16 "Assembly"
- In Paul R. Yoder, "Opto-Mechanical Systems Design," Marcel Dekker Inc., New York and Basel, 2006.

8.8 References

1. B. Braunecker, B. Reiss et al., "Production of Monolithical Optics," 4. International IWF Colloquium 1998, Institut für Werkzeugmaschinen und Fertigung, Technical University Zurich, p. 101.
2. B. Braunecker, J.R. Rogers, B. Gächter, "Optical sensor for determining the angle of inclination," Patent WO 97/45701 A1/EP0901 608 B1.

Chapter 9

Future Trends

9.1 Introduction

"...Prognoses are difficult, particularly if they concern the future..."

Karl Valentin (1882–1948); Bavarian poet and pessimist

9.2 Preliminary Remarks

We mentioned in the Sec. 1.1 the need to distinguish between two characteristic market segments: the mass market of consumer products dominated by Asian companies and the market for a smaller series of higher quality products made in Europe and the United States. In the case of consumer products, very specific process solutions are developed that prevent a profound understanding of the basic technologies. We therefore focus in this compendium on products for the latter market segment, where a large number of application fields exist. This leads to many optical variants and in consequence, to many different individually optimized process solutions. However, it allows us to get a deep insight into modern production methods and to understand the motivation drivers behind the optical technologies involved. Thus, in short, the market is divided as follows:

- *Asia*: high volume, low cost, just good enough quality, small product range, large number of product variants, very special equipment and technology.
- *Europe* and *USA*: small series, high price, and best possible quality, wide product range, highly flexible equipment (for economical reasons), wide range of technologies, high grade of automation.

Requirements for high quality and highly flexible equipment often lead to an intrinsic discrepancy: expenses for investment grow rapidly and sometimes limit the best possible result in quality and price. On the other hand, optical workshops need a certain variety of product range for sufficient utilization.

Another area of discrepancy lies in the conflict between flexibility and automation. Often the automation of processes leads to restrictions in flexibility.

In conclusion, a clear focus on the product range, good communication with the customer, close cooperation with the equipment manufacturer, and reliable control over the technologies are the keys to success. The comments in the Secs. 9.3 to 9.7 are extracts we draw from the authors' contributions, from contacts and interviews with them, but also from our personal knowledge and experience.

9.3 Applications

The widespread use of aspheres is best illustrated by the table in Sec. 3.7, typical for European and U.S. industries. From the corresponding contributions in Chapter 11, we note a large diversity of appropriate processing methods to optimally produce the particular asphere under consideration. We also saw that, in the majority of application cases, aspheres will only be accepted if their production costs justify the advantage we gain. We heard a factor of 2 as a challenging target, that an asphere could cost at a maximum, over an equivalent spherical lens.

Considering the contributions in Part II, we note that all authors are quite optimistic about the progress in producing aspheres. The reasons to integrate aspheres are manifold and mentioned several times, but clearly the better optical performance and the reduced number of elements are the most frequently named drivers.

9.4 Materials

Today, one focus of material development activities concerns "low-T_g" glasses, which are used with precision-molding technologies. The main driver is the consumer market for optical devices, for example, digital still cameras or cell phone cameras, which need high optical performance for high lens production volumes at reasonable production costs. With aspherical low-T_g-glass lenses, these demands can be fulfilled. The overall number of lenses in an optical system can be reduced while keeping the same optical performance and reducing the lens system size, and at the same time enabling the low-cost mass production process of precision molding to be used. In the near future, for all optical positions of standard glasses in the Abbe diagram, low-T_g glasses will be available.

Other future developments of optical materials will try to overcome the other disadvantages of the existing materials. Looking into the Abbe diagram, there are trends to extend the different material regions. For instance, polymers will try to enter the glass area through the development of new material systems or by modification of existing systems with nanoparticles (Fig. 9.1).

Efforts are being undertaken to generate glasses within the crystal n_d–v_d area. But there are clear limitations set by the crystallization stability. Only in combination with new melting and hot-forming techniques, which might combine high melting temperatures with fast cooling (e.g., noncontact laser melting), an extension of the glass area to higher refractive indices and lower dispersions is achievable. Direct pressing of aspheres from the melt could open a way to produce low-cost aspheres using non-low-T_g glasses.

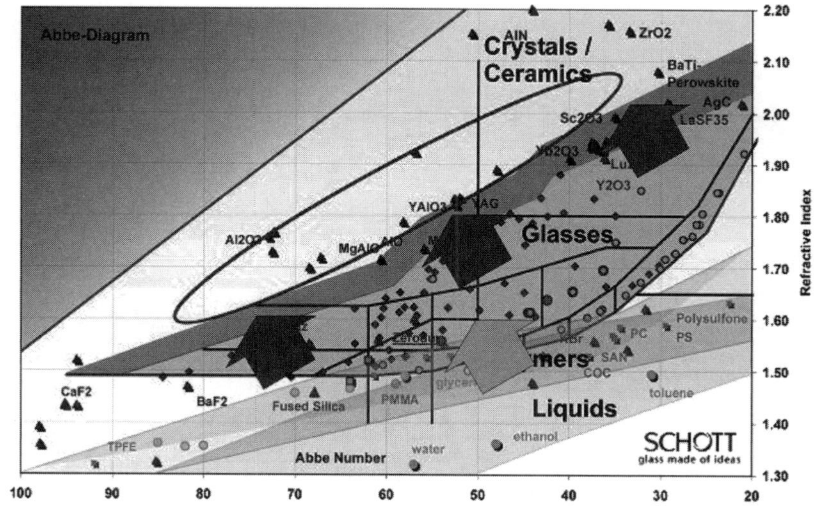

Figure 9.1 Development trends, n_d–v_d.

Crystals are clearly limited in their applications because of cost issues in primary production as well as on the secondary processing side. General new approaches with respect to material systems and production technologies, which have an additional impact on aspherical lenses, cannot be foreseen at the moment. However, a new class of optical materials, "polycrystalline ceramics," have emerged over recent years. Optoceramics will try to offer optical positions in the crystal field at lower costs. The material production costs for these typical sintering processes can be significantly lower when compared with crystal growth. The reasons are lower process temperatures as well as the possibility for near net shape forming. Nevertheless, the high grinding and polishing costs will remain unaltered. The gain of optical performance must be weighed with the higher overall costs and compared to those of glasses and polymers.

With respect to polychromatic applications, it will be of the highest importance to have material pairs with the "right" anormal partial dispersion available to achieve apochromates. Two target areas, as marked in Fig. 9.2, are of great interest and can be potentially addressed by glasses or ceramics. In these areas today, no materials are commercially available, but experimental results have already shown the feasibility for achieving these optical positions.

Development trends can also be seen in the direction of polycrystalline ceramics with high transmission. Here, YAG, Y_2O_3, AlON, and Spinel ceramics are under development. The original drivers of development are active laser applications and windows with high thermal and mechanical stability.

Apart from higher refractive indices at lower dispersions values and favorable anormal partial dispersion values, the high UV and IR transmissions of optoceramics are interesting. This stimulates R&D work for 193-nm microlithography or for military IR applications.

In the IR area, significant development has been made in developing IR transmission glasses or glass ceramics that can be used in a precision molding

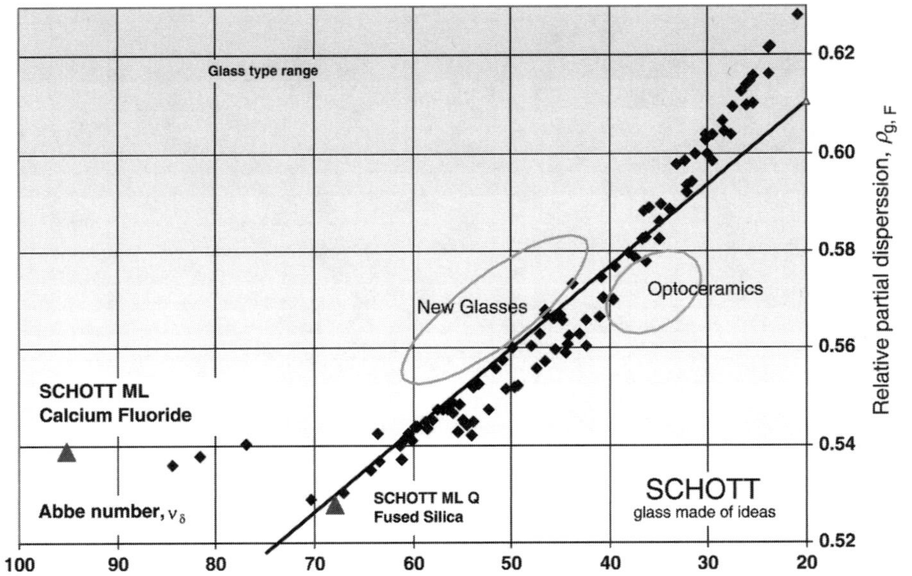

Figure 9.2 Overview of relative partial dispersion.

process. In the future, materials will be developed that will be processed with more cost-effective grinding and polishing technologies even in high-quality applications.

9.5 Processing Technologies and Metrology

High-accuracy processing requires a strong integration of metrology methods. Therefore, it is appropriate to combine future trends for processing technologies and metrology.

It has been shown that aspherical surfaces would be used more frequently if their manufacturing costs could be reduced to about twice the costs of a spherical surface. We are still far off this. The cost factor is very important in general, but it is critical for mass consumer products like camera objectives and optics for projectors. The application of novel aspherical elements as well as new approaches in design and production technology and developments in in-process metrology will help to reduce cost.

Aspheres will be needed in many new prospering business fields, for example in astronomy or in lithography (today and for the next generation of objectives, working at a wavelength of 13 nm). In both cases, reflective systems with aspherical components are under development.

The economical fabrication of aspheres of high and medium quality was and will be the main technology driver. Two main trends are identified. Replication techniques in plastic materials will allow consumer optics to be produced in batch sizes larger than 10^4. The trend is moving towards multifunctional optical elements with an aspherical first surface and a diffractive element on the other. Additionally, the fabricated elements will include a complete mounting structure to allow them to be fixed by clips in the optical device.

Concerning precise glass replicas, we note trends towards new materials with a larger variety of refractive indices and the development of more sophisticated tools.

The classical fabrication process of generating and polishing will be used for batch sizes smaller than 10^4. For the future, we expect the three-step process—generating surface, smoothing/polishing, and local correction—still to be valid. For local correction, CCP and MRF technologies are well established. For higher requirements, IBF is employed. All future developments for local correction methods must exceed this benchmark. Typical patterns (mm range, see Fig. 2.7) are often generated by the actual equipment. A promising approach is an overlay of a statistical movement (noise) on the deterministic tool motion. Investigations of the replacement of hydraulic movement by a direct electrical drive look successful and will continue.

The trend is moving towards deterministic grinding with low subsurface damage and an *in situ* metrology to shorten process time. For polishing, the trend is towards fast CCP processes that allow reproducible process conditions. For the far future, a further development might not be unrealistic: the CCP process in the future has to remove subsurface damage as well as correct the final surface figure to specification without a third process step. Development will go on to develop tools with good long-term stability. The main focus will be on the description of the tool function, for example, with an amplitude–wavelength plot and an algorithm that allows us to improve shape and roughness in one step.

There is still a lack of superpolishing technologies (<0.2 nm) beyond pitch polishing. Pitch polishing is more a global polishing method and is not really appropriate for local correction polishing.

9.5.1 Integrated process–metrology

One of the most prominent cost drivers for nearly all production technologies today is the insufficient integration of metrology in the manufacturing process. Measurements of nanometres need special precautions: granite slabs, constant temperature and humidity, no vibrations, clean surfaces and especially an external reference system to properly align the asphere under test. Optical methods in particular need uncontaminated surfaces to achieve accuracy. To integrate cleaning steps into the workflow without introducing new error sources is rather difficult. It is the goal of many activities to control the actual surface form permanently during the shaping process. Then, processing time and costs are expected to drop significantly.

For the grinding process, roughness and form can be measured by stylus instruments. Alternatively, confocal principal and even white light interferometry can be used in the future. Pointwise measurement principles need to be integrated into the form-generating equipment.

For form measurement of polished surfaces, interferometry is the preferred method today. Interferometry is not yet easily integrable into the processing device.

Therefore, measurement principles could be adapted to polished surfaces, although the measurement process is slow because scanning is needed.

9.5.2 Null optics

Despite the progress in in-line measurements, an off-line inspection is still necessary. The advantage of in-line metrology improvements will reduce the frequency of off-line measurements.

We mentioned earlier that polished aspherical surfaces when measured by interferometers need a Null optics as a wavefront compensator. Computer-generated holograms (CGH) are ideally suited but still too time consuming and too expensive. Furthermore, their integration together with the asphere under test in the interferometer is difficult but is relaxed by use of additionally introduced holograms for positioning and centering. Furthermore, calibration is more difficult. It was shown recently that the calibration of aspherical surfaces requires seven measurements to avoid symmetric and antisymmetric errors (for spherical surfaces only three measurements are required). For centering and calibration, a multihologram technique will be used in the future.

Significantly improved short-term availability and price reduction will enhance the application of CGHs.

9.5.3 Alternative metrology methods

Null lens techniques used for testing aspherical surfaces by interferometry are well established, but too expensive for a low number of elements or prototypes. An alternative method is "stitching." Stitching integrates partially recorded, overlapping interferograms. This method is time consuming and requires, in general, very precise mechanical movement (lateral, angular, or along the optical axis) of the inspected lens or measurement setup. An overlap of a minimum of 30% is usually required. Highly precise defocusing allows analysis of different fringe patterns consecutively.

Methods using flexible adaptive optics are making progress in development and could find future implementation to measure aspherical optical elements in production.

To reduce the sensitivity of interferometric testing, a wavelength in the infrared such as $\lambda = 10.6\,\mu\text{m}$ can be used, but the resolution is also reduced, and the detector arrays are not as appropriate as for visible light. Therefore, a multiwavelength technique is another attractive method. By coherent mixing of several wavelengths, a larger artificial wavelength is synthesized to reduce the number of fringes, to increase their contrast, and to avoid ambiguity problems. Accurate measurement needs to be carried out with a single wavelength.

9.5.4 Hybrid technologies

In the future, hybrid techniques will gain in importance. Thin layers of photoresist or polymer on spherical glass surfaces shape spherical lenses to aspheres. Furthermore, simultaneously added diffractive structures on the surface allow correction of chromatic and thermal aberrations. Stray light could limit its application. The advantages are low cost and good thermal stability due to the glass body.

9.5.5 Adaptive systems

For astronomical telescopes, deformable aspherical mirrors are already used, and some new principles are under development. Their purpose is to eliminate residual system aberrations when moving the telescopes or to dynamically compensate image blurring by air turbulence. Closed-loop systems with wavefront sensors adjust the flexible mirror surface to compensate the measured wavefront aberrations. Recently, thin membrane mirrors have been introduced. A reflecting, only 1-μm-thick Si-nitride membrane, coated with Al, is stretched over an array of electrodes. Its bending is controlled by applying an electric field between the membrane and electrodes. The largest diameter at present is about 50 mm. More applications are thinkable if the slope of the surface deformation could be increased. Its use in interferometers to test aspherical surfaces is important.

9.5.6 Free-form surfaces

Free-form aspheres will attract more attention to squeeze systematic wavefront errors down to zero or allow components with larger tolerance values while keeping the specifications. As an example we will show a compensator plate in the space telescope of Leica/Contraves (Sec. 11.8). To fabricate such an element with reduced symmetry makes in-line metrology necessary [1].

9.5.7 Liquid lenses

The trend to miniaturize optical systems will continue. If optical systems need variable magnifications (zoom optics), then new nonmechanical means to change the focal length must be developed. New adaptive optics like liquid crystals or liquid lenses are mature enough to be used in mobile phones for automatic focusing and soon for zooming. VariOptics™ lenses are plastic devices containing two liquids. One liquid is based on a water-soluble formulation, and the other is oil-based. The nonpolar oil layer is in contact with a positive electrode, and the polar water-based solution is negatively biased. The curved interface between the oil and the water layers acts as a lens, with variable curvature depending on the voltage applied. The speed of the variation is already of the order of 100 ms. These new technologies may therefore overcome some of the problems we have in adaptive optics and optical metrology for testing aspherical surfaces.

9.5.8 Simulation and modeling

The general trend will be to use computer power for simulation of the process within the machine before performing any hardware job, so that an operator without any background can produce precision aspheres within a few days.

Since the earliest times, mankind knew that rigid bodies of concave and convex spherical shape are generated by pressing two plane plates at each other and performing extended but random movements. This leads, when averaged over time, to a constant mass ablation all over the body, which is a self-consistent and self-stabilizing process. Hobby astronomers grind and polish their high-quality mirrors even today by this archaic but proved method.

However, to make aspherical or even free-form surfaces, one has to locally modify the mass ablation rate. The problem is the continuation of the local operation over the whole surface. The limited precision of the tooling (for example, their finite kinematic and dynamic accuracy for guidance) leads to surface errors. One can easily show that correcting surface errors with the same tool favors only artifacts of a higher spatial frequency. This leads finally to the famous "orange skin," a phenomenon feared by lens makers.

It was emphasized in Chapter 5 "Processing" that the entirety of all involved parameters and their physical interaction must be understood. The Preston equation in its consequence (Sec. 5.1.3) describes the dependence of the ablation rate on load pressure, viscosity, tool motion, lens temperature, cooling rate, glass heat conductivity, capacity, and so on. Obviously, we need to determine the right amount of mass per local area per time to control all parameters with the necessary precision.

We conclude that a flexible, reliable, and error-minimized production of non-spherical surface shapes needs a clear understanding and modeling of the ablation physics together with high-end technology tools for the ablation process. Strong computing power is required to analyze the measurements of the actual surface shape, to predict corrections, and to guide the process.

While we have focused so far on mass removal, the same holds for the opposite, the thin layer coating, which is a mass-adding process of comparable complexity. We see the following similarity between both processes. Reliability and cost-efficiency require that "analysis, prediction, and guidance" must run in real time within the *same reference coordinate system*, consequently on the same machine. Although this is already the case for coating (obviously driven by the need to operate in vacuum chambers), more efforts have to be undertaken regarding the surface-shaping processes. In particular, the in-line integration of metrology may still be improved.

The dramatic progress of numerical controlled machines concerning reliability, precision, and cost efficiency will set the trend in optics. We expect in the future an "all in one" process (Fig. 9.3). The goal is to perform a closed loop of polishing and shape measurement of the lens under test within the same coordinate system. This

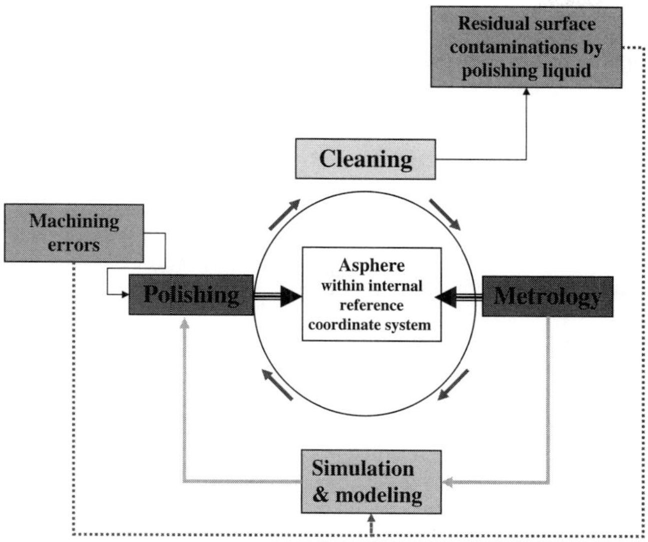

Figure 9.3 Computer-controlled interaction of integrated processing technology and metrology.

needs to detect all systematical and stochastical errors, like mechanical axial jitters, but also the remaining surface contaminations left by the polishing liquid after the integrated cleaning step. Only then is a reliable modeling and, in consequence, a trustable preadjustment of the polishing tool possible.

We also expect this development on the *system level*, including assembly. This could lead to a stronger use of monoliths, where optical multifunctionality is laid down in one physical item, containing all mechanical interfaces for easy build-in in cameras, and so on.

Thus, our final and provoking question: Will the single lens of Sec. 2.1.1 be replaced in the future by a glass monolithical *system* with intelligent shaped and coated boundaries? We think quite soon, at least for many sensoric applications.

9.6 Coating Technologies

- New coatings and coating technologies are permanently driven by new developments in optics.
- Classical and energetic evaporation will remain the most common deposition technique.
- Sputtering becomes more important for optics on larger and flat areas.
- Advanced process control and monitoring strategies will allow low-cost mass production of classical and complex inhomogeneous coatings.
- Virtual deposition will reduce coating costs and increase coating performance.

- Optical coatings, like 1D photonic crystal, will spread into two and three dimensions by merging with microstructuring techniques.
- Basic understanding of coating formation as a self-organized atom-by-atom process at the picometer level, following the bottom-up principle in nature, will allow shaping multifunctional properties.

Both photons and electrons show a completely analogous behavior when subject to a periodic potential. Photonic crystals, at first as 1D optical coatings, will meet electronic crystals to form new optoelectronic devices.

9.7 Assembly

9.7.1 Automatization

We have worked out in Chapter 8 the true meaning of assembly, that is, to introduce compensators. They allow us to relax considerably the fabrication tolerances of the optical elements. This leads obviously to cost reductions; but on the other hand, it causes new labor costs, a classical conflict! Because demand for better optical performance at lower costs will grow in the future, more efficient assembly methods must be studied. We hear from many companies that fully automatic robot mounting, performed under computer control and actively guided by wavefront sensors, is considered to be the most promising method, especially when aspheres are involved.

9.7.2 Cements and glues

Another question to be answered by Assembly is how to maintain the position and orientation of any optical element over the whole lifetime of the instrument? What progress concerning cementing and glueing can be expected?

Cementing of two lenses as in an achromat has to ensure the correct optical transmittance and mechanical stability over time. Ageing of the cement should not introduce any mechanical stress on either glass bodies. New cements will improve elasticity and transmittance, especially in the UV region, and minimize outgassing. Glueing describes the fixation of a lens or lens group in the mechanical structure. It has to ensure the mechanical position and orientation over time. Additionally, the elasticity of the glue has to protect the lens element from mechanical shocks and vibrations and should not be altered by ageing.

The chemical consistency of glues and cements, their handling, and their pre-aging are part of company proprietary knowledge, based on long experience. Also, one of the most-important aspects of traditional optics (and thus for aspheres) is determining how the gluing "spots" are defined to avoid any mechanical stress on the lens body while keeping the position accuracy.

9.7.3 Flexures

In the future, we also see trends that combine glueing with elastic mechanical structures, called flexures. Flexures are masterpieces of mechanical engineering; they can be designed and machined to keep a lens in an exact orientation and lateral position, but they move the lens along the optical axis if mechanical shocks occur. The lens returns to the old position with a high degree of accuracy. We know of rather heavy lenses (some kg) for UV wafer inspection, which are moved along the optical axis, about some microns range at the rather high frequency of 20 Hz for a focusing scan. The lens is held by flexures that keep lateral decentering below 50 nm.

9.7.4 Complete processes

We expect an extension of the new technologies, like the robot assembly method of TRIMO from sensorics to endoscopy, but also to systems with components of a larger diameter. The same holds for monolithic elements, especially if several monoliths are moved for zooming and focusing.

9.7.5 Monolithic optics

Monoliths carry lenses and mirrors. Any postcontrol of their position and orientation is irrelevant because it is unchangeable. However, in cases where the optical quality is out of spec and one has to identify the location of wavefront degradation, a tomographic method using a heterodyne technique was proposed and will be implemented in the future [1].

Centering, mounting, and adjustment of aspherical elements require more technological efforts. New mounting procedures need to be studied. Robot mounting, with its inherent potential to miniaturize systems, but maintain accuracy, operating with speed and achieving large flexibility, will be implemented to a higher degree. The TRIMO assembly method (Sec. 16.4), optimized for micro-optics in photonic applications, illustrates this trend.

9.8 Reference

1. B. Braunecker, B. Reiss et al., "Production of Monolithical Optics," 4. International IWF Colloquium 1998, Institut für Werkzeugmaschinen und Fertigung, Technical University Zurich, p. 101.

Chapter 10

Mathematical Formulation

To create aspherical surfaces, a common approach uses polynomials:

$$z(x, y) = (A_1x^2 + A_2x^4 + A_3x^6 + \cdots) + (B_1y^2 + B_2y^4 + B_3y^6 + \cdots)$$
$$+ (C_1x^1 + C_2x^3 + C_3x^5 + \cdots) + (D_1y^1 + D_2y^3 + D_3y^5 + \cdots).$$

Linear terms, which represent a tilt of the basic surface, as well as terms with odd exponents, are usually omitted in practice.

10.1 Surfaces of Second-Order (Quadrics)

The quadratic terms $z(x,y) = A_1x^2 + B_1y^2$, taken for themselves, are contained in the surfaces of second order. They are also called quadrics in the 3D space (Quadric 3D) and have some importance for optics. These surfaces are aspherics themselves and therefore will be used as a basis for aspherics. Note that the sphere is also a surface of second order.

They have the property that the intersection of any plane with them creates curves, which are called cone sections: ellipses, parabolas, hyperbolas, straight lines, and points. Straight lines and points are called degenerate cone sections in this context.

The surfaces of second order are generally determined through the equation

$$F(x, y, z) = a_{11} + 2a_{12}x + 2a_{13}y + 2a_{14}z + 2a_{23}xy + 2a_{24}xz + 2a_{34}yz$$
$$+ a_{22}x^2 + a_{33}y^2 + a_{44}z^2 = 0,$$

as result of the multiplication of a vector $\mathbf{u}^T = (1, x, y, z)$ with a symmetrical 4×4 matrix $\mathbf{A}$, that is,

$$F(x, y, z) = \mathbf{u}^T\mathbf{A}\mathbf{u} = 0,$$

where $\mathbf{A}$ is the general matrix of the surfaces of second order.

The main axis transformation of the general matrix **A** leads to a matrix **H**, in which all elements are zero except the elements on the main diagonal h_{11}, h_{22}, h_{33}, h_{44}, and the element h_{14}.

By introducing the following conditions,

- h_{11} is either 0 or -1
- h_{14} is either 0 or 1
- if $h_{14} = 1$, then $h_{44} = 0$ and $h_{11} = 0$
- if $h_{44} = 0$, then $h_{22} \leq h_{33}$, else $h_{22} \leq h_{33}$ and $h_{33} \leq h_{44}$,

the surfaces of second order are definitely located in a Cartesian coordinate system in the Euclidean space. The matrix $\boldsymbol{H}$ can therefore be called the generating matrix of surfaces of second order.

By introducing the geometrically significant semiaxes a, b, and c, the matrix elements can be written:

$$h_{22} = 1/a^2, \qquad h_{33} = 1/b^2, \qquad h_{44} = 1/c^2,$$

where a, b are either real or imaginary, and c is always real. The product

$$\mathrm{N}(x, y, z) = \mathbf{u}^{\mathrm{T}}\mathbf{H}\mathbf{u} = h_{22}x^2 + h_{33}y^2 + h_{44}z^2 + 2h_{14}z + h_{11} = 0$$

leads to the normal forms of second order surfaces;

$$\text{Case 1: } \mathrm{N1}(x, y, z) = x^2/a^2 + y^2/b^2 + z^2/c^2 = 0,$$

that is, h_{11} and $h_{14} = 0$ leads to second order conical surfaces;

$$\text{Case 2: } \mathrm{N2}(x, y, z) = x^2/a^2 + y^2/b^2 + z^2/c^2 - 1 = 0,$$

that is, $h_{11} = -1$ and $h_{14} = 0$ leads to second order centered surfaces;

$$\text{Case 3: } \mathrm{N3}(x, y, z) = x^2/a^2 + y^2/b^2 + 2z = 0,$$

that is, $h_{14} = 1$ and $h_{44} = 0$ leads to second order parabolic surfaces.

10.2 Basic Equation by ISO 10110—Part 12

In ISO 10110—Part 12, the basic equation for surfaces of second order is

$$z = f(x, y) = \frac{\dfrac{x^2}{R_x} + \dfrac{y^2}{R_y}}{1 + \sqrt{1 - (1 + \kappa_x)\left(\dfrac{x}{R_x}\right)^2 - (1 + \kappa_y)\left(\dfrac{y}{R_y}\right)^2}}.$$

It is derived from the centered surfaces of second order by the following procedure.

1. Transform into the vertex

$$F(x, y, z) = \mathbf{u}^{\mathrm{T}}\mathbf{T}^{\mathrm{T}}\mathbf{A}\mathbf{T}\mathbf{u} = 0, \qquad \frac{x^2}{a^2} + \frac{y^2}{b^2} + \frac{(z-c)^2}{c^2} - 1 = 0,$$

   using the T matrix

$$T = \begin{bmatrix} 1 & 0 & 0 & 0 \\ t_x & 1 & 0 & 0 \\ t_y & 0 & 1 & 0 \\ t_z & 0 & 0 & 1 \end{bmatrix}$$

   with $t_z = -c$ and $t_x, t_y = 0$.

2. Rearrange for z

$$\frac{z}{c} = 1 - \sqrt{1 - \left(\frac{x^2}{a^2} + \frac{y^2}{b^2}\right)}.$$

   The positive sign of the square root is omitted, as it is meaningless in optics. Multiplying both sides by the term $1 + \sqrt{}$ immediately leads to

$$z = c\frac{\dfrac{x^2}{a^2} + \dfrac{y^2}{b^2}}{1 + \sqrt{1 - \dfrac{x^2}{a^2} - \dfrac{y^2}{b^2}}}.$$

3. Introduce the vertex radii R_x, R_y

$$\frac{a^2}{c} = R_x, \qquad \frac{b^2}{c} = R_y.$$

4. Introduce the conical constants κ_x, κ_y

$$\kappa_x = \frac{a^2}{c^2} - 1, \qquad \kappa_y = \frac{b^2}{c^2} - 1.$$

5. Replace the semiaxes expressions by the vertex radii and conical constants, which results in the equation given in ISO 10110—Part 12 (see above).

The derivation from the parabolic surfaces of second order is somewhat different, because the semiaxis c is missing.

1. Because the parabolic surfaces are already in the vertex position, immediately rearrange for z:

$$\frac{x^2}{a^2} + \frac{y^2}{b^2} + 2z = 0, \qquad z = -\frac{1}{2}\left(\frac{x^2}{a^2} + \frac{y^2}{b^2}\right).$$

2. Introduce the vertex radii R_x, R_y,

$$-a^2 = R_x, \qquad -b^2 = R_y.$$

3. Introduce the conical constants κ_x, κ_y,

$$\kappa_x = \frac{1}{a^2}, \qquad \kappa_y = \frac{1}{b^2}.$$

4. Replace the semiaxes expressions by the vertex radii and conical constants.

To our knowledge, a derivation from the conical surfaces of second order, as stated in ISO 10110—Part 12, is not possible.

10.2.1 Modifications

The basic equation may be modified by adding a polynomial of higher order to the polynomial mentioned previously (Sec. 2.2.2.1).

Part II

Experts' Contributions

Chapter 11

Applications

11.1 Illuminations

A. Jacobsen

11.1.1 Digital projectors and rear-projection TVs

Light output and light distribution over a screen was already a very important issue for slide and film projectors when it became even more critical with the emergence of digital projectors and rear-projection TVs over the past 15 years. Such projectors use liquid crystal displays (LCDs) or Texas Instruments digital micromirror devices (DMDs) to modulate image information onto the light beam, which is then finally projected to a screen. Especially during the early times, transmissivity of the displays was very low. Also, the filaments of halogen lamps or the arcs of metal–halide discharge lamps were rather bulky and did not match the dimensions and accepted apertures of the displays. In addition, spectral uniformity over the field became an issue, especially in industrial applications like business presentations and process visualization. Special measures were implemented to efficiently collect the emitted radiation from the light source and to homogenize spectral and brightness uniformity over the screen. For color images, complex optical relay systems were established to efficiently split and recombine the light into three color channels: red, green, and blue (Fig. 11.1).

New light sources with significantly smaller arc gaps and higher luminous efficacy in conjunction with the ongoing introduction of smaller imager displays enabled the development of more compact projectors (Fig. 11.2) with continuously higher light output and lower power consumption. While maintaining brightness and spectral performance, miniaturization and cost optimization became a driving force for optical system design. This allows the introduction of front and rear projection engines for large-screen consumer projection TVs, but again it reinforces the need for more compact and cost-efficient optical systems. Recent advances in light emitting diode (LED) development enable battery-powered, ultra-compact pocket projectors, which soon will be applied in consumer rear-projection TVs.

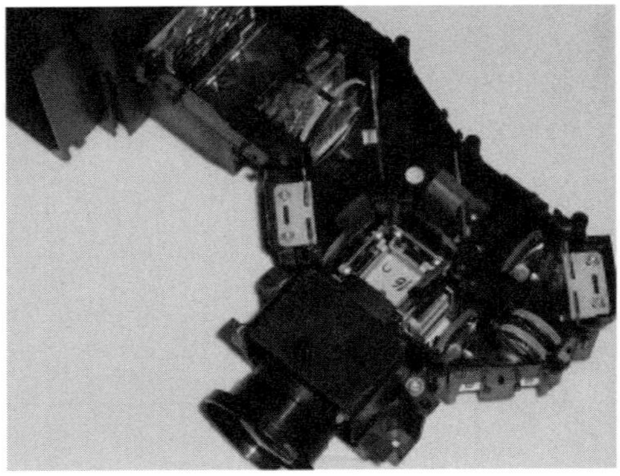

Figure 11.1 Illumination and relay system of a digital LCD projector (Source: OpSys Proj. Consulting, Schöffengrund, Germany).

Figure 11.2 Digital light processing–based compact digital projector (Source: Resolution AS, Drammen, Norway).

Low light output compared to discharge lamps and the different characteristics of LEDs will fortify design requirements for efficiency and light distribution in the illumination systems of projectors.

Application fields for digital projectors cover a wide range of products, from compact projectors for business and conference room projection, projection systems for industrial applications (Fig. 11.3), projectors for home cinema and entertainment, and rear-projection TVs to high-brightness large-venue and digital cinema projectors.

11.1.2 Automotive headlighting

During the mid-1980s, car manufacturers started to put more emphasis on the design aspects and aerodynamic shape of car front ends. Until then, car headlights used parabolic mirrors to collect light. Structured cover glasses then spread the light onto the road. Flat front ends and space restrictions led to the invention of projection headlights using PE (polyellipsoid) reflectors to collect the light and to project and distribute it onto the street with multifunctional single-element projection lenses. Introduced by BMW, this technology made its way into the cars of

Figure 11.3 Control room multiscreen projection (Source: Barco Control Rooms GmbH, Karlsruhe, Germany).

Figure 11.4 Aspheres in a PE Headlight with clear cover glass (Source: Docter Optics GmbH, Neustadt, Orla, Germany).

almost all manufacturers, including compact and luxury cars, trucks, and buses. Implementation of high-intensity discharge (HID) lamps increased demands to accurately image and distribute the light onto the street within applicable safety standards. Like no other industry, car manufacturers drive technology developments into commercial products but also push costs down towards its limit. Latest developments include Super PE projection systems with all-clear cover glasses (Fig. 11.4) and adaptive headlight systems that automatically orient themselves to the prevailing road and visibility conditions (Fig. 11.5).

11.1.3 Optical systems

11.1.3.1 Illumination and relay systems

Figure 11.1 shows the typical illumination system layout of a 3-Chip LCD projector. Light from the discharge lamp arc is collected by an ellipsoidal mirror

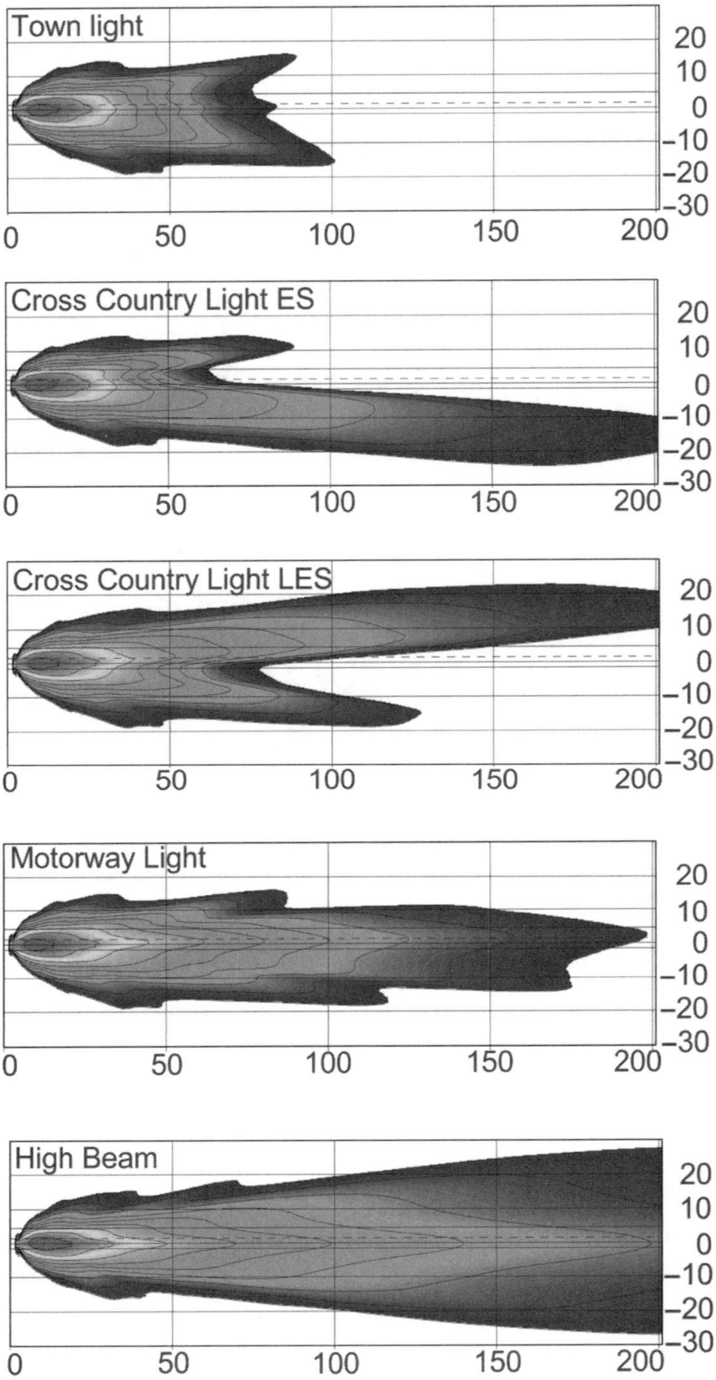

Figure 11.5 Light distributions of Hella's Variox module (Source: Hella KgaA Hueck & Co., Lippstadt, Germany).

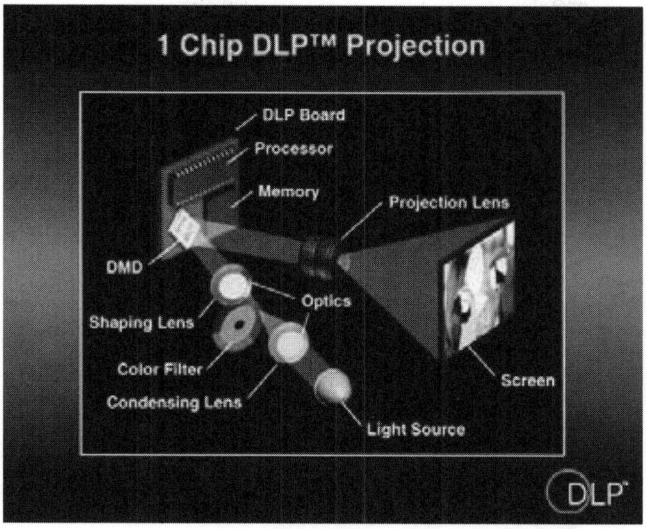

Figure 11.6 Basic layout of a single-chip DLP® projector (Source: Texas Instruments Inc., Plano, USA).

with a dichroic coating to prevent UV and IR from entering the optical system. Two segmented lens elements (integrator plates) differentiate the light beam from the reflector into a large number of rectangular sub-bundles. The second integrator plate, combined with further lens elements and dichroic filters, images the rectangular patterns into the LCD imagers and superimposes them to eliminate spectral and intensity inhomogeneities. In addition, the segmentation of the circular light beam allows squaring the circle to be overcome. Miniaturization of the lamp arc helps increase the light output of projectors by introducing space-saving polarization recovery elements.

In Single-chip DLP® projectors (Fig. 11.6), light is again collected by a dichroic ellipsoidal mirror (sometimes combined with a condensing lens) and is focused into the entrance of an integrator rod that spatially homogenizes the light beam. Refractive and reflective optical elements image the rod exit plane onto the DMD imager with the projection lens toward the screen. Other than in LCD projectors, color information is modulated onto the light beam with a rotating color wheel in synchronization with the DMD imager.

11.1.3.2 Headlight projection system

Figure 11.7 shows side and top views of a Super-PE projection headlight. A multi-ellipsoidal or free-formed reflector captures the light from the halogen lamp filament or the HID lamp arc. The high aperture light passes a horizontal aperture baffle and a single lens element projects the light onto the road. This single-element projection lens has to meet high levels of safety standards including light distribution on the road, accurate and soft transition over the bright–dark edge, and low

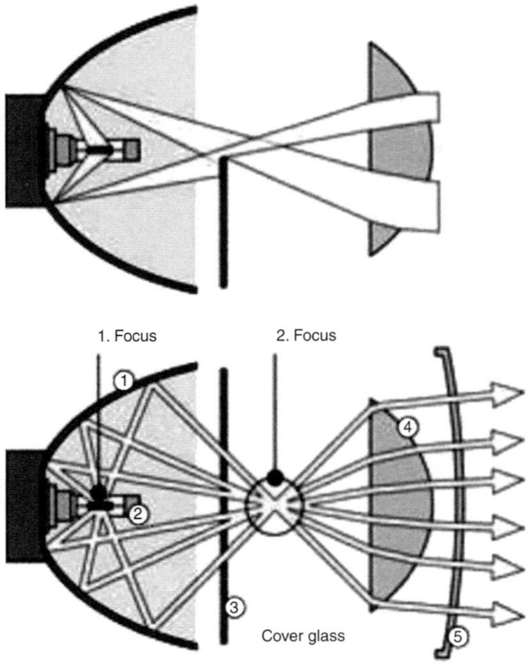

Figure 11.7 Side and top views of a Super-PE projection headlight (Source: Hella KGaA Hueck & Co., Lippstadt, Germany).

colorization along the edge of the lens elements. This is achieved by applying special surface features (modulations, textures, and free-form deformations) onto the lens surface. Movable aperture baffles and mirror elements support the features of adaptive headlight systems.

11.1.4 Design drivers and degree of asphierization

To date, brightness is one major criterion for digital projectors. The use of parabolic, ellipsoidal, and general aspherical lamp reflectors has dramatically improved light collection efficiency. Correction of spherical aberrations in condensing and relay systems helped to optimize light transport and beam shaping in these high aperture systems (F/1.0 and higher). Cost and space saving requirements have also driven designers to reduce the number of optical components while maintaining optical performance. The use of aspherical lens and mirror elements typically leads to excellent compromises. Accuracy requirements are in the range of 2–15 μm global surface deviation from calculated aspherical shapes.

Even in rather simple fog headlights, aspheres have been used as projection lens elements from scratch. Cost-saving requirements have never allowed more than one single lens element; therefore, asymmetries in light distribution, colored edges, and all other requirements listed in the safety standard were designed into and corrected with one, single lens surface. Surface modulations and free-form elements within the aspherical lens surface are implemented and require accuracies within a few micrometers local surface deviation, while the global asphere allows between 10 and 30 μm of deviation.

11.1.5 Process and performance parameter

11.1.5.1 Materials

Manufacturing technologies and cost considerations very strongly influence the selection of glass materials. Most components are manufactured from B270, a technical glass similar to K5 and supplied by Schott AG as blanks and rods suitable for efficient manufacturing processes. In some cases, borosilicate glasses with lower expansion coefficients are used to prevent damage on lens elements from thermal stress. In some very rare cases, demands for a higher refractive index have required use of F2 from Schott. Because of the use in illumination and lighting systems and due to subsequent manufacturing processes, requirements for homogeneity and purity of the glass material are not very high. Primarily, manufacturing processes require a fire-polished surface quality on the glass blanks and rods.

11.1.5.2 Manufacturing and tolerances

Digital projectors and car headlights are manufactured in tens to hundreds of thousands or even millions per type and demand very cost-efficient manufacturing processes for aspherical components (Fig. 11.8). About a handful of companies use the glass blank molding process to press heated glass blanks or rods into precise mold toolings made of steel or ceramics. Careful tempering helps to remove thermal stress in the semi-finished components and prevents thermal shrinkage. In most cases, the second surface of a lens element experiences spherical or plano grinding and polishing, but cost–performance trade-offs are applied. Typical global

Figure 11.8 Molded glass component, integrator plate for digital projector (Source: Moulded Optics GmbH, Schöffengrund, Germany).

Figure 11.9 Tactile 3D measurement (Source: In-Vision Digital Imaging GmbH, Austria).

aspherical shapes must be within an accuracy of 5 and 30 μm; smaller displays and surface modulations require even higher accuracy. Surface roughness is not highly critical, because it supports homogenization of the light beam. Other tolerances and mounting parameters scale accordingly.

11.1.5.3 Quality assurance

Supervision of molding parameters (glass and mold temperature, pressure, temperature profile during annealing) is the most important criterion for product verification. Tactile 3D measurement of molds after regular maintenance intervals and comparison with the molded components (Fig. 11.9) verify process stability. For components in digital projection, most customers request compliance with mechanical specifications to be identified with statistical means, combined with a functional test of the components in test jigs on statistical levels versus reference components ("golden" samples). Light output and distribution on screen are measured and compared. For automotive lens elements, special goniophotometers (Fig. 11.10) are used to verify light and stray light levels at defined coordinates over the field with reference to safety standards.

11.1.6 Outlook

Digital projectors and rear-projection TVs, as well as ellipsoid projection headlights in vehicles, are mass products with strong consumer dedication and have to follow very rigid cost-reduction programs.

Lamp power reduction and LED light sources in digital projectors will accelerate the trend towards plastic injection-molded components. Cost–performance trade-offs lead in the same direction but also might make spherical components be considered. With the ongoing miniaturization of imager displays, all optical components will also shrink in size. Increased accuracy of molded glass elements will

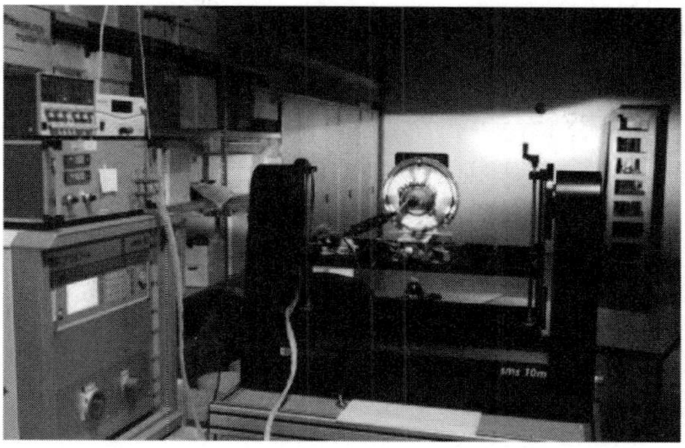

Figure 11.10 Goniophotometer for headlight lens element inspection (Source: Docter Optics GmbH, Neustadt/Orla, Germany).

allow the use of molded aspheres in the imaging lens of wide-angle projectors. This will lead to significant cost savings while maintaining performance and reducing complexity of the projection lens.

Cost improvements in ellipsoidal projection headlights are achieved by process modifications through the use of side molded aspheres, allowing a higher degree of automation. In addition, recent advances in plastic material development and development trends towards use of high-brightness LEDs in car headlights will enable new design considerations in car front ends.

11.1.7 References

1. I. Menzel and U. Rohlfing, "Unter Druck—Blankpressen komplexer optischer Komponenten," *F&M Feinwerktechnik, Mikrotechnik, Mikroelektronik*, Vol. 4, pp. 18–22 (2000).
2. U. Weichmann, et al. "Lightsources for small-etendue applications: a comparison of xenon and UHP lamps," *Proc. SPIE Int. Soc. Opt. Eng.*, Vol. 5740, pp. 13 (2005).
3. M. Duelli, et al. "Integrator rod with polarization recycling functionality," *SID Symposium Digest*, Vol. 33, pp. 1078 (2002).
4. M.H. Keuper, et al. "RGB LED Illuminator for Pocket-Sized Projectors," *SID Symposium Digest*, Vol. 35, pp. 943 (2004).
5. Dr. Ing. P. Christiani GmbH & Co. KG, Technisches Institut für Aus- und Weiterbildung, Konstanz, Germany: http://www.kfztech.de/kfztechnik/elo/licht/licht.htm.
6. H. Bauer, *Bosch Kraftfahrzeugtechnik—Lichttechnik und Scheibenreinigung am Kraftfahrzeug* (2002).
7. Hella KGaA Hueck & Co.: Technical Information—Lighting Headlamps; Lippstadt, Germany; http://www.hella.com/produktion/HellaCOM/WebSite/Misc Content/ Download/AutoIndustry/Light/LI_Headlamps.pdf.
8. Motor Vehicle Lighting Council: Adaptive Front Lighting Facts Sheet; Research Triangle Park, NC, USA; http://www.mvlc.info/pdf/AFS_Fact_Sheet.pdf.

11.2 Micro-Optic Cylindrical Aspherical Fast Axis Collimator for High Power Diode Laser

M. Forrer, D. Kura, M. Meeder, E. Langenbach

11.2.1 Application fields

Fast-axis aspherical laser collimators have found numerous applications in high-power diode laser systems. The laser radiation is used indirectly as a pump source for solid-state laser systems as well as for direct applications of the laser-diode emission in industrial processing, defense, space exploration, and medicine. High-power diode lasers will continue to economically penetrate application fields traditionally occupied by conventional laser systems with more than several Watts available from a single diode laser emitter or emitter arrays at IR and visible wavelengths. The high brightness exhibited by the diode-laser source can only be transferred to the application area by a very high-quality collimation of the emitted radiation. This starts at the very first lens after the laser facet, where the highest NA of the optical imaging system is needed.

In particular, in applications with an array of several diode laser emitters on a single substrate along a common emission facet (known as a "bar"), a cylindrical collimation system common to all emitters proves most useful. An optical imaging system of the highest quality is needed for single-mode lasers and arrays thereof. Future application fields for fast-axis aspherical laser collimators include the use of the diode laser in integrated external cavity systems for wavelength and mode stabilization.

11.2.2 Optical systems

High-power diode lasers are built from edge-emitting Fabry–Perot cavities on p–n-junctions in semiconductor crystals. The edge-emitting facets are formed by

Figure 11.11 Plano aspherical fast-axis collimators of different focal lengths.

cleaving resonator mirrors along crystal lattice surfaces and coating them. The lateral dimensions of the laser cavity are determined by the epitaxial processing of the base wafer material. The wave-guiding properties of the laser cavity provide varying resonator functions such as single-mode emitter, broad-area emitter, or amplification by trapezoid geometry. All of these laser cavities are also available in parallel structures aligned at a defined pitch in an array on a common semiconductor substrate known as a diode laser bar. They have an almost single-mode Gaussian laser emission in the axis vertical to the p–n interface, commonly known as the fast axis (a fast, spot-size increase in free space due to the larger divergence), and different degrees of mode content in the horizontal axis along the p–n interface, known as the slow axis.

The very high quality of the diode laser emission in the fast axis is advantageous in different coupling and application schemes. Collimation and imaging with rotationally symmetric optics is useful only when the spot-size ellipticity or the prevailing astigmatism is not of any importance to the application, or when enough space is available for subsequent anamorphotic magnification or correction. Cylindrical collimation and imaging optics, however, offer the advantage of simultaneous anamorphotic magnification and correction for astigmatism of the diode laser.

Due to the very small aperture of the diode laser waveguide in the fast axis, collimation in this direction is usually achieved with cylindrical lenses with an NA in the order of NA > 0.8. Optical solutions examined for this purpose are numerous. They range from spherical plano-convex lens systems with various high refractive indexes, over combined systems made from different plano-convex cylinders and hybrid lens systems with diffractive or refractive aspherical correction layers, aspherical plano-convex lenses or spherical or plano-parallel graded index lenses (Fig. 11.11).

11.2.3 Process and performance parameters

The performance of the different optical systems can be theoretically compared in terms of the residual wavefront distortion (Fig. 11.12). Each has specific advantages for certain applications. Nevertheless, regarding the requirement for optimum quality and the concurrent demand for economic production processes, the optical design of the aspherical plano-convex system is the best performing and is therefore used for collimation and imaging purposes in various applications.

The aspherical design parameters used for the second surface of the plano-convex system depend on the desired focal length. However, for actual systems, the aspherical constants up to the third order are by far sufficient to reach the diffraction limit.

The main drawback with the cylindrical collimation systems is found in their imperfect optical performance with respect to the oblique rays emitted in directions between the fast and slow axes.

Most important for high-brightness applications is an optimum surface quality and a performing broadband antireflection coating. For specific applications

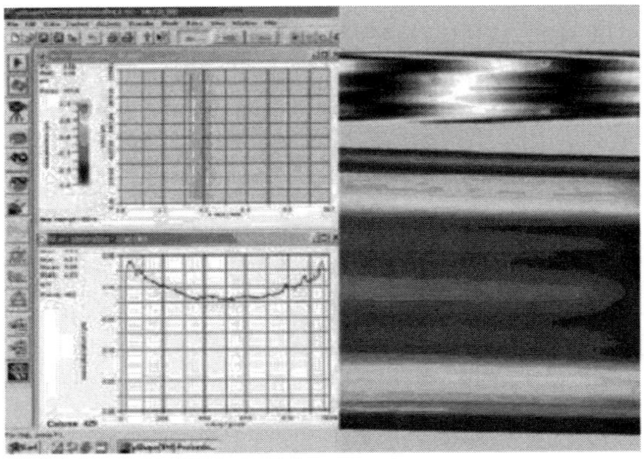

Figure 11.12 Interferometric measurement of cylindrical FAC with FISBA's μPhase® 2 and μShape™ measurement and analysis software.

targeting only one wavelength, the coating can even be optimized to reach values of less than 0.1% residual reflection loss per surface.

The focal length can be produced in a range from as low as 0.1 mm to 2.5 mm with NA > 0.8. In this range, each focal length requires a specific aspherical design (Fig. 11.13). The production technologies available for production range from etching, grinding, and polishing or wafer-scale fabrication together with appropriate coating and miniaturization processes.

11.2.4 Materials

The materials used for the fabrication of the aspherical fast-axis collimators are chosen from the available high-index glass types, with a focus on glass with a low dispersion coefficient when the lenses are to be applied with diode lasers at a spectral range in the near-IR band. Before being processed, the glass is checked for

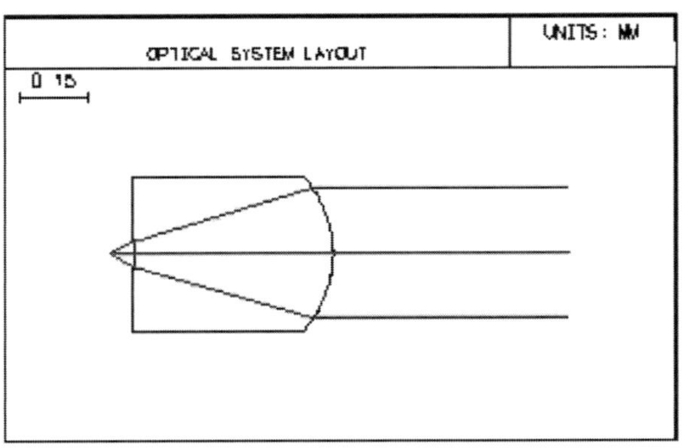

Figure 11.13 Optical design layout for a plano-aspherical FAC with $f = 300$ micrometers.

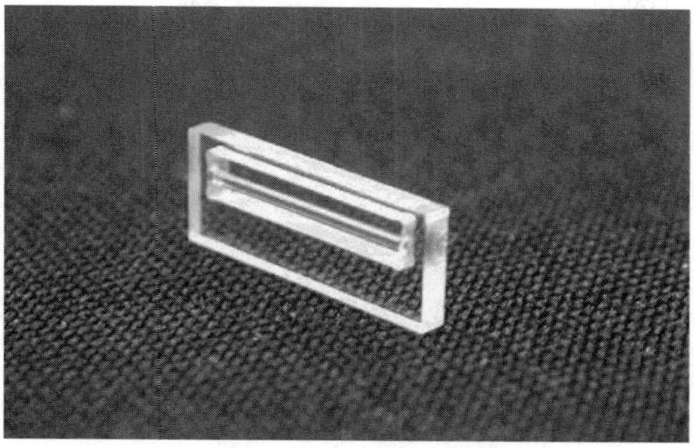

Figure 11.14 Mechanical design layout for a FAC with $f = 300$ micrometers and FAC mount (width of the holder 12 mm).

homogeneity of the refractive index. Only very high index materials ($n > 2.3$) are able to minimize the spherical aberration to acceptable values when focusing with spherical fast-axis collimators.

11.2.5 Manufacturing and tolerances

Special factors affect the relation between the effective focal length and the back focal length, which is the distance between the laser facet and the first optical surface. The back focal length should be as small as possible; assembly, however, requires a security margin of some 30–100 micrometers. Straightness tolerances of the optical focus line set special manufacturing demands in the different directions of the laser emission in the fast axis as well as in the beam direction.

It is most important, finally, that the active surfaces have to be kept absolutely free of any absorbing residuals leading to thermal heat-up and eventually to the destruction of the lens. A typical list of lens manufacturing tolerances is given in Table 11.1.

Table 11.1 Typical manufacturing tolerances for a plano-aspherical cylindrical laser collimator.

	Value	Typical tolerance	Unit	Remark
Effective focal length	0.1–2.5	0.005–0.05	mm	EFL
Residual divergence	<1.5	0.1	mrad	@EFL = 1.0 mm
Numerical aperture	>0.8	0.05	NA	
Antireflection coating transmission	>98%	0.5%		780–1000 nm
Back focal length	0.01–0.15	0.05	mm	

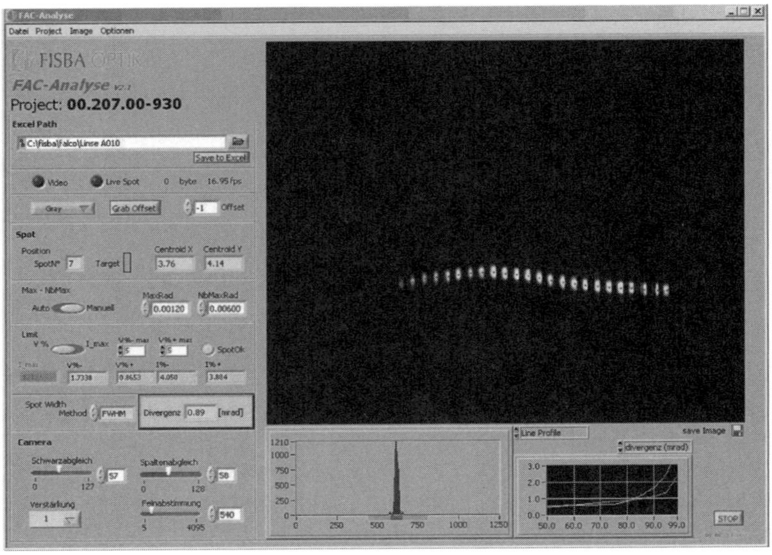

Figure 11.15 FAC performance evaluation in imaging mode with a high-power diode laser. The imaging is realized with an anamorphotic system imaging each individual emitter of a diode array in collimation position of the FAC. The gray-scale image is evaluated for the vertical distortion of the combined FAC and diode laser (SMILE), as well as for the individual imaging quality of every emitter. The lower graph insert shows the vertical intensity distribution of a single emitter.

The handling of micro-optic cylindrical lenses is critical for collimators with a focal length below ~0.5 mm. This is typically solved with specific mounts and assembly structures also used for joining the lenses with the diode lasers (Fig. 11.14).

11.2.6 Quality control

To verify performance, the lenses can be measured interferometrically with FISBA's μPhase® 2 Twyman–Green digital interferometer. The mega-pixel resolution of this CCD-based evaluation allows for full qualification along the cylindrical axis in a single interferometric measurement (Fig. 11.12).

Additional direct testing concerns the optical performance in cooperation with diode laser emission. In these tests, either the residual power in the imaged pedestals or transmission through variable slit widths is utilized for a qualification of the performance (Fig. 11.15).

11.2.7 Comments and outlook

The future driver of FAC performance is the availability of low-cost and optimized standard designs for specific industrial applications. The residual wavefront distortion as well as the transmission properties can still be optimized. Further development trends are given with a requirement for shorter focal length as well as the integration of specific and optimized assembly structures that will allow for processes adapted to low-cost industrial assembly of micro-optic components and high-power diode lasers.

11.2.8 Reference

P. Loosen and H. Opower, U. Brauch, "High-Power Diode Lasers for Direct Applications," in *High Power Diode Lasers*, Diehl, R. (ed.), Springer Verlag, Berlin (2000).

11.3 Photo-Optics

C. Horneber, P. Karbe

11.3.1 Application fields

Aspherical lenses are often used in camera lenses. These photographic lenses cover a large field of application. They are used for journalistic reports, sport photography, as well as portrait, animal, and nature photography. All these applications demand quite different lenses. For this reason, the range from wide-angle lenses to telephoto lenses is available, varying from small and compact sizes to large and high-aperture lenses. The application of aspherical lenses enables the manufacturers to satisfy the increasingly stringent requirements.

11.3.2 Optical systems

In this section, two photo lenses for different application areas are depicted. The LEICA Summilux-M 1:1.4/50 mm ASPH® (Fig. 11.16) already has extraordinary imaging performance at full aperture. In this camera lens, a molded aspherical lens is used. One domain of this lens is "available light" photography, where the photographer does not want to use a flash light, even in difficult light situations. Using aspherical lenses, it is possible to combine a compact design and high imaging quality.

A typical lens for universal purposes is the LEICA Vario-Elmarit-R 1:2. 8–4.5/28–90 mm ASPH® (Fig. 11.17), which is characterized by a high-speed aperture and very good imaging quality throughout the entire zoom range. In this lens, one molded and one polished aspherical lens are used.

11.3.3 Design driver and degree of aspherization

Often the receiver (film, CCD-chip) limits the imaging chain object $\rightarrow$ camera lens $\rightarrow$ receiver and therefore defines the necessary quality of the camera lens.

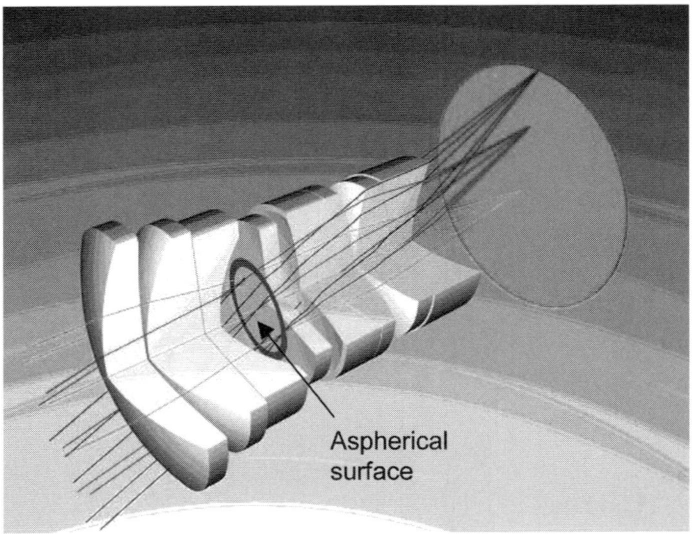

Figure 11.16 Cross section through the LEICA Summilux-M 1:1.4/50 mm ASPH® at full aperture.

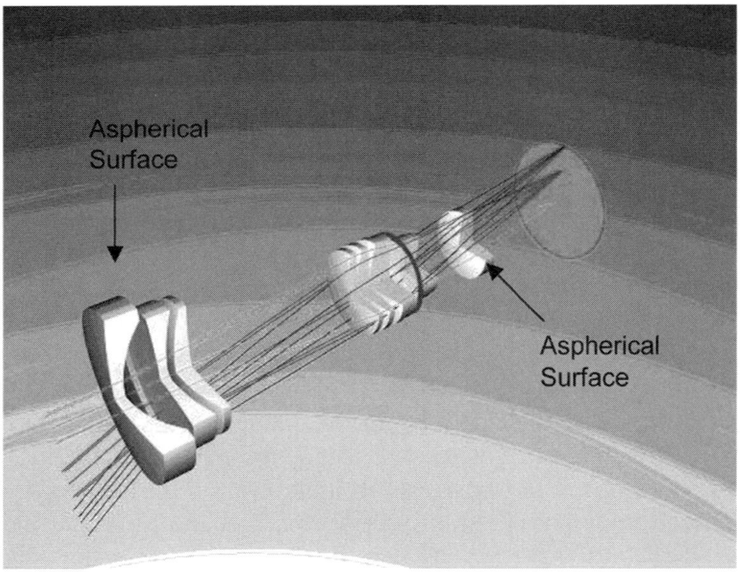

Figure 11.17 Cross section through the LEICA Vario-Elmarit-R 1:2.8–4.5/28–90 mm ASPH® at focal length 28 mm and full aperture.

Because of the continuous increase in resolution of both analog films and digital CCDs, the demands on the camera lens increase, too. At the same time, the imaging quality should increase while the size of the camera lens should decrease.

To satisfy all these requirements, the application of aspherical lenses is essential. First, the aspherical lens is used to correct monochromatic errors such as spherical aberration and distortion. Also, the aspherical lens can be used to minimize coma and astigmatism. The imaging error of the aspherical lens is determined by the position of the aspherical lens within the camera lens. Close to the aperture stop it can be particularly used to correct spherical aberration. One example of this is illustrated by the Summilux-M 1:1.4/50 mm ASPH®. The aspherical lens is positioned immediately behind the aperture stop and minimizes the spherical aberration; the main influence is on the spherical aberration for inclined bunches in the sagittal plane.

The more the aspherical lens moves away from the aperture stop, the more it influences distortion. This effect is used in the Vario-Elmarit-R 1:2.8–4.5/28–90 mm ASPH®, where the aspherical surface is the first surface of the system. It mainly corrects the distortion in the wide-angle position. The second aspherical surface is on the last surface of the system. It is located between the aperture stop and the imaging plane and therefore is used for more than one correction task.

11.3.4 Progress and performance parameters

11.3.4.1 Materials

For the production of aspherical lenses, two manufacturing processes are mainly used: molding and polishing. Which process is used mainly depends on the shape of the aspherical lens and the accepted tolerances.

Molded aspherical lenses are much cheaper in high numbers of pieces than polished lenses. Therefore, one has to pay with some disadvantages. The range of optical glasses is restricted to only a few different glasses that can be molded. As yet, diameters of more than 30 mm cannot be manufactured with the desired quality. Furthermore, the radius of curvature of the spherical surface is more restricted than when using polished lenses. Due to the molding process, one has to tolerate a larger wedge angle between the front and back sides of the lens.

On the other hand, polished aspheres can be produced even in tiny numbers with acceptable financial effort. They can be up to 200 mm in diameter, and many different optical glasses can be used. Because of the manufacturing effort, polished aspherical surfaces are many times more expensive than molded surfaces. In return, they usually have a higher optical quality.

11.3.4.2 Manufacturing and tolerances

Due to the ongoing miniaturization of imaging sensors, the focal length of photographic lenses must be accordingly shorter. One disadvantage of this evolution is the fact that corresponding tolerances decrease as well. Nowadays, it is common to tolerate the radial shift of a spherical lens to only 0.01 mm and the tilt of such a lens to 0.5 arcmin. For aspherical lenses, these tolerance values may not be small enough; an aspherical surface has up to six degrees of freedom because of its continuously varying curvature, compared to only one degree of freedom for a spherical surface. With six degrees of freedom, aspherical surfaces can be designed exactly for a specific correction purpose. Consequently, the whole camera lens is very sensitive to misalignment of the aspherical lens.

The demanded tolerances cannot be reached with common lens barrels. Therefore, so-called shift elements are required. These elements consist of lenses or lens groups that can be shifted perpendicularly to the optical axis of the camera lens. During shifting of these elements, the imaging quality is measured continuously to determine the optimal position of the shift element. In this way, it is possible to reposition the element with a precision of only a few micrometers.

11.3.4.3 Quality assurance

The position of each lens is controlled during the manufacturing process. In particular, one has to pay attention to the exact position of the aspherical surfaces. The manufactured camera lenses are subject to 100% control. This is mainly done with modulation transfer function (MTF)-measuring devices. Through special elements that can be shifted along the optical axis and the abovementioned shift elements that are more perpendicular to the optical axis, it is possible to level out small misalignments.

11.3.5 Comments and outlook

Future developments will allow the production of precise and low-cost aspherical lenses. Therefore, new fields of application will be opened up for camera lenses. At present, camera lenses for mobile phones consist of several aspherical lenses, because lenses of a small size are obligatory. The trend to compactness and light weight will be transferred to high-quality camera lenses in the future.

11.3.6 Further reading

L. Kölsch and P. Karbe, "Der siegeszug der asphäre oder der durchbruch der asphärentechnologie", *Photonik*, Vol. 3, pp. 54–57, AT-Fachverlag GmbH, Stuttgart (2003).

11.4 Aspheres for Large Format Lenses

H. Ebbesmeier

11.4.1 Application of aspherical lenses for camera lens systems

High-performance wide-angle lenses for conventional analog and digital photography, in particular wide-angle zoom systems, are today unthinkable without

Figure 11.18 Super-Symmar Aspheric 5.6/150 XL.

aspherical surfaces. The aims of using aspherical surfaces are the realization of compact designs, cost reduction by reducing the number of lenses, and, of course, to improve resolution and performance. In retrofocus systems, distortion can be reduced significantly by an aspherical surface situated on one of the front lenses. In high-aperture lens systems, aspherical lenses have to be located near the aperture in order to reduce spherical aberration.

The manufacturing method—pressed, ground, and polished, or hybrid technology—depends on the shape, the dimensions, and the required tolerances of the respective aspherical lens. The required quantity is a further significant factor.

11.4.2 Application of aspherical lenses for large, wide-angle systems

Several years ago, a number of wide-angle systems with one aspherical surface and an aperture angle of 105 deg were developed for different large-format cameras. These lens systems (Super-Symmar Aspheric XL), with focal lengths between 80 and 210 mm, replaced the six-lens Super-Angulon lenses with f from 120 to 210 mm.

The aims and results are described in the following sections with regard to the Super-Symmar Aspheric XL 5.6/150 mm lens (Fig. 11.18).

11.4.3 The task

A further aim was to improve the performance and to increase the aperture size from 1:8 to 1:5.6. For small- or medium-format cameras, a large flange focal distance of the lenses is required, because the mirror used in such cameras needs a certain space. This means that in the wide-angle range, asymmetrical systems—retrofocus-systems—have to be used.

For large-format cameras, relatively symmetric lenses like the dual Gauss-lenses can be used, where distortion and color aberrations can be corrected easily.

Existing systems offer distortion values of less than 1% for a field angle of more than 100 deg.

Commercially available wide-angle systems for large-format cameras (such as Super-Angulon and Grandagon lenses) offer excellent performance for long object distances and apertures from 16 to 22. However, for close-up photography (table-top), lateral chromatic aberrations increase and the resolution deteriorates.

The aim of the development of the new system was a compact and cost-efficient solution with larger maximum aperture and better reproduction performance, particularly when realizing short taking distances.

11.4.4 The result

A system with an asymmetric refractive power sequence was chosen as a possible solution compared with the Super-Angulon lens. This made it possible to achieve a substantial reduction in diameter of the rear element.

Compared with the Super-Angulon 8/165, the length of the Super-Symmar Aspheric 5.6/150, which covers the same image circle, was reduced from 135.5 mm to 80 mm, although the aperture diameter was increased (Fig. 11.19). The weight could be reduced from 1605 g to 740 g.

The larger residual error due to the asymmetric configuration versus a symmetric solution, particularly for distortion, coma, and lateral chromatic aberrations, can only be compensated by using aspherical surfaces (Fig. 11.20). In the system described here, this surface is located on the second side of the fifth lens.

In addition, it was possible to achieve an initial aperture of 1:5.6 without increasing the number of lenses and the diameters. In the past, seven- or eight-lens systems had to be used for f-numbers 4.5 and 5.6. With its high resolution for long-distance photographs, at working aperture, the Super-Angulon Aspheric is, like the Standard Super-Angulon, close to an ideal system.

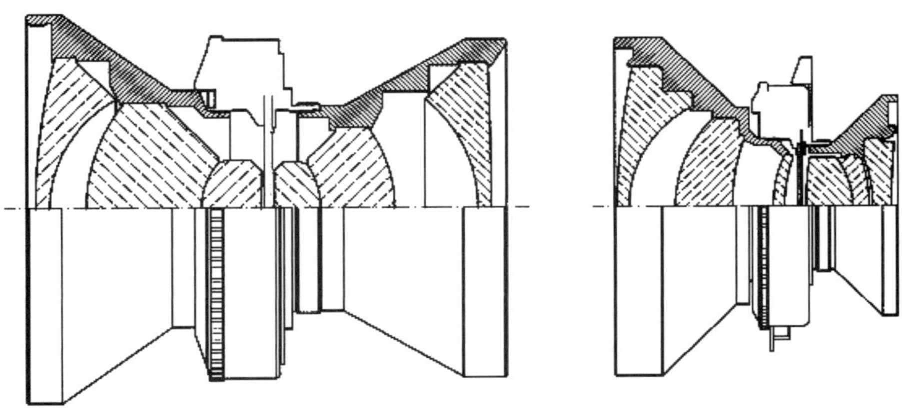

Figure 11.19 Super-Angulon 8/165 and Super-Symmar Aspheric 5.6/150 XL.

Figure 11.20 Lateral chromatic aberration in the edge.

For close-up photographs, it offers significantly improved performance compared with conventional systems. The lateral chromatic aberrations were reduced significantly for these photographing situations, that is, by about 50% compared with conventional lenses over the whole distance range. Of course, this also leads to higher values for the modulation transfer function.

11.4.5 Production: manufacturing process

So far, pressed aspherical glass lenses are only available in a limited number of glass types, and diameters and shapes are also restricted. Furthermore, tool costs are very high and unacceptable for the small production runs associated with large-format photography.

The manufacturing techniques are usually unable to meet the required surface precision tolerances and the requirements in terms of dispersion and refractive index deviation (e.g., change in refractive index of the glass during cooling for pressed lenses compared with the original value). Ground and polished aspherical glass lenses were very expensive to produce in the past.

11.4.6 Precision and measuring equipment

Only the high-precision CNC-grinding and polishing machines, developed in recent years, have made it possible to produce aspherical glass lenses with the required small tolerances for a reasonable price. These machines enable small-area grinding and polishing of aspherical lens surfaces. Any deviations from the required shape remaining after polishing are measured interferometrically using computer-generated holograms (Fig. 11.21). The information is fed back to the polishing machines, and the process is repeated until an acceptable residual error has been achieved. This process is able to achieve deviations from the required surface shape of less than 0.5 μm, both for global and local errors. The quality of the holograms used obviously has to be very high. Interferometric verification is also essential.

Figure 11.21 Surface testing.

11.4.7 Future perspectives

Because the sales figures for analog large-format cameras and systems are declining, certainly no further new systems with aspherical lenses can be expected in the future. High development costs and expensive holograms can only be justified for certain quantities and/or over long periods. However, aspherical surfaces can definitely be expected to be used in lenses for special digital camera backs. Chips with formats of 24×36 mm up to 38.8×50 mm and a pixel pitch of less than 9 μm are currently being used for this type of application. The lens systems are usually installed in electronic shutters, currently available in size 0 (Schneider with 23 mm aperture diameter) and 1 (Rollei with 34 mm aperture diameter).

For using small depths of field in image composition, larger apertures than 1:4.0 would be desirable both for wide-angle lenses and for long focal lengths. For telephoto lenses, a large aperture cannot be realized at the moment due to the limited aperture diameter of the electronic shutters.

As soon as a new electronic shutter system is available, surely lenses with aspherical elements will be developed to achieve compact systems with high performance.

11.5 Aspherical Projection Lenses for UV- and EUV-Lithography

H. Feldmann, H.-J. Mann, W. Ulrich

11.5.1 Introduction

Advanced dioptric projection lenses from Carl Zeiss are used in some of the world's most advanced deep UV projection lithography systems. These lenses provide a resolution of better than 100 nm across the entire field of view, with a level of aberration control that maximizes critical dimension uniformity and lithographic process latitude. These dioptric projection lenses are currently being used for critical layer device patterning for a wide array of complex logic, memory, and application-specific integrated circuits.

The continuing trend towards higher integration density of microelectronic circuits requires steadily decreasing feature sizes. One of the fundamental requirements for lithography with a resolution of 100 nm and below is the development of high-performance optical designs for projection lenses. Specifically, it is demonstrated that aspheres can be used effectively to reduce the volume and cost of full field hyper-NA projection lenses for dry and immersion lithography.

The use of light in the extreme ultraviolet (EUV) spectral region combined with purely catoptric projection lenses will extend the resolution limit below a 32 nm feature size. Polishing of aspherical mirror surfaces with subnanometer accuracy is one of the most important enabling technologies for building such systems.

11.5.2 Optical lithography at the edge of Raleigh's law

The manufacturing of integrated circuits with smaller and smaller features demands leading-edge projection lenses with specifications that nobody had considered possible a few years ago. The Rayleigh resolution R of a lithographic printing system is expressed as $R = k_1\lambda/\text{NA}$, where k_1 is a process-dependent factor, λ is the wavelength of illumination, and NA is the numerical aperture. The process-dependent k_1 factor takes into account several factors, such as partial coherence and the influence of resolution-enhancement techniques like off-axis illumination and phase shift masks. Wavelength scaling, numerical aperture scaling, and k_1 process optimization have all been used to improve resolution. For example, a projection lens with an NA of 0.70 operating at 193 nm can achieve a resolution of 100 nm in resist, assuming a k_1 factor of 0.36.

Sematech's International Technology Roadmap for Semiconductors (ITRS) shows that 100 nm design rules are achieved today using either 248 nm or 193 nm technology. These high-NA tools, like the Twinscan XT:1250 (Fig. 11.22) produced by ASML (http://www.asml.com), are almost exclusively supported by dioptric projection lens technology in a step and scan mode. According to Moore's law, the leading-edge performance of integrated circuits doubles every 18 months without any increase in manufacturing cost. Recently, the actual rate of development has even exceeded the prediction of this golden rule.

11.5.3 Aspheres for compact high-NA lenses

Designing projection lenses with the highest NAs for microlithography combines several difficulties of lens design. Field sizes comparable to those of photographic lenses have to be achieved at NAs of high-aperture microscope objectives

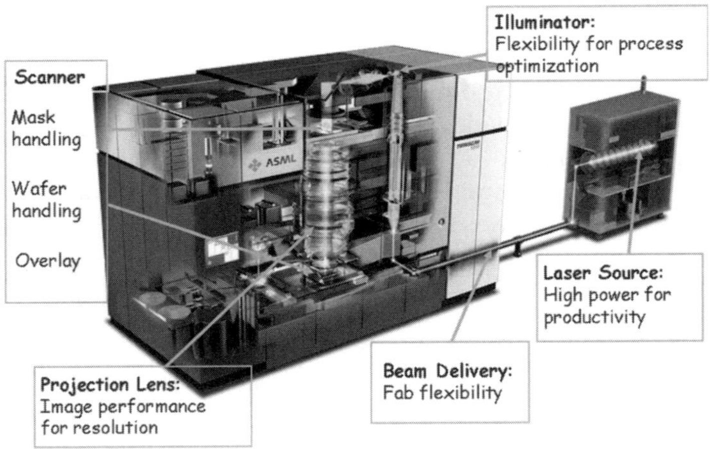

Figure 11.22 State-of-the-art ASML Twinscan XT: 1250.

while maintaining near-perfect wavefront correction. Compared to the standard definition of a diffraction limited lens, where the wavefront rms aberration is smaller than $\lambda/14$, lithographic lenses have to be better corrected by more than one order of magnitude. With the new Starlith 1400, Carl Zeiss SMT AG (http://www.smt.zeiss.com) has established new standards in 193 nm lithography. These lenses enable the fabrication of microchips with structure widths of 65 nm and below (approximately 1/1000 of the diameter of a human hair), and this strongly requires lens designs that approach a zero-aberration condition over the whole field. Of course, these lenses are totally free of distortion, and they also have to be manufacturable in quantity. Figure 11.23 visualizes how design modifications, and particularly the use of aspheres, have enabled the increase in NA for

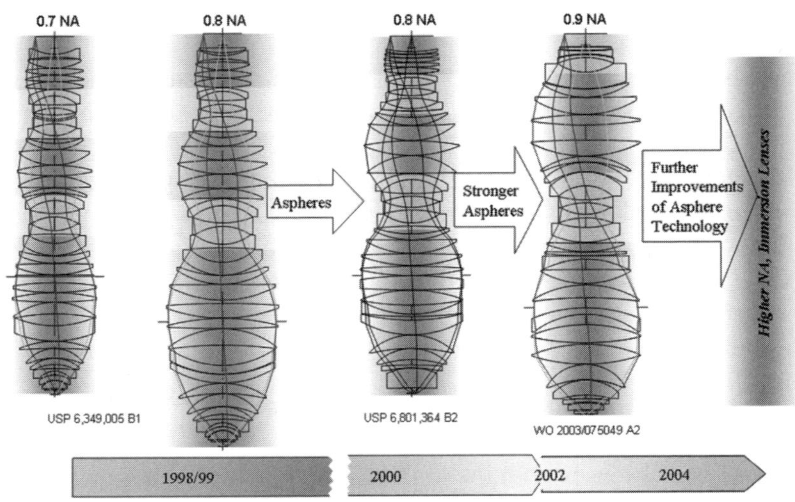

Figure 11.23 Compact Hyper-NA lens designs using asphere technology.

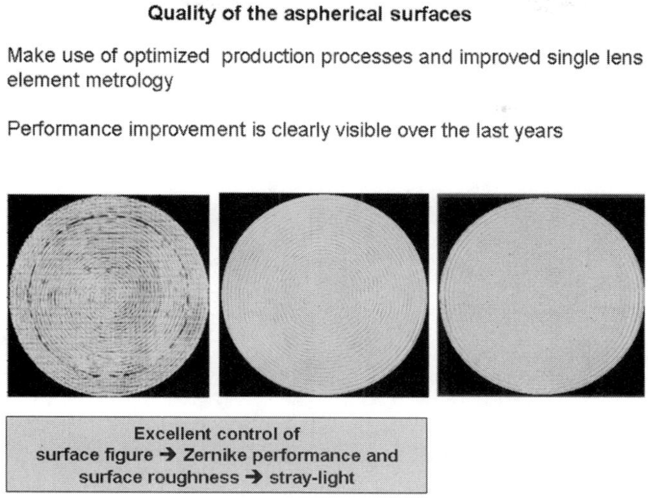

Figure 11.24 Asphere experience.

DUV lenses from 0.7 to 0.9 with a moderate growth in lens volume [1–3]. This was a huge success in lens design and very helpful in realizing these optics, because appropriate DUV-materials such as quartz and CaF_2 are only available in restricted quantities and at very high costs. The total blank mass of such lenses is about 500–1000 times higher compared to a sophisticated fast-speed camera lens, and the price per kg of ArF-material is 10–50 times higher compared to the high-index optical glass used in modern camera lenses.

However, the demands for accuracy in the aspheres also grew dramatically as they were used more frequently in lens design. Optimized production processes and improved single lens element metrology has led to excellent control of surface figure and surface roughness over recent years (Fig. 11.24).

11.5.4 Immersion lithography

The NAs of lithographic lenses have almost reached the physical limit of NA [1]. Driven by Moore's law, the use of an immersion liquid to increase the NA further was recently introduced to the semiconductor industry. For a given NA, there is a substantial drop in lens complexity due to immersion for a given NA [4]. However, Fig. 11.25 shows also that even the use of strong aspheres cannot prevent the explosion of volume for NA $>$ 1.1. Here, new design means, such as aspherical catadioptric concepts, are required. The optical designers at Carl Zeiss have optimized multi-axial designs with folding mirrors [5] as well as the so-called inline designs with just one common optical axis for all lens elements [6], but the latter one is much more preferable for some important reasons from the customer's point of view. In particular, only the inline design concept allows the customers to use the same reticles as today.

By introducing catadioptric immersion concepts, lens volume could be reduced dramatically. Simultaneously, we have improved the lens design performance below an aberration level of 0.5 nm.

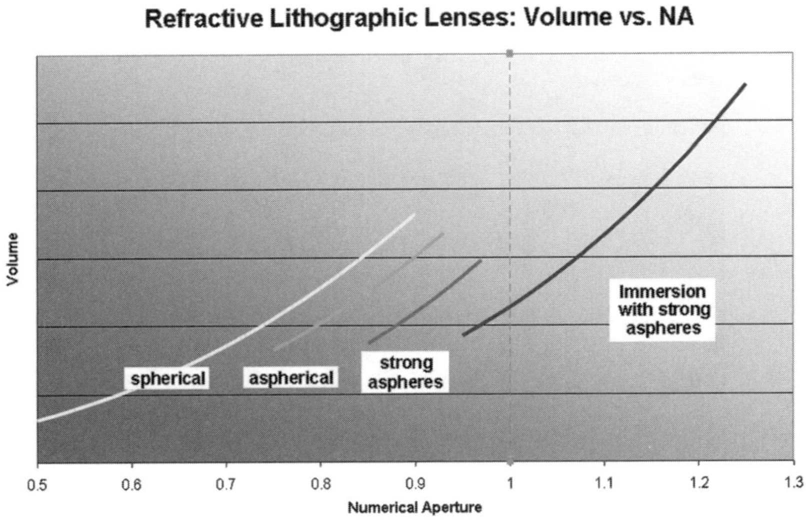

Figure 11.25 Aspheres enable low-mass designs.

11.5.5 EUV lithography

Extending the roadmap of lithographic methods to ever higher resolution, one arrives at the point where the NA cannot be enlarged any further. This is due to the theoretical limit of 1.0 NA in air or about 1.5 NA in immersion fluids. An even higher NA depends on the availability of high refractive index fluids and lens material. Also, a transition to a shorter wavelength (EUV) cannot be accomplished by a refractive lens due to absorption of the EUV radiation in the bulk material. One method to overcome this difficulty uses reflective optics consisting of multilayer mirrors. Due to the absorption of these mirrors (~70% maximum reflectivity), there is a strong demand to use as few of them as possible in the design. This reduces the number of degrees of freedom for the design process and results in an upper bound for the etendue of the lens. However, additional degrees of freedom are provided through the use of aspherical mirrors.

The Micro Exposure Tool (MET) is the first commercialized EUV system and is used as a process development tool by chip manufacturers. The projection lens of the MET consists of two aspherical mirrors made of Zerodur, a material with an extremely small thermal expansion coefficient. The illumination system includes an aspherical collector. Because shape preservation during light exposure is less important for the illumination system, the mirrors can be made of nickel, silicon, or aluminum. The most promising candidates for the EUV production tools are ring field scanner designs. Off-axis ring fields are necessary to avoid obscuration effects, which are inherent in on-axis systems like the MET tool. Figure 11.26 illustrates the design concepts of both EUV tools together with their key specifications. All mirrors in the projection lenses have an aspherical shape, enabling a wavefront correction beyond the diffraction limit.

Optical fabrication of aspherical mirrors is one of the most challenging key technologies necessary to achieve the extremely tight system specifications. A

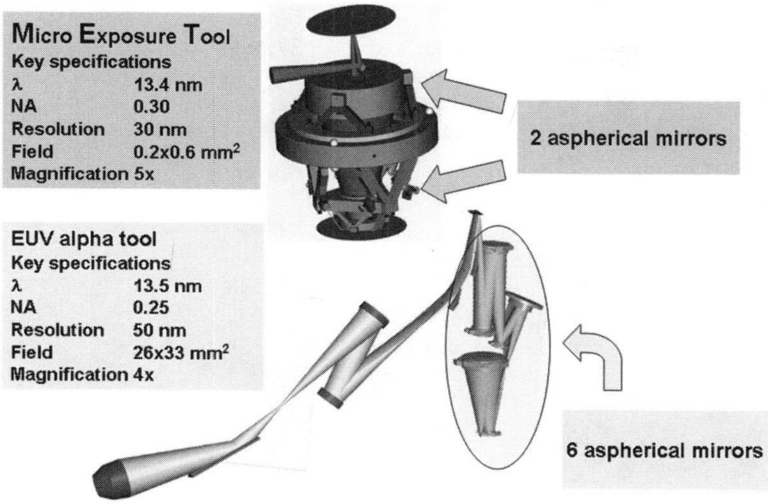

Figure 11.26 EUV systems at Carl Zeiss SMT AG: MET and EUV alpha tool.

convenient description of surface quality is provided by the power spectral density function (PSD). The PSD of an EUV mirror is divided into three different regions corresponding to their impact on optical system performance; the long spatial frequency region is called "figure" and causes wavefront aberrations, the mid spatial frequency region (MSFR) is responsible for stray light in the image field, and the high spatial frequency region (HSFR) scatters light out of the optical system, resulting in transmission loss. The surface specifications are derived by integration of the PSD and are expressed in terms of a root mean square (rms) value for each of the three different regions. A typical value for these rms values is 0.25 nm (Fig. 11.27). This surface quality specification is illustrated by the following

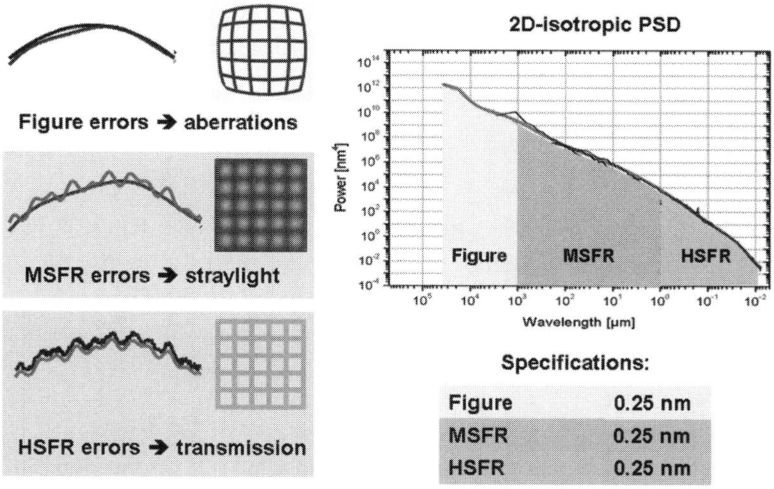

Figure 11.27 Requirements for optics fabrication of aspherical mirrors.

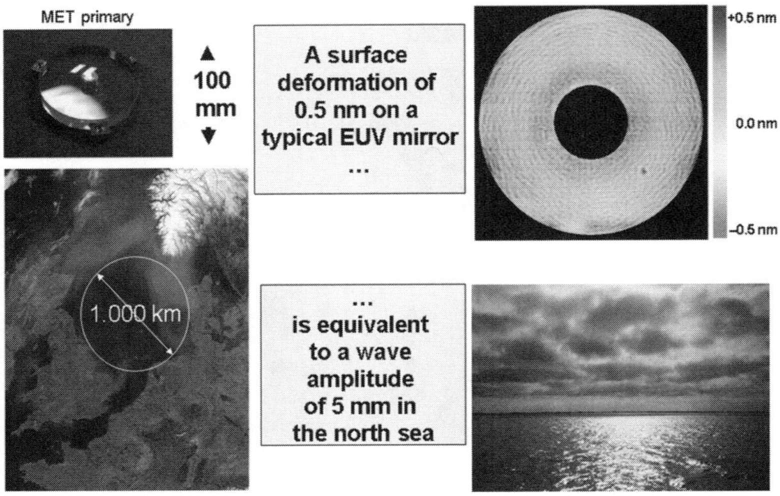

Figure 11.28 Optical requirements for EUV mirrors: a comparison.

relation between mirror size and mirror surface deformation: scaling up one typical MET mirror to the size of the North Sea, a surface deformation of 0.5 nm is scaled into a wave amplitude of only 5 mm (Fig. 11.28).

11.5.6 Outlook

In order to follow Moore's law, optical designers have to optimize new optical designs for projection lenses with NAs higher than 1.10 to an aberration level of a few Angström. In parallel, EUV technology requires further improvements in overall transmission as well as in resolution. Leading-edge asphere technology will continue as the key technology for leading-edge optical systems in lithography over the next ten years. Larger diameters of lens elements and mirror elements, aspherical deformations of several millimeters, and even better surface quality are some of the challenges for the future. And again, this technology has to be available for cost-effective serial production.

11.5.7 Acknowledgments

We would like to thank Winfried Kaiser, Gerd Fürter, Peter Kürz, Erik Sohmen, and Reiner Garreis for valuable discussions, supporting, and reviewing this contribution.

11.5.8 References

1. K.-H. Schuster, "Projection objective for microlithography", US Patent 6,801,364 B2.
2. H.-J. Rostalski, R. Hudyma and W. Ulrich, "Refractive projection objective," WO 2003/075049 A2.

3. W. Ulrich, H.-J. Rostalski and R. Hudyma, "Development of dioptric projection lenses for deep ultraviolet lithography at Carl Zeiss," *J. Microlithography, Microfabrication and Microsystems*, Vol. 3(1), pp. 87–96 (2004).
4. A. Dodoc and A. Epple, "Katadioptrisches Projektionsobjektiv mit Zwischenbildern," Patent pending.
5. CZ SMT AG, Patents pending.
6. http://www.smt.zeiss.com.
7. http://www.asml.com.
8. K.-H. Schuster, et al., "Microlithographic reduction objective, projection exposure equipment and process", US Patent 6,349,005 B1.

11.6 Large-Format Lenses for Aerial Surveying

M. Biber, B. Braunecker

11.6.1 Application fields

11.6.1.1 Airborne photogrammetry

Airborne photogrammetric cameras take pictures of the earth's surface, flying between 2 and 6 km above ground. The images are either recorded on film or registered by a digital sensor. Professional systems use large-sensor formats to collect a maximum of data in order to reduce the costs of the flight. Consequently, film-based cameras like the Leica RC30 record on 9-in. square film, while its digital "pendant," the Leica ADS40, uses three CCD lines with 24,000 pixels for panchromatic (black/white) recording. The sensor lines are oriented normal to the flight direction and are looking down to the ground in forward, nadir, and backward directions for stereo recording (pushbroom principle). The airplane is controlled and guided by GPS satellites (Fig. 11.29a).

Depending on the actual altitude and speed, the same ground pixel is registered by the forward and backward line with a delay time of several seconds. During this time interval, the common reference coordinate system needed for the stereo evaluation is lost due to the erratic roll, pitch, and yaw motions of the airplane. To overcome this problem, the camera is mounted on a heavy, gyro-controlled mechanical platform and is equipped with navigation sensors that record the residual 3D linear and 3D rotational acceleration values. The navigation data are taken to extract the 3D topographic data out of the recorded-image data. Due to the huge amount of data, the rectification calculations are performed after flight, on the ground. Application fields for airborne cameras include topographic mapping, urban planning, and traffic monitoring.

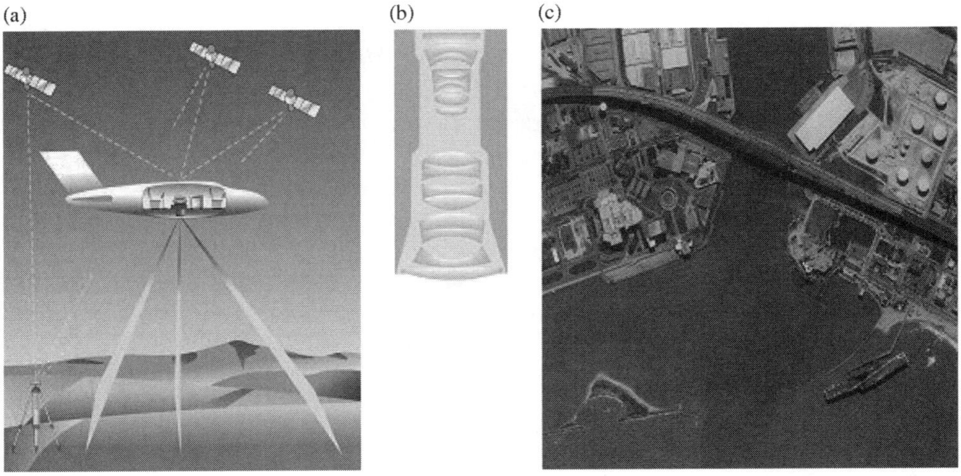

Figure 11.29 (a) Three-line pushbroom mode; (b) ADS40; (c) RGB picture.

11.6.1.2 Remote sensing

In the case of the ADS40, four additional CCD lines are placed in the focal plane at the half forward looking position. They carry spectral filters for the blue, green, red, and IR spectral range. The interference filters are comprised of stacks of about 60 thin layers for obtaining a rectangular bandpass transmission curve with spectral edges of 2 nm. The CCD lines for the red, green, and blue bands (RGB) are integrated on the same chip for stability reasons, with a separation distance of about 2 mm in the flight direction. This distance, however, is already large enough to limit the recording accuracy because of the erratic airplane motions. To avoid the cumbersome algorithmical treatment mentioned above, a hardware solution was found: all four color CCD lines look through a beam combiner prism down to earth on the same "white" ground line.

All these efforts allow *quantitative* spectral intensity measurements to be obtained for remote-sensing applications (Fig. 11.29c). However, it also increases the data rate; up to 1 Terabyte of polychromatic intensity data is recorded during a two-hour flight. Again, the relevant topographic and spectral information must be extracted by digital post-processing on the ground. Typical application cases are environmental control, forestry monitoring, agricultural classifications, and disaster recovery.

11.6.2 Optical systems

11.6.2.1 Film cameras

Figure 11.30 shows Leica's 15/4 UAGS (Ultra Aviogon Super) lens with a focal length of 150 mm and an F-number of 4. Note the mechanical adapter frame for the 9-in. square film. This format corresponds to a maximum field angle of ±45

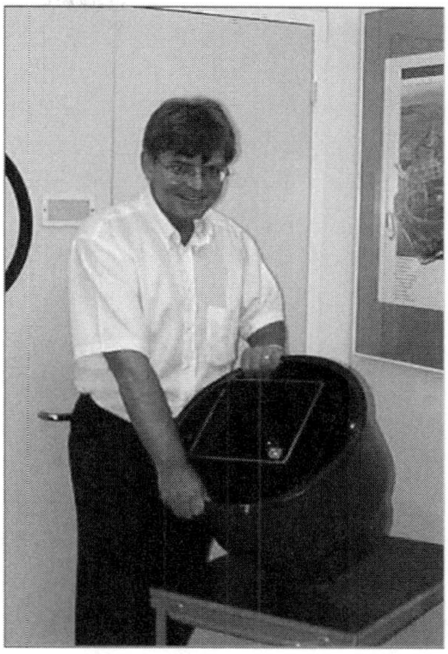

Figure 11.30 15/4 UAGS.

deg. When looking under this maximum field angle on the lens, we still see a round exit pupil, similar to a cat's pupil, indicating the homogenous intensity distribution across the film area. The homogeneity is a nondisputable customer requirement for best use of the film dynamics. The fight against the $\cos^4$-law, that is, against the intensity drop across the field, is one of the challenges for the designer. Film lenses are metric lenses where the geometrical distortion is kept below 0.001% over the full 9-in. format, that is, below 2 μm.

11.6.2.2 Digital cameras

Figure 11.29(b) and Fig. 11.31 show lenses designed for digital sensors. The specifications are similar to those of the film lens except for more relaxed optical distortion values, which can be eliminated by software. What is different to film lenses is that

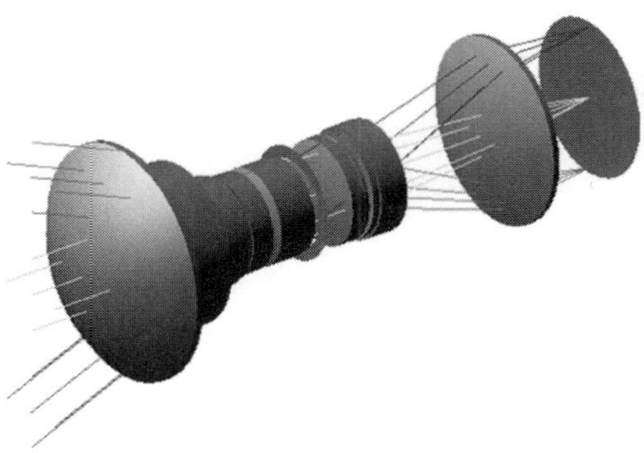

Figure 11.31 Aspheric 12/2.8.

all light bundles have to hit the sensor lines under 90 deg. One reason for this is the angular dependence of the interference filters mentioned above, which would lead in the case of a non-normal incidence to color shifts or more severe to unacceptable spectral information loss in quantitative remote-sensing applications. To avoid this, digital cameras need telecentrical lenses where the exit pupil is placed to infinity. This type of lens is difficult to correct and often requires a large construction length.

11.6.3 Design drivers and degree of asphérization

11.6.3.1 Aspherization

Today, all large-format photogrammetric lenses in our designs are spherical. The use of aspheres was not considered for cost reasons. Also, there was no actual need for them, because the prime performance criteria in photogrammetry are a high spatial contrast and resolution in a broad spectral range, that is, the color correction from 420 nm to 900 nm across the wide angular field of 90 deg. This depends more on the proper selection of the glass material than on the surface shapes. Therefore, any aspherization of the smaller inner lenses near the aperture stop position was not necessary. However, aspherizing at least one of the larger outer lenses would allow facilitation of the correction of distortion, field curvature, field homogeneity, and, in the case of digital cameras, telecentricity, with fewer lenses, and therefore would reduce weight and costs. Figure 11.31 shows a design study with an aspherical lens element before the sensor plane, which is corrected for F-number 2.8.

11.6.3.2 Athermalization

Airborne lenses must operate under severe environmental conditions, such as flying heights of up to 6 km. One requirement is to maintain image quality and registration between $-30°C$ and $+50°C$. To this purpose, optomechanical solutions for the stationary thermal case, where all lenses are at the same temperature, exist [2]. But the important nonstationary situation, where, for example, a warm lens is brought into a colder environment, is more difficult to handle. Due to the large lens diameters, radial thermal gradients inside the glass material occur, causing field curvature. To overcome this thermal aberration, a large depth of focus must be designed [3].

11.6.4 Process and performance parameters

11.6.4.1 Materials

We mentioned the required high quality of the glass material. Due to the large dimensions, our key suppliers are Schott and Ohara. The glass blocks for each lens are delivered with the melting data for the refraction indices at different wavelengths. Sometimes, interferometrical measurements in transmission of the whole glass block are performed to identify the regions of best optical homogeneity.

11.6.4.2 Manufacturing and tolerances

Using actual melting data, the design is reoptimized, causing small changes in the lens thickness values. These correction data are forwarded to the optical workshop, which later reports the actual measurement data back to the design office to calculate the definitive air distances between the lenses. Typical production tolerances are ± 0.010 mm for the lens thickness, ± 15 arcsec for the lens wedge angles, surface shape errors of $3/1(0.5)$ according to the ISO-Norm, ± 0.005 mm for the air gaps between the lenses, and centering tolerances of some ± 10 arcsec. Extremely good process control is asked from the manufacturer, SwissOptic AG.

11.6.4.3 Quality assurance

A special "vertical goniometer" was developed to verify and calibrate the geometrical performance of the completely assembled camera. The camera looks downward into a small collimator with a test pattern. By precisely moving the collimator in both lateral directions and by slightly defocusing the test pattern, a complete 3D test flight can be simulated. From the recorded pattern images, the optical transfer function (OTF) and the distortion values are determined. The fully automated procedure is performed for the three panchromatic and the four spectral CCD lines across field, in and normal to the flight direction (Fig. 11.32). Finally, all CCD pixels are radiometrically calibrated by measuring the dark current and the sensor current dependent on the illumination level. For this purpose, the complete assembled camera looks into a large illumination sphere (Fig. 11.33).

11.6.5 Comments and outlook

Digital recording in airborne photogrammetry has replaced the traditional film technology. Besides the online data processing capability, advantages include the much higher dynamic range of the electronic sensors. In the case of the ADS40, the image intensity data are quantized by 12 to 14 bits, compared to about 6 to 7 bits resolution for photofilm. Another significant advantage is that the spectral information can be easily adapted to the customer needs by simply changing the filter sets. Finally, we mention the spatial resolution, where the digital technique is equivalent to film recording. However, because the film resolution is limited by the unavoidable wet development and scan-in process, the digital technique is permanently under improvement with new sensor materials and sensor architectures, like

Figure 11.32 Geometrical calibration.

Figure 11.33 Radiometrical calibration.

'staggered' arrangements [2]. Lens aspherization was not yet realized for cost reasons but would be necessary to further improve the light efficiency (F-number) and the geometrical mapping characteristics of image homogeneity, distortion, and telecentricity. Thus, we carefully watch for any progress in the manufacturing of aspherical lenses.

11.6.6 References

1. A. Eckardt, B. Braunecker and R. Sandau, "Performance of the imaging system in the LH Systems—ADS40 Airborne Digital Sensor," in *International Archives of Photogrammetry and Remote Sensing*, Vol. 33, part B1, pp. 104–109, Amsterdam, Netherlands (2000).
2. R. Sandau, *Digitale Luftbildkamera, Handbuch für die Praxis*, Wichmann Verlag (2005).
3. B. Braunecker and B. Aebischer, "Thermal-Management grossformatiger Optiksysteme," http://www.dgao-proceedings.de/archiv/105_chronologisch_d.php, http://www.leica-geosystems.com/corporate/en/ndef/lgs_4045.htm

11.7 Mirror Telescope for Space Communication

E. Fischer

11.7.1 Application fields: optical link between satellites for data communication

It is widely accepted that, in the future, about 10% of Internet traffic will be transmitted via satellites. The unprecedented growth of the Internet poses a continuing challenge for satellite companies trying to react to a changing customer mix. Customers nowadays are seeking individualized access to communication links and not traditional services such as sending bundles of phone calls or television over long distances and providing broadcast services. However, satellites have historically been designed to cover distance and not to efficiently deliver computer connections over relatively compact areas.

Future high-speed data exchange over large distances will doubtless be dominated by optical technologies, using fiber and optical free-space connections between stations. In particular, a global network of satellites is considered to be the most flexible and most efficient technique for communication purposes between continents. Even if the worldwide communication crisis in recent years has dampered many activities, it has led to a critical review of all technologies involved. But new applications like the transfer of large 3D images in real time are of growing importance and will need very powerful point-to-point and point-to-multipoint communication channels. This will cause a renaissance of satellite-to-satellite but also satellite-to-groundstation communication technologies.

11.7.2 Optical free-space communication systems

11.7.2.1 Advantage of optics versus radio frequency (RF)

The relatively narrow beamwidth of free-space optical communication beams is one of its primary advantages over RF systems. This is because the beam divergence of an RF beam or an optical beam is proportional to λ/D, where λ is the transmitted wavelength and D the transmitter aperture diameter. Due to the significantly shorter wavelength optics, much higher "antenna gains" are achieved and can thus project higher intensity at the receiver for the same power at the transmitter. However, the narrow beamwidth also causes the main technical challenge for optical communication systems in space. The receiver must acquire the slender transmit beam, and the optical connection must be maintained with sub-arcsecond accuracy throughout communication.

11.7.2.2 Afocal telescope

The data exchange by optical means between two satellites or between a satellite and a groundstation requires optical transceivers, so called optical terminals, at both link ends. The sender terminal emits the information to be transmitted via a laser beam towards the counter station. This forward datastream beam is either intensity or phase modulated. In the reverse direction, the terminal simultaneously receives the very weak backward data signals from the counter station. An important part of each optical terminal is the telescope, often also called the optical antenna. The telescope has to fulfill two requirements: (1) the diameter of its entrance pupil must be reasonably large so as to collect enough backward light, and (2) the emitted forward wavefront must be diffraction limited to concentrate enough laser light at the counter station.

The telescope used in the Contraves Space mid-range optical terminals is an all-mirror system with an afocal magnification of 10× and an entrance pupil diameter of 135 mm.

11.7.3 Design drivers and degree of aspherization

For energetic reasons, the transmitter beam system, that is, the telescope, has to be designed to achieve a communication beam divergence of 2–3 arcseconds. On the other hand, the acquisition cone of uncertainty resulting from satellite position and attitude uncertainties as well as from thermal-induced satellite platform deformations is about 0.6 deg. The uncertainty cone is therefore the design driver for the terminal's telescope and internal beam relay/imaging optics field of views (FOVs). The tight transmitter beam also causes a steep intensity decrease on the receiver side in the case of transmitter mispointing. The tolerable 4 sigma mispointing loss is usually limited to 3 dB in link budgets. This means that the mispointing angle of the telescope line of sight (LOS) has to be actively controlled to values less or

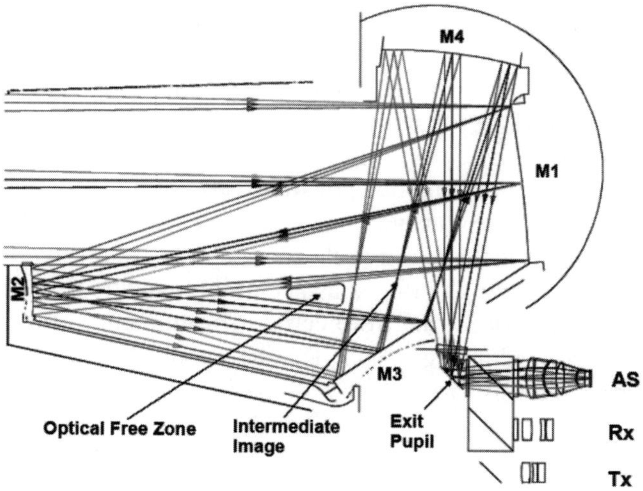

Figure 11.34 Afocal all-mirror free-space optical intersatellite communication telescope.

equal to 0.2 arcseconds. This LOS stability requirement is the driver for the tolerable optical wavefront error (WFE) of the receiver optics. In order to achieve the required tracking performance, it is mandatory that not only the transmitter optics is well diffraction limited, but also the receiver optics. Furthermore, it is required that the shape of the received beam spot falling onto the tracking receiver stays constant over the whole tracking FOV.

Low-mass, compact packaging size and rigidity against space environmental loads (launch vibrations, separation shock, zero-gravity operation, thermal gradients, and severe radiation impacts) favor an all-mirror solution (Fig. 11.34). The emitted wavefront must be diffraction limited without central obscuration, causing mirrors M1, M2, and M4 in the figure to be 'off-axis' aspheres. The mirror M3 is used as a plane folding mirror. To reduce the manufacturing complexity, the aspherical shape of all mirrors is only conical. The telescope was designed by Leica-Geosystems AG, Heerbrugg/Switzerland.

11.7.4 Process and performance parameters

11.7.4.1 Materials

From various materials that have been investigated, the favorite choice that emerged was Zerodur, from Schott Glas Werke. Zerodur exhibits the best combination of properties with regard to thermal expansion and long-term stability, even under high-energy radiation loads. The disadvantage of a glass-ceramic material is the fact that it is not a typical structural material and that the effort to machine such structures bears a considerable risk of failure (via brittleness, micro-cracking, and so on).

11.7.4.2 Manufacturing and tolerances

Lightweight telescope mirrors were manufactured and tested by SwissOptic AG, Heerbrugg, Switzerland. A computer numerically controlled (CNC) grinding machine was used to come close to the "mother aspheric mirror" shape with an accuracy of 1 μm. After finishing this polishing process, the required off-axis mirror segments are cut out. The rear sides of all such telescope mirrors are lightweight. The minimum thickness in the center is chosen according to the manufacturing limits. To avoid any print through effects, a special internal stress relaxation process for the mirrors was developed. Interferometric measurement results for mirror M2 in Fig. 11.34 are depicted in Fig. 11.35. The measured value of 0.074 rms fringes at 546.1 nm corresponds to a rms value of $\lambda/52.6$ ($\lambda = 1064$ nm).

11.7.4.3 Telescope structure, integration, and verification

In the course of optics design and tolerance analysis, it emerges that the key opto-mechanical requirements of the telescope (mirror positions <0.010 mm; tilt angles ~10 arcsec) will be very hard to meet over the full range of environmental loads. A number of options have been considered during the development cycle, including metallic structures and glass-ceramic-based structures. Finally, a metallic

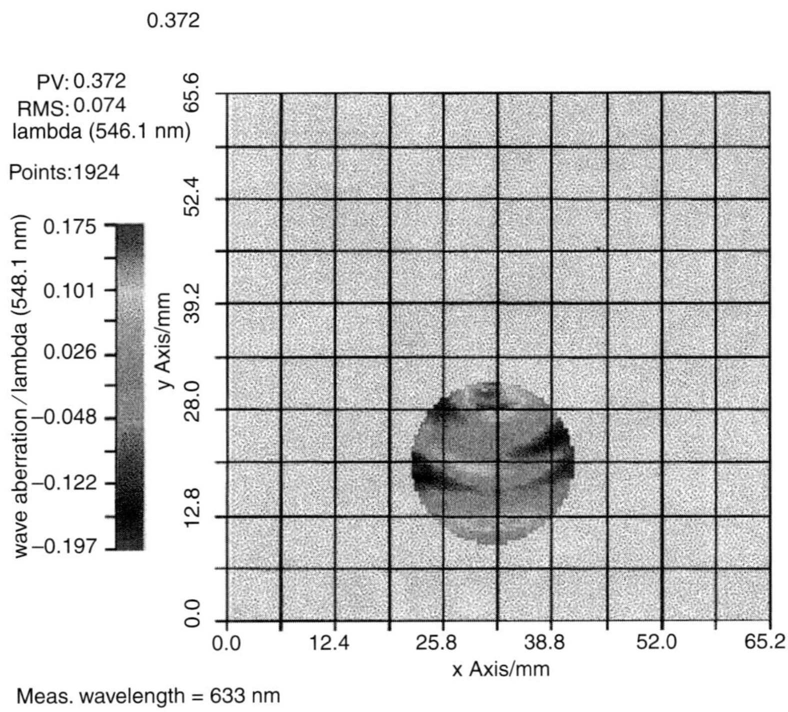

Figure 11.35 Interferometric measurement result for mirror M2.

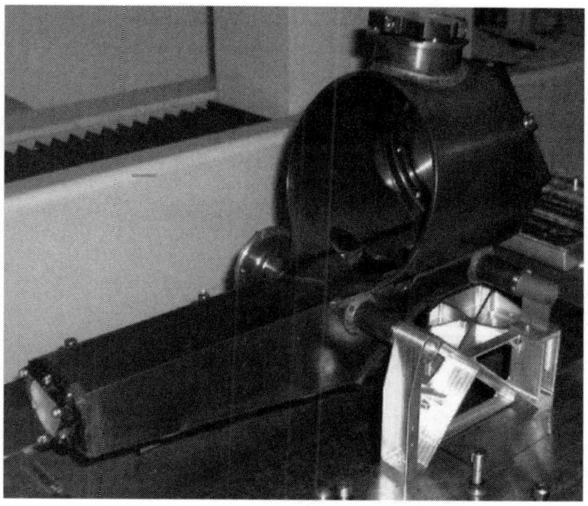

Figure 11.36 Telescope mounted in integration rig.

telescope structure design in combination with a segmented active temperature control was selected. Contraves Space AG, Zürich, was responsible for the design and analysis of the telescope structure and the mirror cells. Finally, the Engineering Model of the telescope was integrated and verified by Contraves Space (Fig. 11.36).

11.7.5 Quality assurance

The interferometric telescope performance measurements were performed with a modified Fizeau-type interferometer from FISBA Optik. The measurements were carried out in a double-pass configuration. With the temperature control system of the telescope switched off, the measured wavefront deviation from a plane surface was $\lambda/12$ rms@ $\lambda = 1064$ nm. By switching the temperature control system of the telescope on, it was possible to improve the alignment of the optical system significantly. This improvement was achieved by applying segmented heating at strategic locations on the telescope structure. The wavefront error achieved was finally $\lambda/22$ rms@ $\lambda = 1064$ nm (Fig. 11.37 and Table 11.2).

Table 11.2 Comparison of required performance for various telescope models with achieved engineering model performance.

	Requirements for flight model	Requirements for engineering model	Achieved with engineering model
rms wavefront error	$\lambda/25$ (@ 1064 nm)	$\lambda/8$ (@ 1064 nm)	$\lambda/22$ (@ 1064 nm)*
Line of sight error	<100 μrad	<100 μrad	<25 μrad

*Temperature control system operated.

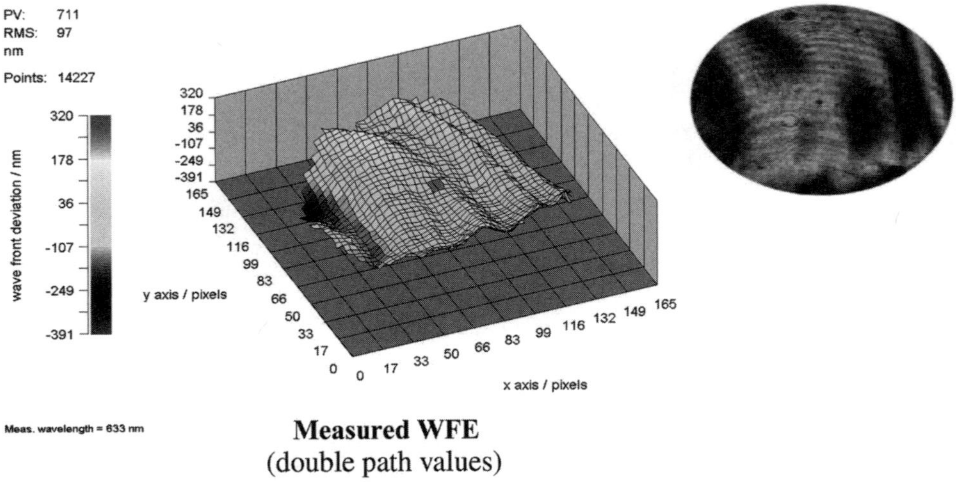

Figure 11.37 Interferometric measurement of telescope engineering model without operating the thermal control system.

11.7.6 Comments and outlook

It would be advantageous if all mirrors could be directly grinded and polished as off-axis segments of the bulk material. Although this technology needs strong development efforts, it is clearly the only way to reduce production costs.

Design studies have been performed to shape the plane mirror M3 with a free-form aspherical correction. The results of a design study are presented in "Free-form Correction Plate for Telescopes", Sec. 11.8, by Leica Geosystems.

11.7.7 Reference

1. G.C. Baister, et al., "The ISLFE terminal development project—results from the engineering breadboard model," *20th AIAA International Communication Satellite Systems Conference*, 12–15 May (2002).

11.8 Free-form Correction Plate for Telescopes

M. Biber, B. Braunecker

11.8.1 Application fields

11.8.1.1 Performance improvement of mirror telescopes

Mirror telescopes are mostly used for astronomical observations, but also, as shown in Sec. 11.7, for space communication purposes. The afocal all-mirror Leica design consists of two parabolic mirrors M1 and M4, and a hyperbolic mirror M2. They are mounted as off-axis segments to avoid beam obscuration and thus to minimize the intensity loss of the emitted laser beam to the opposite terminal over distances of up to 60,000 km. Because mirror M3 is a plane folding mirror, all four elements can be easily manufactured by state-of-the-art technology. Contraves Space verified by experiment that the specifications set for the wavefront quality and the angular field distortion were reached.

We will show next how to further improve the image quality and the distortion. The idea is to polish some free-form surface deformations on the flat mirror M3 or on elements near to M3.

11.8.1.2 Optical system

The optical performance of any optical system can be improved in general by adding more components to the system. This, however, increases the system complexity and may lead to narrower fabrication and assembly tolerances for all optical elements. Therefore, we decided to increase the aspherization degree of one of the four elements. As a quality measure, we also had to consider (besides the wave front flatness in the pupils) the image registration, that is, the angular field distortion. This function is specified, because the star pattern inside the field of view is registered by a 2D sensor and immediately compared with preprogrammed star configurations to obtain the actual terminal orientation (Fig. 11.38).

11.8.2 Design drivers and degree of aspherization

The correction of wavefronts in optical systems requires Schmidt-plates, which must be located near pupil planes, while the control of distortion errors should occur near the image planes. The *simultaneous* correction of both effects consequently needs an optical element positioned *between* an image plane and a pupil plane. This position must also be mechanically accessible. Figure 11.38 indicates that the plane

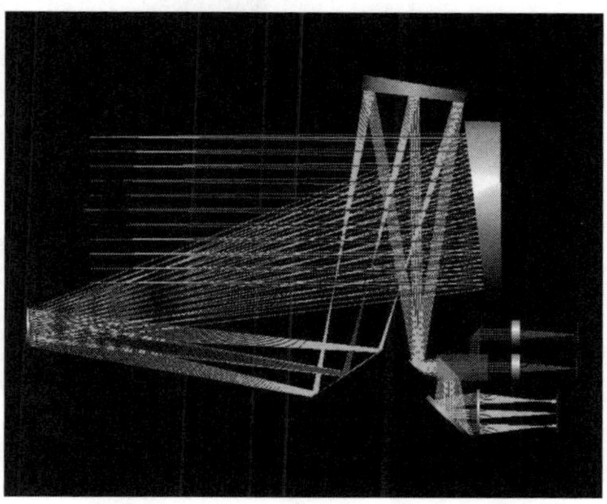

Figure 11.38 Afocal space telescope with plane folding mirror M3.

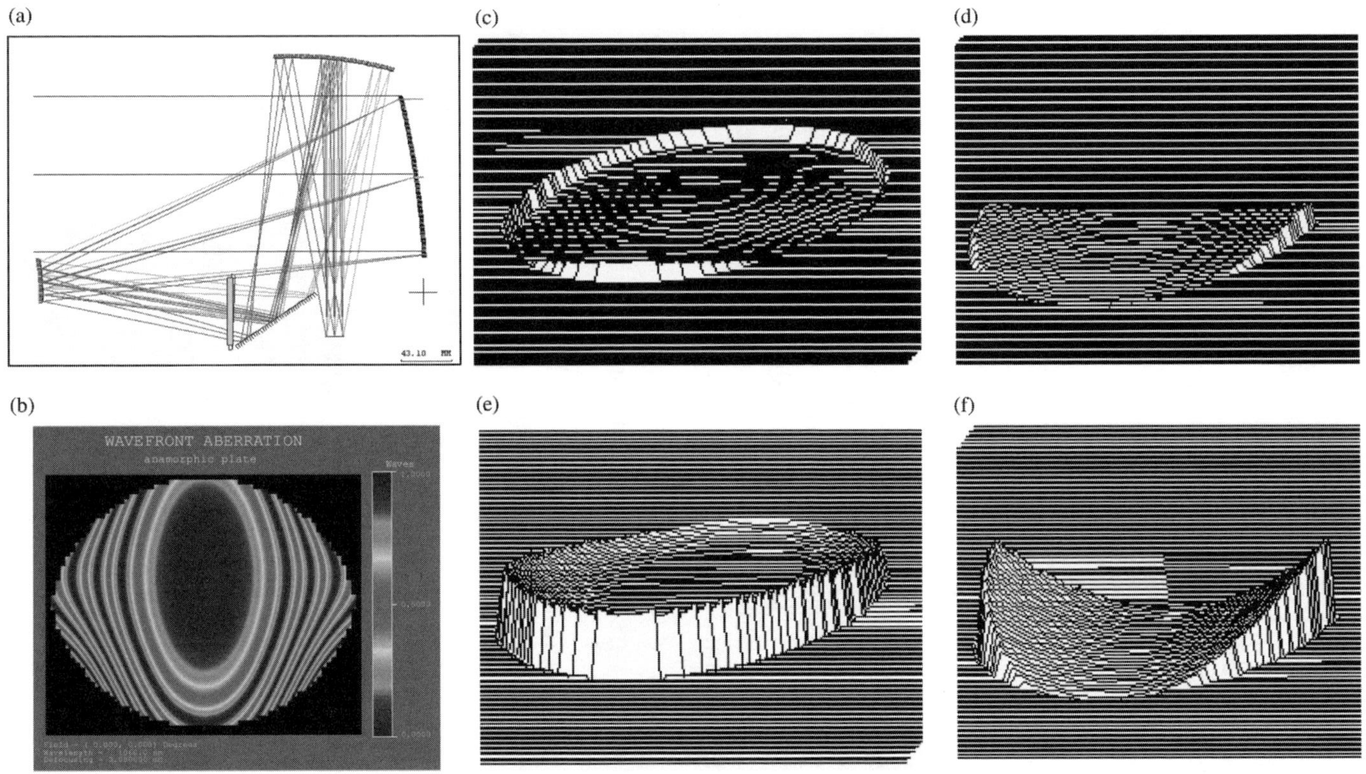

Figure 11.39 (a) The location of a refractive corrector plate before M3. (b) Its surface deviation from the ideal plane. (c, d) The wavefront in the exit pupil for the on-axis and the full field case of ± 1.0 deg with the correction plate. (e, f) For comparison purposes, the same situations without correction.

Table 11.3 Summary of results.

Wavefront quality @ $\lambda = 1064\,\text{nm}$	On axis	Full field
No correction	rms $\lambda/33$	rms $\lambda/7$
With correction	rms $\lambda/70$	rms $\lambda/20$

folding mirror M3 is ideally suited due to its location between the entry pupil at M1 and the intermediate image plane between M3 and the eyepiece M4. Note that the layout is a Kepler telescope.

The correction function could be polished either directly on mirror M3 itself, or on a separate glass plate placed just before or after M3. We show design results obtained with a thin transparent glass plate before M3. For real applications, the direct mirror correction would be preferred in order to keep the achromatism of an all-mirror telescope.

11.8.3 Process and performance parameters

Because mirrors M1, M2, and M4 are slightly tilted around the x-axis (i.e., the axis perpendicular to the paper plane in Fig. 11.39), the surface deformation function for a corrector element near M3 is symmetric only in x. It is a free-form asphere in the y-direction, and its manufacturing needs CNC-driven grinding and polishing tools. These technologies exist today, but are still expensive.

The quality improvements we present in the following are results of a design study. The results are summarized in Table 11.3.

11.8.4 Comments and outlook

The improvement of mirror telescopes by inserting a special correcting plate is an old, well-proven technology, initiated by the early works of Bernhard Schmidt around 1930 to correct spherical aberration. He polished the surface correction shape in a rather ingenious way; by first bending the glass plate by applying vacuum at the one side and then polishing it as a normal plane plate. After relaxing the vacuum, the surface of the unbended plate had the right shape needed to correct the aberration. In our case, we apply modern CNC technologies to grind and polish a surface deformation. Even if the polishing process is still a challenge at the moment, and expensive, the performance improvements are so impressive that further steps in maturization of robot-driven polishing are to be expected.

11.8.5 Reference

1. Patent: "Method for correcting optical wavefront errors and optical systems, such as a telescope, produced accordingly," US 6,426,834.

Chapter 12

Materials

12.1 Low-T_g Glass ($n_d < 1.6$, $v_d > 65$)

B. Hladik, S. Ritter

12.1.1 Intended purpose of the glass

Preforms (precision gobs) for precision glass molding.

12.1.2 Glass types[1]

Glass type	Manufacturer	Glass family
P-PK53	Schott AG	Phosphate glass
P-SK57	Schott AG	Phosphate glass
N-FK5	Schott AG	Phosphate glass
N-FK51A	Schott AG	Fluoro phosphate glass
N-PK52A	Schott AG	Fluoro phosphate glass
N-PK51	Schott AG	Fluoro phosphate glass
K-CaFK95	Sumita	Fluoro phosphate glass
K-PG325	Sumita	Fluoro phosphate glass
K-PBK40	Sumita	Borosilicate glass
K-PG375	Sumita	Phosphate glass
K-GFK70	Sumita	Fluoro phosphate glass
K-GFK68	Sumita	Fluoro phosphate glass
M-BaCD12	Hoya	na
M-BACD5N	Hoya	na
M-PCD51	Hoya	na
L-BSL7	Ohara	na
L-PHL2	Ohara	na
L-PHL1	Ohara	na
L-BAL42	Ohara	na
L-BAL35	Ohara	na
P-SK12	Hikari	na
P-SK5	Hikari	na

na, not available.

[1] Only a selection of glass types.

12.1.3 Optical properties

Glass type	Refractive index, n_d	Abbe number, ν_d	Partial dispersion, $P_{g,F}$	Deviation of relative dispersion, $\Delta P_{g,F}$	Color code ($\lambda 80/\lambda 5$)	Stress optical coefficient, K (10^{-6} mm^2/N)
P-PK53	1.52707	65.91	0.5408	0.0084	36/31	2.06
P-SK57	1.58700	59.60	0.5412	−0.0024	34/31	2.17
N-FK5	1.48749	70.41	0.5290	0.0036	30/27	2.91
N-FK51A	1.48656	84.47	0.5359	0.0342	34/28	0.7
N-PK52A	1.49700	81.61	0.5377	0.0311	34/28	0.67
N-PK51	1.52855	76.98	0.5401	0.0258	34/29	0.54
K-CaFK95	1.434245	95	0.534	0.0385	33/28	na
K-PG325	1.50670	70.5	0.538	0.0064	34/30	na
K-PBK40	1.51760	63.5	0.534	−0.0085	33/29	na
K-PG375	1.54250	62.9	0.544	0.0010	36/31	na
K-GFK70	1.56907	71.3	0.545	0.0145	34/28	na
K-GFK68	1.59240	68.3	0.546	0.0105	34/28	na
M-BaCD12	1.58313	59.46	0.5404	−0.0008	35/29	na
M-BaCD5N	1.58913	61.25	0.5373	−0.0007	35/30	na
M-PCD51	1.59201	67.02	0.5357	0.0081	35/28	na
L-BSL7	1.51633	64.1	0.5333	−0.0045	33/30	2.93
L-PHL2	1.55880	62.6	0.5430	0.0027	34/30	2.99
L-PHL1	1.56455	60.8	0.5457	0.0026	34/31	3.29
L-BAL42	1.583126	59.4	0.5423	−0.0031	34/29	na
L-BAL35	1.589130	61.2	0.5382	−0.0043	35/30	2.29
P-SK12	1.58313	59.46	na	na	34/29	na
P-SK5	1.58913	61.25	na	na	35/30	na

na, not available.

12.1.4 Mechanical properties

Glass type	Density, ρ (g/cm^3)	Knoop hardness, $HK_{0.1/20}$	Young's modulus, E (10^8 N/m^2)	Poisson's ratio, μ
P-PK53	2.83	335	593.7	0.271
P-SK57	3.01	535	926.6	0.249
N-FK5	2.45	520	620	0.232
N-FK51A	3.68	345	810	0.293
N-PK52A	3.75	355	710	0.298
N-PK51	3.96	400	740	0.295
K-CaFK95	3.54	331	718	0.287
K-PG325	3.00	352	642	0.265
K-PBK40	2.39	615	799	0.229
K-PG375	2.90	368	625	0.252

(Continued)

Glass type	Density, ρ (g/cm^3)	Knoop hardness, $HK_{0.1/20}$	Young's modulus, E (10^8 N/m^2)	Poisson's ratio, μ
K-GFK70	4.41	332	663	0.295
K-GFK68	4.51	368	683	0.308
M-BaCD12	3.01	575	896	0.252
M-BaCD5N	2.82	600	na	na
M-PCD51	3.14	440	885	0.271
L-BSL7	2.38	560	793	0.214
L-PHL2	3.03	370	645	0.272
L-PHL1	3.18	350	589	0.280
L-BAL42	3.05	590	891	0.247
L-BAL35	2.82	630	1008	0.252
P-SK12	2.80	530	na	na
P-SK5	2.93	588	na	na

na, not available.

12.1.5 Chemical properties

Glass type	Alkali resistance class (AR) (acc. ISO 10629)	Acid resistance class (SR) (acc. ISO 8424)	Acid resistance class (SR) (Powder)	Climatic resistance class (CR) (acc. ISO/WD13384)	Water resistance (WR) (acc. JOGIS)
P-PK53	4.3	51.0	na	2	1
P-SK57	2	52.3	na	4	1
N-FK5	2	4	na	2	4
N-FK51A	2.2	52.3	na	1	1
N-PK52A	3.3	52.3	na	1	1
N-PK51	3.3	51.2	na	2	1
K-CaFK95	na	na	4	na	2
K-PG325	na	na	5	na	1
K-PBK40	na	na	1	na	1
K-PG375	na	na	3	na	1
K-GFK70	na	na	1	na	1
K-GFK68	na	na	1	na	1
M-BaCD12	na	na	3	na	1
M-BaCD5N	na	na	4	na	2
M-PCD51	na	na	4	na	1
L-BSL7	na	1.0	na	na	2
L-PHL2	na	51.1	na	na	1
L-PHL1	na	53.3	na	na	1
L-BAL42	na	5.2	na	na	2
L-BAL35	na	52.2	na	na	2
P-SK12	na	na	2	na	1
P-SK5	na	na	4	na	2

Link to http://www.schott.com/optics_devices/english/download/ (SR ISO).

Link to http://www.hoyaoptics.com/pdf/OpticalGlass.pdf (SR Powder).

Note, methods of determination of chemical stability are different for different manufacturers (acc. ISO or acc. JOGIS).

na, not available.

12.1.6 Thermal properties

Glass type	Thermal expansion, CTE (−30°C/70°C) ($10^{-6}\times 1/K$)	Thermal expansion, CTE (100°C/300°C) ($10^{-6}\times 1/K$)	Thermal expansion, CTE (20°C/300°C) ($10^{-6}\times 1/K$)	Transformation temperature, T_g (°C)	Softening point (°C)	Thermal capacity, C_p (J/(g ∗ K))	Thermal conductivity, λ (W/mK)
P-PK53	13.31	na	15.97	383	453	0.77	0.64
P-SK57	7.23	na	8.9	493	593	0.76	1.01
N-FK5	9.2	na	10.0	466	672	0.808	0.925
N-FK51A	12.74	na	14.81	464	527	0.69	0.76
N-PK52A	13.01	na	15.0	467	538	0.67	0.73
N-PK51	12.7	na	14.4	487	568	0.62	0.65
K-CaFK95	na	16.7	na	431	na	na	na
K-PG325	na	17.3	na	285	na	na	na
K-PBK40	na	7.3	na	501	na	na	na
K-PG375	na	16.9	na	343	na	na	na
K-GFK70	na	15.6	na	475	na	na	na
K-GFK68	na	15.2	na	512	na	na	na
M-BaCD12	7.0	8.8	na	500	540	na	na
M-BaCD5N	6.8	8.9	na	515	545	na	na
M-PCD51	9.3	11.6	na	499	541	0.709	na
L-BSL7	5.8	7.1	na	498	630	na	1.169
L-PHL2	9.9	13.0	na	381	440	na	0.683
L-PHL1	10.5	14.0	na	347	408	na	0.627
L-BAL42	7.2	8.8	na	506	607	na	1.028
L-BAL35	6.6	8.1	na	527	619	na	1.126
P-SK12	na	8.6	na	512	na	na	na
P-SK5	na	8.4	na	520	na	na	na

na, not available.

12.1.7 Applications and limitations

Current application: Precision glass molding.
Potential application and outlook: Also as "standard" optical glass.
Limitations: Medium chemical resistance and medium crystallization stability.

12.1.8 Further reading

1. H. Bac and N. Neuroth, *The Properties of Optical Glass,* Springer-Verlag, Berlin (1998).
2. S. Hirota, "Precision molded glass lenses in TAF-series and FD · FDS-series," *New Glass*, Vol. 17, No. 2, pp. 53–55 (2002).
3. S. Uehara, "Low T_g optical glass for precision press molding," *Materials Integration*, Vol. 16, No. 8, pp. 41–48 (2003).
4. S. Omi, "Precision molding technology of glass lens," *New Glass,* Vol. 19, No. 1, pp. 65–68 (2004).

12.1.9 Links

- http://www.schott.com/optics_devices/english/download/
- http://www.sumita-opt.co.jp/en/optical.htm
- http://www.hoyaoptics.com/specialty_glass/optical.htm
- http://www.oharacorp.com/swf/catalog.html
- http://www.hikariglass.com/downloads.htm
- http://www.hikariglass.com/prod04.htm

12.1.10 Research and development

See http://www.schott.com/ft/english.

12.2 Low-T_g Glass ($1.6 < n_d < 1.9, 40 < v_d < 65$)

B. Hladik, S. Ritter

12.2.1 Intended purpose of the glass

Preforms (precision gobs) for precision glass molding.

12.2.2 Glass types[1]

Glass type	Manufacturer	Glass family
P-SF8	Schott AG	Silicate glass
K-VC79	Sumita	Silicate glass
K-LaFK60	Sumita	Lanthanum borate glass
K-CD45	Sumita	Silicate glass
M-BACD15	Hoya	na
M-LaC130	Hoya	na
M-TAF101	Hoya	na
L-TIM28	Ohara	Silicate glass
L-LAM69	Ohara	na
L-LAM60	Ohara	na
P-SK16	Hikari	na
P-LaK13	Hikari	na

na, not available.

12.2.3 Optical properties

Glass type	Refractive index, n_d	Abbe number, ν_d	Partial dispersion, $P_{g,F}$	Deviation of relative partial dispersion, $\Delta P_{g,F}$	Color code ($\lambda 80/\lambda 5$)	Stress optical coefficient, $K\,(10^{-6}\,\text{mm}^2/\text{N})$
P-SF8	1.68893	31.1	tbd	tbd	tbd	tbd
K-VC79	1.60907	57.8	0.543	−0.0076	34/29	na
K-LaFK60	1.63246	63.8	0.541	−0.0006	38/28	na
K-CD45	1.69320	33.7	0.591	0.0044	41/35	na
M-BACD15	1.62263	58.16	0.5390	−0.0045	35/28	na
M-LaC130	1.69350	53.20	0.5465	−0.0060	37/29	na
M-TAF101	1.76802	49.2	0.5515	−0.0081	40/32	na
L-TIM28	1.68893	31.1	0.5986	0.0074	40/36	na
L-LAM69	1.73077	40.5	0.5728	−0.0031	41/34	2.03
L-LAM60	1.74320	49.3	0.5529	−0.0088	37/31	1.83
P-SK16	1.62041	60.29	na	na	35/31	na
P-LaK13	1.69350	53.20	na	na	37/30	na

na, not available; tbd, to be determined.

[1]Only a selection of glass types.

12.2.4 Mechanical properties

Glass type	Density, ρ (g/cm^3)	Knoop hardness, $HK_{0.1/20}$	Young's modulus, E (10^8 N/m^2)	Poisson's ratio, μ (W/(m $*$ K))
P-SF8	tbd	tbd	tbd	tbd
K-VC79	3.09	599	1008	0.256
K-LaFK60	4.32	547	997	0.288
K-CD45	3.12	562	903	0.260
M-BACD15	3.02	760	1049	0.267
M-LaC130	3.52	650	1135	0.29
M-TAF101	4.56	749	1205	0.298
L-TIM28	2.88	530	845	0.254
L-LAM69	3.24	630	1133	0.273
L-LAM60	4.2	620	1147	0.289
P-SK16	3.36	556	na	na
P-LaK13	3.66	590	na	na

na, not available; tbd, to be determined.

12.2.5 Chemical properties

Glass type	Alkali resistance class (AR) (acc. ISO 10629)	Acid resistance class (SR) (acc. ISO 8424)	Acid resistance class (SR) (Powder)	Climatic resistance class (CR) (acc. ISO/WD13384)	Water resistance (WR) (acc. JOGIS)
P-SF8	tbd	tbd	tbd	tbd	tbd
K-VC79	na	na	4	na	1
K-LaFK60	na	na	5	na	1
K-CD45	na	na	1	na	2
M-BACD15	na	na	4	na	2
M-LaC130	na	na	4	na	1
M-TAF101	na	na	4	na	1
L-TIM28	na	1.0	na	na	2
L-LAM69	na	52.2	na	na	1
L-LAM60	na	51.2	na	na	1
P-SK16	na	na	5	na	3
P-LaK13	na	na	4	na	2

Link to http://www.schott.com/optics_devices/english/download/ (SR ISO).
Link to http://www.hoyaoptics.com/pdf/OpticalGlass.pdf (SR Powder).
Note, methods of determination of chemical stability are different for different manufacturers (acc. ISO or acc. JOGIS).
na, not available; tbd, to be determined.

12.2.6 Thermal properties

Glass type	Thermal expansion, CTE (−30°C/70°C) ($10^{-6} \times 1/K$)	Thermal expansion, CTE (100°C/300°C) ($10^{-6} \times 1/K$)	Thermal expansion, CTE (20°C/300°C) ($10^{-6} \times 1/K$)	Transformation temperature, T_g (°C)	Softening point (°C)	Thermal capacity, C_p (J/(g * K))	Thermal conductivity, λ (W/mK)
P-SF8	tbd	tbd	tbd	tbd	tbd	tbd	tbd
K-VC79	na	9.3	na	516	na	na	na
K-LaFK60	na	11.4	na	485	na	na	na
K-CD45	na	11.0	na	456	na	na	na
M-BACD15	6.3	8.2	na	530	570	0.742	1.040
M-LaC130	6.9	8.5	na	500	560	na	na
M-TAF101	5.9	7.4	na	605	690	0.446	0.780
L-TIM28	10.1	13.0	na	504	582	na	1.020
L-LAM69	8.6	10.5	na	497	574	na	1.114
L-LAM60	7.4	9.2	na	541	623	na	0.876
P-SK16	na	8.6	na	548	na	na	na
P-LaK13	na	8.3	na	544	na	na	na

tbd, to be determined; na, not available.

12.2.7 Applications and limitations

Current application: Precision glass molding.
Potential application and outlook: Its potential application and outlook is as "standard" optical glass.
Limitations: Medium chemical resistance and medium crystallization stability.

12.2.8 Further reading

1. H. Bach and N. Neuroth, *The Properties of Optical Glass,* Springer-Verlag, Berlin (1998).
2. S. Hirota, "Precision molded glass lenses in TAF-series and FD–FDS-series," *New Glass*, Vol. 17, No. 2, pp. 53–55 (2002).
3. S. Uehara, "Low T_g optical glass for precision press molding," *Materials Integration*, Vol. 16, No. 8, pp. 41–48 (2003).
4. S. Omi, "Precision molding technology of glass lens," *New Glass*, Vol. 19, No. 1, pp. 65–68 (2004).

12.2.9 Links

- http://www.schott.com/optics_devices/english/download/
- http://www.sumita-opt.co.jp/en/optical.htm
- http://www.hoyaoptics.com/specialty_glass/optical.htm
- http://www.oharacorp.com/swf/catalog.html
- http://www.hikariglass.com/downloads.htm
- http://www.hikariglass.com/prod04.htm

12.2.10 Research and development

See http://www.schott.com/ft/english

12.3 Low-T_g Glass ($1.8 < n_d$, $30 > v_d$)

B. Hladik, S. Ritter

12.3.1 Intended purpose of the glass

Preforms (precision gobs) for precision glass molding.

12.3.2 Glass types[1]

Glass type	Manufacturer	Glass family
P-LASF47	Schott AG	Lanthanum borate glass
P-SF67	Schott AG	Niobium phosphate glass
P-SF68	Schott AG	Bismuth glass
K-VC89	Sumita	Lanthanum borate glass
K-PSFn3	Sumita	Niobium phosphate glass
K-PSFn1	Sumita	Niobium phosphate glass
M-FD60	Hoya	na
M-NbFD82	Hoya	na
M-FDS910	Hoya	na
L-LAH53	Ohara	na
L-LAH81	Ohara	na
L-LAH83	Ohara	na
P-LaSF021	Hikari	na
SFS01	Hikari	na

na, not available.

12.3.3 Optical properties

Glass type	Refractive index, n_d	Abbe number, v_d	Partial dispersion, $P_{g,F}$	Deviation of relative dispersion, $\Delta P_{g,F}$	Color code ($\lambda 80/\lambda 5$)	Stress optical coefficient, $K\,(10^{-6}\,\mathrm{mm}^2/\mathrm{N})$
P-LaSF47	1.80610	40.9	0.5671	−0.0079	39/33	2.39
P-SF67	1.90680	21.4	0.6334	0.0256	*48/39	2.96
P-SF68	>2.0	20.7	tbd	tbd	tbd	tbd
K-VC89	1.81000	41.0	0.567	−0.0084	40/34	na
K-PSFn3	1.83917	23.9	0.622	0.0206	*43/38	na
K-PSFn1	1.90680	21.2	0.631	0.0262	*49/39	na
M-FD60	1.80486	24.74	0.6220	0.0183	47/37	2.92
M-NBFD82	1.81474	37.03	0.5815	−0.0002	42/35	1.95
M-FDS910	1.82114	24.06	0.6236	0.0187	50/38	2.99
L-LAH53	1.80610	40.9	0.5688	−0.0065	40/34	na
L-LAH81	1.80610	40.4	0.5704	−0.0057	41/34	2.17
L-LAH83	1.86400	40.6	0.5669	−0.0089	*37/31	1.61
P-LaSF021	1.85026	32.28	na	na	41/36	na
SFS01	1.92286	21.3	na	na	na	na

tbd, to be determined; na, not available.

[1]Only a selection of glass types.

12.3.4 Mechanical properties

Glass type	Density, $\rho\,(g/cm^3)$	Knoop hardness, $HK_{0.1/20}$	Young's modulus, $E\ (10^8\ N/m^2)$	Poisson's ratio, μ (W/(m*K))
P-LaSF47	4.54	620	1195.6	0.298
P-SF67	4.24	440	900	0.248
P-SF68	tbd	tbd	tbd	tbd
K-VC89	4.75	644	1124	0.290
K-PSFn3	3.9102	409	883	0.256
K-PSFn1	4.1334	444	886	0.233
M-FD60	3.54	470	905	0.26
M-NbFD82	4.34	615	1149	0.29
M-FDS910	3.69	440	893	0.256
L-LAH53	4.49	660	1151	0.298
L-LAH81	4.53	640	1127	0.300
L-LAH83	5.29	680	1271	0.299
P-LaSF021	4.31	na	na	na
SFS01	6.05	268	na	na

tbd, to be determined; na, not available.

12.3.5 Chemical properties

Glass type	Alkali resistance class (AR) (acc. ISO10629)	Acid resistance class (SR) (acc. ISO 8424)	Acid resistance class (SR) (Powder)	Climatic resistance class (CR) (acc. ISO/WD13384)	Water resistance (WR) (acc. JOGIS)
P-LaSF47	1	51.4	3	1	1
P-SF67	1.3	1	1	1	1
P-SF68	tbd	tbd	tbd	tbd	tbd
K-VC89	na	na	3	na	1
K-PSFn3	na	na	1	na	1
K-PSFn1	na	na	1	na	1
M-FD60	na	na	1	na	1
M-NbFD82	na	na	4	na	1
M-FDS910	na	na	1	na	1
L-LAH53	na	51.2	na	na	1
L-LAH81	na	51.2	na	na	1
L-LAH83	na	3.2	na	na	1
P-LaSF021	na	na	3	na	2
SFS01	na	na	4	na	1

Link to http://www.hoyaoptics.com/pdf/OpticalGlass.pdf (SR Powder).
Link to http://www.Schott.com/optics_devices/english/download/ (SR ISO).
Note, methods of determination of chemical stability are different for different manufacturers (acc. ISO or acc. JOGIS).
na, not available; tbd, to be determined.

12.3.6 Thermal properties

Glass type	Thermal expansion, CTE (−30°C/70°C) ($10^{-6} \times 1/K$)	Thermal expansion, CTE (100°C/300°C) ($10^{-6} \times 1/K$)	Thermal expansion, CTE (20°C/300°C) ($10^{-6} \times 1/K$)	Transformation temperature, T_g (°C)	Softening point (°C)	Thermal capacity, C_p (J/(g * K))	Thermal conductivity, λ (W/mK)
P-LaSF47	6.04	na	7.27	530	627	0.55	0.85
P-SF67	6.2	na	7.4	539	663	0.53	0.790
P-SF68	tbd	na	tbd	<500	tbd	tbd	tbd
K-VC89	na	8.3	na	528	na	na	na
K-PSFn3	na	11.8	na	477	na	na	na
K-PSFn1	na	10.2	na	498	na	na	na
M-FD60	9.3	11.9	na	470	520	0.61	0.866
M-NbFD82	5.9	7.7	na	550	590	na	na
M-FDS910	9.5	12.1	na	455	505	0.648	0.92
L-LAH53	5.9	7.2	na	574	646	na	0.862
L-LAH81	5.8	7.3	na	566	645	na	0.86
L-LAH83	6.8	8.0	na	608	713	na	0.894
P-LaSF021	na	8.1	na	569	na	na	na
SFS01	na	9.5	na	402	na	na	na

tbd, to be determined; na, not available.

12.3.7 Applications and limitations

Current application: Precision glass molding.
Potential application and outlook: Also "standard" optical glass.
Limitations: Medium chemical resistance and medium crystallization stability.

12.3.8 Further reading

1. H. Bach and N. Neuroth, *The Properties of Optical Glass*, Springer-Verlag (1998).
2. S. Hirota, "Precision molded glass lenses in TAF-series and FD• FDS-series," *New Glass*, Vol. 17, No. 2, pp. 53–55 (2002).
3. S. Uehara, "Low T_g Optical Glass for Precision Press Molding," *Materials Integration*, Vol. 16(8), 41–48 (2003).
4. S. Omi, "Precision molding technology of glass lens," *New Glass*, vol. 19(1), 65–68 (2004).

12.3.9 Links

- http://www.schott.com/optics_devices/english/download/
- http://www.sumita-opt.co.jp/en/optical.htm
- http://www.hoyaoptics.com/specialty_glass/optical.htm
- http://www.oharacorp.com/swf/catalog.html
- http://www.hikariglass.com/downloads.htm
- http://www.hikariglass.com/prod04.htm

12.3.10 Research and development

See http://www.schott.com/ft/english.

12.4 UV-Transmitting Glasses

P. Hartmann

12.4.1 Intended purpose of the glass

Illumination and imaging lenses for i-line (365 nm) microlithography with very high requirements with respect to homogeneity, striae, refractive index, Abbe number, and transmittance at 365 nm including test equipment UV-(fluorescence)-spectroscopy.

12.4.2 Glass types

Glass type	Manufacturer	Glass family
FK5HT	Schott	Fluoro Crown
BK7HT	Schott	Boro Crown
K5HT	Schott	Crown
K7HT	Schott	Crown
LLF1HT	Schott	Lead Silicate Extra Light Flint
LLF6HT	Schott	Lead Silicate Extra Light Flint
LF5HT	Schott	Lead Silicate Extra Light Flint
LF6HT	Schott	Lead Silicate Extra Light Flint
F2HT	Schott	Lead Silicate Extra Light Flint
F8HT	Schott	Lead Silicate Flint
F14HT	Schott	Lead Silicate Extra Light Flint
S-FPL51Y	Ohara	Fluoro Crown
S-FSL5Y	Ohara	Fluoro Crown
BSL7Y	Ohara	Boro Crown
BAL15Y	Ohara	Barium Silicate Crown
BAL35Y	Ohara	Barium Silicate Crown
BSM51Y	Ohara	Barium Silicate Crown
PBL1Y	Ohara	Lead Silicate Extra Light Flint
PBL6Y	Ohara	Lead Silicate Extra Light Flint
PBL25Y	Ohara	Lead Silicate Extra Light Flint
PBL26Y	Ohara	Lead Silicate Extra Light Flint
PBM2Y	Ohara	Lead Silicate Flint
PBM8Y	Ohara	Lead Silicate Extra Light Flint
PBM18Y	Ohara	Lead Silicate Flint

12.4.3 Optical properties

Glass type	Refractive index, n_d	Abbe number, ν_d	Partial dispersion, $\Delta P_{g,F}$	365 nm; 10 mm	Stress optical coefficient, K (10^{-6} mm^2/N)
FK5HT	1.48749	70.41	0.0036	0.999	na
BK7HT	1.51680	64.17	−0.0009	0.998	na
K5HT	1.52249	59.48	0.0000	0.995	na
K7HT	1.51112	60.41	0.0000	0.995	na
LLF1HT	1.54814	45.75	−0.0009	0.996	na
LLF6HT	1.53172	48.76	−0.0013	0.998	na
LF5HT	1.58144	40.85	−0.0003	0.995	na
LF6HT	1.56732	42.85	−0.0007	0.995	na
F2HT	1.62004	36.37	0.0002	0.986	na
F8HT	1.59551	39.18	−0.0006	0.991	na
F14HT	1.60140	38.23	−0.0006	0.990	na

(*Continued*)

Glass type	Refractive index, n_d	Abbe number, ν_d	Partial dispersion, $\Delta P_{g,F}$	365 nm; 10 mm	Stress optical coefficient, K (10^{-6} mm^2/N)
S-FPL51Y	1.49847	81.14	0.1462	0.997	na
S-FSL5Y	1.48915	70.36	0.0331	0.999	na
BSL7Y	1.51825	64.24	0.0014	0.998	na
BAL15Y	1.55897	58.68	−0.0073	0.994	na
BAL35Y	1.59143	61.23	−0.0064	0.996	na
BSM51Y	1.60548	60.65	−0.0073	0.995	na
PBL1Y	1.55098	45.83	−0.0113	0.997	na
PBL6Y	1.53430	48.95	−0.0087	0.998	na
PBL25Y	1.58482	40.77	−0.0081	0.995	na
PBL26Y	1.57047	42.86	−0.0074	0.996	na
PBM2Y	1.62409	36.27	−0.0011	0.986	na
PBM8Y	1.59911	39.26	−0.0060	0.991	na
PBM18Y	1.59915	38.77	−0.0099	0.993	na

Link to definition of optical properties: Schott Technical Informations at http://www.Schott.com/optics_devices/english/download/ and http://www.oharacorp.com/swf/og.html.
na, not available.

12.4.4 Mechanical properties

Glass type	Density, ρ (g/cm^3)	Knoop hardness, $HK_{0.1/20}$	Young's modulus, E (10^8 N/m^2)	Poisson's ratio, μ	Thermal stress φ_w (N/m^{2}K) $* 10^{-6}$
FK5HT	2.45	520	62	0.232	703.65
BK7HT	2.51	610	82	0.206	733.25
K5HT	2.59	530	71	0.224	750.26
K7HT	2.53	520	69	0.214	737.40
LLF1HT	2.94	450	60	0.208	613.64
LLF6HT	2.81	470	63	0.203	592.85
LF5HT	3.22	450	59	0.223	690.99
LF6HT	3.11	390	60	0.217	na
F2HT	3.61	420	57	0.220	599.23
F8HT	3.38	450	60	0.222	632.39
F14HT	3.44	430	58	0.216	584.44
S-FPL51Y	3.66	380	716	0.302	13,950.72
S-FSL5Y	2.46	530	622	0.229	7,180.03
BSL7Y	2.50	570	811	0.207	6,954.35
BAL15Y	2.90	560	783	0.236	7,789.01
BAL35Y	3.23	550	881	0.244	6,642.46
BSM51Y	3.36	570	901	0.256	7,629.44
PBL1Y	2.93	430	616	0.212	6,566.50
PBL26Y	3.10	420	589	0.220	6,720.64
PBM2Y	3.61	420	571	0.223	6,319.95
PBL6Y	2.79	450	605	0.205	6,316.35
PBL25Y	3.23	460	585	0.219	6,516.65

(Continued)

Glass type	Density, ρ (g/cm^3)	Knoop hardness, $HK_{0.1/20}$	Young's modulus, E (10^8 N/m^2)	Poisson's ratio, μ	Thermal stress φ_w (N/m^2K) $* 10^{-6}$
PBM8Y	3.36	400	588	0.222	6,424.16
PBM18Y	3.37	410	598	0.223	6,772.72

Link to definition of mechanical properties: Schott Technical Informations at http://www.schott.com/optics_devices/english/download/ and http://www.oharacorp.com/swf/og.html.
na, not available.

12.4.5 Chemical properties

Glass type	Alkali resistance class (AR) (acc. ISO 10629)	Acid resistance class (SR) (acc. ISO 8424)	Climatic resistance class (CR) (acc. ISO WD 13384)	Phosphate resistance (PR) (acc. ISO 9689)
FK5HT	2	4	2	2.3
BK7HT	2	1	2	2.3
K5HT	1	1	1	1
K7HT	1	2	3	2.3
LLF1HT	2	1	1	1
LLF6HT	2.3	1	2	na
LF5HT	2.3	1	2	2
LF6HT	na	na	na	na
F2HT	2.3	1	1	1.3
F8HT	2.3	1	1	2.3
F14HT	2.2	1	1	1
		Weathering resistance, Ohara		
S-FPL51Y	na	51.0	2	4.2
S-FSL5Y	na	3.0	2	2.0
BSL7Y	na	1.0	1	1.0
BAL15Y	na	1.2	1	1.0
BAL35Y	na	4.2	2	1.0
BSM51Y	na	51.2	3	2.2
PBL1Y	na	1.0	1	1.0
PBL6Y	na	1.0	1	1.0
PBL25Y	na	1.0	1	2.0
PBL26Y	na	1.0	1	2.0
PBM2Y	na	1.0	1	2.0
PBM8Y	na	1.0	1	2.0
PBM18Y	na	1.0	2	2.0

Link to definition of chemical properties: Schott Technical Informations at http://www.schott.com/optics_devices/english/download/ and http:// www.oharacorp.com/swf/og.html.
Note, methods of determination of chemical stability are different for different manufacturers.
na, not available.

12.4.6 Thermal properties

Glass type	Thermal expansion, CTE (−30°C/70°C) ($10^{-6} \times 1/K$)	Temperature coefficients of refractive index (dn_e/dT) (+20; +40°C)(10^{-6}/K)	Transformation temperature, T_g (°C)	Softening point, $T_{10}7.6$ (°C)	Heat capacity, C_p (J/gK)	Thermal conductivity, λ (W/mK)	Thermal diffusivity, κ $\lambda/C^*_p\rho$ (10^{-6} m^2/s)
FK5HT	9.2	−1.0	464	672	0.808	0.925	0.467
BK7HT	7.1	3.0	559	719	0.858	1.114	0.517
K5HT	8.2	2.1	543	720	0.783	0.950	0.468
K7HT	8.4	1.6	513	712	na	na	na
LLF1HT	8.1	2.9	431	628	0.65	na	na
LLF6HT	7.5	3.7	422	627	0.700	na	na
LF5HT	9.1	2.0	419	585	0.657	0.866	0.409
LF6HT	na	0.0	431	na	na	na	na
F2HT	8.2	4.4	432	593	0.557	0.780	0.388
F8HT	8.2	4.0	446	609	0.600	0.980	0.483
F14HT	7.9	4.6	438	na	0.560	na	na
S-FPL51Y	13.6	−6.9	448	na	na	0.780	na
S-FSL5Y	8.9	−0.7	na	676	na	1.007	na
BSL7Y	6.8	3.2	577	714	na	1.182	na
BAL15Y	7.6	3.0	507	642	na	1.000	na
BAL35Y	5.7	4.0	590	697	na	0.991	na
BSM51Y	6.3	3.3	585	684	na	0.961	na
PBL1Y	8.4	3.0	442	592	na	0.960	na
PBL6Y	8.3	2.9	453	637	na	1.016	na
PBL25Y	8.7	3.3	440	590	na	0.899	na
PBL26Y	8.9	2.4	432	591	na	0.912	na
PBM2Y	8.7	4.2	432	584	na	0.814	na
PBM8Y	8.5	3.6	445	590	na	0.878	na
PBM18Y	8.8	4.1	441	565	na	0.865	na

Link to definition of thermal properties: Schott Technical Information at http://www.schott.com/optics_devices/english/download/ and http://www.oharacorp.com/swf/og.html.
na, not available.

12.4.7 Form of delivery

Glass type	Manufacturer	Precision gobs
FK5HT	Schott	Cut blanks
BK7HT	Schott	Cut blanks
K5HT	Schott	Cut blanks
K7HT	Schott	Cut blanks
LLF1HT	Schott	Cut blanks
LLF6HT	Schott	Cut blanks
LF5HT	Schott	Cut blanks
LF6HT	Schott	Cut blanks
F2HT	Schott	Cut blanks
F8HT	Schott	Cut blanks
F14HT	Schott	Cut blanks
S-FPL51Y	Ohara	na
S-FSL5Y	Ohara	na
BSL7Y	Ohara	na
BAL15Y	Ohara	na
BAL35Y	Ohara	na
BSM51Y	Ohara	na
PBL1Y	Ohara	na
PBL6Y	Ohara	na
PBL25Y	Ohara	na
PBL26Y	Ohara	na
PBM2Y	Ohara	na
PBM8Y	Ohara	na
PBM18Y	Ohara	na

na, not available.

12.4.8 Applications and limitations

Current applications: Special quality grades of normal optical glass types for i-line lenses.
Potential application and outlook: All applications that require improved UV transmittance compared to “standard” optical glass.
Limitations: Decreased transmittance towards shorter wavelengths and availability in high-transmittance and high-homogeneity grades presupposing dedicated melting runs.

12.4.9 Further reading

1. H. Bach and Krause, *Optical Glass*, Springer-Verlag, Berlin (1996).

12.4.10 Links

- http://www.schott.com/lithotec/english/
- http://www.oharacorp.com/optical_glass/i-Line.html

12.4.11 Research and development

i-line glasses are state-of-the-art material, leading-edge transmitting microlithography now uses fused silica and CaF_2 lenses.

12.5 Fused Silica

J. Alkemper

12.5.1 Intended purpose of the glass

Applications that require high transmission (in particular, in the UV and IR ranges), high radiation damage resistance, temperature resistance, chemical durability, low thermal expansion, and/or ultra-low impurity content.

12.5.2 Glass types

Glass type	Manufacturer (selection)
Electrical fused quartz (Type I)	Heraeus, Tosoh, Shin-Etsu
Flame fused quartz (Type II)	Heraeus, Tosoh, Shin-Etsu
Fused silica (Type III, "wet," flame hydrolysis)	Schott, Corning, Heraeus, Shin-Etsu, Asahi, Tosoh
"Dry" fused silica (Type IIIb, dried soot)	Heraeus, Corning, Asahi, Shin-Etsu
"Dry" fused silica (Type IV, plasma oxidation)	Heraeus

12.5.3 Optical properties

Glass type	Refractive index, n_d	Abbe number, ν_d	Partial dispersion, $P_{g,F}$	Internal transmission (10 nm) λ_1 (200 nm)	λ_2 (300 nm)	λ_3 (500 nm)	λ_4 (1000 nm)	λ_5 (1400 nm)	Stress optical coefficient, K (10^{-6} mm^2/N)
Electrical fused quartz (Type I)	na	na	na	<0.40	0.99	0.99	0.99	0.99	
Flame fused quartz (Type II)	1.4586	67.6	na	<0.40	0.998	0.998	0.995	<0.95	
Fused silica (Type III)	1.45844	67.9	0.5277	>0.99	0.998	0.998	0.995	<0.95	3.4
Dry fused silica (Type IIIb)	1.4585	68.2	na	>0.99	0.998	0.998	0.998	0.998	
Dry fused silica (Type IV)	na	na	na	<0.98	0.998	0.998	0.998	0.998	

na, not available.

12.5.4 Mechanical properties

Glass type	Density, ρ (g/cm^3)	Knoop hardness, $HK_{0.1/20}$	Young's modulus, E (10^8 N/m^2)	Poisson's ratio, μ	Thermal stress, ϕw (N/m^{2}K) $* 10^{-6}$
Electrical fused quartz (Type I)	2.203	580	725	0.17	445.48
Flame fused quartz (Type II)	2.203	580	725	0.17	445.48
Fused silica (Type III)	2.201	580	720	0.17	468.43

12.5.5 Chemical properties

Solution at 95°C	Duration of test	Weight loss (mg/cm^3)
5% HCl	24 h	<0.01
5% NaOH	6 h	0.45
H_2O deionized	24 h	0.01
10% HF (at 25°C)	20 min	0.23

http://www.corning.com/semiconductoroptics/products_services/semiconductor_optics/.

12.5.6 Thermal properties

Glass type	Thermal expansion, CTE (−30°C/70°C) ($10^{-6} \times 1/K$)	Thermal expansion, CTE (100°C–300°C) ($10^{-6} \times 1/K$)	Thermal expansion, CTE (20°C–300°C) ($10^{-6} \times 1/K$)	Temperature coefficients of refractive index, dn/dt (632.8 nm $* 10^{-6}$.+20/+40°C)	Transformation temperature, T_g (°C)	Softening point, T_s (°C)	Thermal capacity, C_p (J/gK)	Thermal conductivity, λ (W/mK)	Thermal diffusivity, κ ($\lambda/C_p * \rho$) (10^{-6} m^2/s)
Electrical fused quartz (Type I)	0.51	0.58	0.59		1220	1125	0.77	1.38	0.8106
Flame fused quartz (Type II)	0.51	0.58	0.59	10.4	1160	1070	0.77	1.38	0.8106
Fused silica (Type III)	0.54	0.58	0.59	9.7	1080	980	0.79	1.31	0.7572

12.5.7 Form of delivery

Glass type	Manufacturer
Electrical fused quartz (Type I)	Heraeus, Tosoh, Shin-Etsu
Flame fused quartz (Type II)	Heraeus, Tosoh, Shin-Etsu
Fused silica (Type III, "wet," flame hydrolysis)	Schott, Corning, Heraeus, Shin-Etsu, Asahi, Tosoh
"Dry" fused silica (Type IIIb, dried soot)	Heraeus, Corning, Asahi, Shin-Etsu
"Dry" fused silica (Type IV, plasma oxidation)	Heraeus

12.5.8 Applications and limitations

Current applications: UV optics (including microlithography, LCD lithography), IR optics, optical fibers (telecom, datacom, light guides), laser optics, chemical production equipment (e.g., crucibles, window material for reactors, substrates), lamp industry, glass wafers for high-temperature p-Si Thin Film Transistor (TFT) displays.
Potential application and outlook: Microstructured optical elements.
Limitations: High melting temperature, leading to complicated and cost-intensive production and shaping processes.

12.5.9 Further reading

1. R. Brückner, "Silicon dioxide," *Encyclopedia of Applied Physics,* Vol. 18, pp. 101–131 (1997).

12.5.10 Links

- http://www.schott.com/lithotec/english/
- http://www.heraeus-quarzglas.com
- http://www.tosohquartz.com/main.html
- http://www.corning.com/semiconductoroptics/products_services/semiconductor_optics/hpfs.asp
- http://www.sqp.co.jp/e/
- http://www.agc.co.jp/english/quartz/index.html

12.5.11 Research and development

1. C.M. Smith and L.A. Moore, "Fused silica for 157 nm transmittance," *Proc. SPIE*, Vol. 3676, pp. 834–841 (2000).
2. Y. Ikuta, S. Kikugawa, A. Masui, N. Shimodaira, S. Yoshizawa, and M. Hirano, "The new silica glass for 157 nm lithography," *Proc. SPIE*, Vol. 3676, pp. 827–833 (2000).

12.6 Optical Polymers

A. Laschitsch

12.6.1 Intended purpose of the polymer

Optical applications for transparent lightweight materials, such as displays, light-guides, screens, lenses, lens arrays, (sun) glasses, and lighting such as lamps (automotive and exterior).

12.6.2 Types of polymer

Trade name/type	Manufacturer	Polymer family
HW55	Röhm GmbH	PMMA-copolymer
LURAN®	BASF AG	Poly(styrene-co-acrylonitrile) (SAN)
MAKROLON®, LEXAN®	Bayer AG	Polycarbonate (PC)
PLEXIGLAS®, LUCITE®	Röhm GmbH	Polymethyl methacrylate
PLEXIGLAS RESIST® (data in tables from molding compound zk4BR)	Röhm GmbH	(PMMA)
PLEXIMID® 8817	Röhm GmbH	Polyimide (PMMI)
RADEL® R-5000	Solvay	Polyphenylsulfon (PPSU)
STYRON, POLYSTYRENE	BASF AG	Polystyrene (PS)

(Continued)

Trade name/type	Manufacturer	Polymer family
TOPAS® 5013 S-04	Ticona	Cyclo-olefin copolymer (COC)
TROGAMID CX	Degussa GmbH/HPP	Polyamide (PA)
TROGAMID T	Degussa GmbH/HPP	Polyamide (PA)
ZEONEX 480®	Zeon Chemicals	Cyclo-olefin polymer (COP)

HPP, High Performance Polymers.

12.6.3 Optical properties

Polymer	Refractive index, n_d (ISO 489)	Abbe number, ν_d	Haze (%) (ASTM D1003)	Transmission		
				τ_{D65}(%) (DIN 5036)	Internal transmission 350–1100 nm	Internal transmission >2800 nm
HW55	1.512	47	<0.5	91	99.3%	0%
LURAN	1.571	35.3	na	90	99.9%	0%
MAKROLON, LEXAN	1.586	34.5	na	89	99.3%	0%
PLEXIGLAS®	1.492	55	<0.5	92	100.0%	0%
PLEXIGLAS RESIST®	1.498	50	<1.5	92	99.9%	0%
PLEXIMID® 8817	1.541	41	<1.5	90	99.1%	0%
RADEL R-5000 (1.8–3.3 mm)	1.672	18.7	3.1–5.1	74–80	92.7%	0%
STYRON, POLYSTYRENE	1.590	30.9	na	89	99.4%	0%
TOPAS® 5013 S-04	1.535	56.1	<1	91 (2 mm)	99.9%	0%
TROGAMID CX 7323	1.516	52	na	91.7 (2 mm)	100.0%	0%
TROGAMID T5000	1.566	na	na	90 (2 mm)	99.7%	0%
ZEONEX 480R	1.525	56.2	na	91	99.6%	0%

na, not available.

12.6.4 Mechanical properties

Type of polymer	Density (g/m^3)	Young's modulus, E (GPa) (ISO 527)	Charpy impact strength (unnotched), kJ/m^2 (ISO 179)
HW55	1.19	3.6	20
LURAN	1.07	3.7	16
MAKROLON, LEXAN	1.20	2.4	No break
PLEXIGLAS®, LUCITE	1.19	3.2	20
PLEXIGLAS RESIST®	1.18	2.8	25*
PLEXIMID® 8817	1.21	4.5	20
RADEL R-5000 (1.8–3.3 mm)	1.29	2.3	48
STYRON, POLYSTYRENE	1.10	3.2	Not available (varies)

(Continued)

Type of polymer	Density (g/m^3)	Young's modulus, E (GPa) (ISO 527)	Charpy impact strength (unnotched), kJ/m^2 (ISO 179)
TOPAS® 5013 S-04	1.02	3.2	13
TROGAMID CX7323	1.02 (at 23°C)	1.4	No break (23, 0, −30°C)
TROGAMID T5000	1.12 (at 23°C)	2.8	No break (23, 0, −30°C)
ZEONEX 480R	1.01	2.4	Not available (brittle)

*Depends on impact modifier content; can be up to 80 kJ/m^2.

12.6.5 Chemical properties

Type of polymer	Resistant to	Not resistant to
HW55	Bases, low-concentration acids, fatty acids, pure gasoline, glycerine, phosphates	(M)Ethanol, acetone, ether, ketones, carbon hydroxides
LURAN	na	na
MAKROLON, LEXAN	na	na
PLEXIGLAS®	Bases, low-concentration acids, fatty acids, pure gasoline, glycerine, phosphates	(M)Ethanol, acetone, ether, ketones, carbon hydroxides
PLEXIGLAS RESIST®	Bases, low-concentration acids, fatty acids, pure gasoline, glycerine, phosphates	(M)Ethanol, acetone, ether, ketones, carbon hydroxides
PLEXIMID® 8817	Bases, low-concentration acids, fatty acids, pure gasoline, glycerine, phosphates	(M)Ethanol, acetone, ether, ketones, carbon hydroxides
RADEL R-5000 (1.8–3.3 mm)	na	na
STYRON, POLYSTYRENE	na	Benzene
TOPAS 5013 S-04	na	na
TROGAMID CX	Methanol, ethanol, glycerine, acetone, ethyl acetate, aliphatic and aromatic hydrocarbons, gasoline, diesel fuel, hydraulic fluid, lubricating oil, alkali hydroxides, weak acids	Phenols, halogenated hydrocarbons, conc. sulfuric acid
TROGAMID T	Ethyl acetate, aliphatic and aromatic hydrocarbons, gasoline, alkali hydroxides, weak acids	Alcohols, phenols, MEK, conc. sulfuric acid, conc. hydrochloric acid, methylene chloride
ZEONEX 480R	na	na

For a detailed listing, please contact the specific manufacturers of the polymer.
na, not available.

12.6.6 Thermal properties

Type of polymer	CTE (10^{-6}/K) (0–50°C) (ASTM E 831)	VST (B/50) (ISO 306)	T_g* (IEC 10006)	C_p (J/gK) ($T = 23$°C)	λ (W/mK)	κ ($\lambda/C_p * \rho$) (10^{-6} m^2/s)	Operation max. temp. (°C)
HW55	70	119	122	na	na	na	
Luran	70	103	106	1.30	0.17	0.1222	
Makrolon®, Lexan	65	146	146	na	0.20	na	
PLEXIGLAS®, LUCITE	80	108	117	1.459	0.191	0.1100	For all polymers, in general, 25–30 K below Vicat softening temperature
PLEXIGLAS RESIST®	80	102	108	na	na	na	
PLEXIMID® 8817	45	170	172	na	na	na	
Radel R-5000	56	214	220	na	0.35	na	
Styron, Polystyrene	80	84	90	1.28	0.110	0.0781	
TOPAS® 5013 S-04	60	127	na	na	na	na	
TROGAMID CX7323	90	130	140	2.5	0.25	0.098	
TROGAMID T5000	54	150	153	2.2	0.21	0.085	
Zeonex 480R	60	133	na	na	na	na	

*Varies for all polymers with molecular weight.

Abbreviations: CTE, coefficient of thermal expansion; VST, Vicat softening temperature; C_p, specific heat capacity; λ, heat conductivity.

na, not available.

12.6.7 Form of delivery

Type of polymer	Manufacturer	Trade name	Molding compound	Semi-finished sheet
HW55	Röhm GmbH	PLEXIGLAS®	HW55	
SAN	BASF AG	Luran		
PC	Bayer AG, Dow Chemical	Makrolon, Lexan	LQ3147	
PMMA	Röhm GmbH	PLEXIGLAS®, LUCITE	7N, 8N, 7H, 7M (standard grades) zk series HW55	XT 20070 (extruded) GS 233 (cast material) PLEXIGLAS RESIST®
PLEXIGLAS RESIST® (molding compound zk4BR)	Röhm GmbH	PLEXIGLAS®	zk series HW55	XT 20070 (extruded) GS 233 (cast material) PLEXIGLAS RESIST®
PLEXIMID®	Röhm GmbH	PLEXIGLAS® PLEX® grades	PLEX® 8817, 8813	
PPSU	Solvay	Radel	Radel R-5000	
PS	BASF AG	Styrone, Polystyrene	Several grades	
COC	Ticona	TOPAS®	5013S-04	
PA	Degussa GmbH/HPP	TROGAMID	TROGAMID CX7323, TROGAMID T5000	
COP	Zeon Chemicals	ZEONEX 480R, E48R		

HPP, High Performance Polymers.

12.6.8 Applications and limitations

Current applications: Used in lenses, displays in opto-electronic devices, automotive lighting, building materials, sound protection, polymer optical fibers, and lightguides.
Potential application and outlook: Optical storage media and fibers as lightguides.
Limitations: Temperature stability—depending on glass temperature of the specific polymer, which is low compared to inorganic low-molecular glasses; and temperature dependence of optical properties—the refractive index decreases slightly with increasing temperatures, and haze depends on temperature.

12.6.9 Further reading

1. Polymer Handbook
2. Handbook of Chemistry & Physics
3. S. Bäumer (ed.) *Handbook of Plastic Optics*, Wiley-VCH (2005).

12.6.10 Links

- http://www.beuth.de (German DIN-standards for testing of materials)
- http://www.astm.org (American standards for testing of materials)
- http://www.roehm.com (Manufacturer of PLEXIGLAS®)
- http://www.degussa-hpp.com/dl/brochure/trogamid_cx_eng.pdf
- http://www.degussa-hpp.com/dl/brochure/trogamid_cx_optics_eng.pdf (Data sheets TROGAMID®)
- http://www.makrolon.de/BMS/DB-RSC/MakrolonCMSR6.nsf/id/home_en (Data sheets MAKROLON®)
- http://www.geplastics.com/ (Data sheets LEXAN®)
- http://www.solvayadvancedpolymers.com/static/wma/pdf/1/6/0/R5_1_500.pdf
- http://www.kern-gmbh.de/cgi-bin/riweta.cgi?nr=2461&Ing=2 (Data sheets RADEL®)
- http://www.ensinger.ltd.uk/docs/datasheets/TECASON%20P.pdf (PPSU (TECASON) data sheets)
- http://www.plasticsportal.net/wa/plasticsEU~en_GB/portal/styrenicsrc/content/products/styrenics/polystyrol (BASF Polystyrene information)
- http://www.zeon.co.jp/business_e/enterprise/speplast/speplast1_8.html (Data sheets Zeonex®)

12.7 Crystals for UV Optics

J. Korth

12.7.1 Intended purpose of the crystals

Highly UV-transmitting optics and color-corrected optics.

12.7.2 Types of crystals

Type of crystal	Manufacturer	Crystal family/system
BaF_2	Korth	Fluorite/cubic
CaF_2	Canon, Korth, Nikon, Schott Lithotec, St. Goban	Fluorite/cubic
LiF	Korth	Rocksalt/cubic
MgF_2	Korth, St. Goban	Rutile/tetragonal
NaF	Korth	Rocksalt/cubic
SrF_2	Korth	Fluorite/cubic
SiO_2 (α-Quarz)	Korth in Coop., Sawyer and other Companies	Quartz/trigonal-rhombohedral

12.7.3 Optical properties

Type of crystal	n_d (0.58956 μm)	ν_d	Cutoff UV/VIS λ (μm)	Internal transmission* (10 nm)			Cutoff IR, λ (μm)	Stress optical coefficient, K (10^{-6} mm^2/N)
				λ_1 (0.157 μm)	λ_2 (0.193 μm)	λ_3 (0.248 μm)		
BaF_2	1.472	68.202	0.14	0.935	0.997	1	13	na
CaF_2	1.434	94.984	0.12	1	0.998	1	10	na
LiF	1.392	97.283	0.12	0.765	0.906	0.955	6.6	na
MgF_2 o	1.378	106.206	na	0.999	0.997	1	7.7	na
MgF_2 e	1.388	86.592	0.13	na	na	na	na	na
NaF	1.325	88.277	0.13	0.957	0.971	0.982	12	na
SrF_2	1.438	91.582	0.13	0.972	0.992	0.988	12	na
Quarz o	1.544	69.948	0.16	0.821	0.934	0.968	4	na
Quarz e	1.554	71.497	na	na	na	na	na	na

*$T = (1 - R)^2 \times e^{-k \times d} \times e^{-A_{\text{surface}}} = T_{\text{theor}} \times T_{\text{intern}} \times T_{\text{surface}}$, with reflection loss $R = [(n-1)/(n+1)]^2$ (n-refractive index), volume absorption $k \times d$ (d-optical path, k-absorption constant), and surface absorption A_{surface}.

na, not available.

12.7.4 Mechanical properties

Type of crystal	Density (g/cm^3)	Knoop hardness (kg/mm^2)	Cleavage	Young's modulus, E (10^8 N/m^2)	G-module (GN/m^2)	Poisson's ratio, μ	Thermal stress, ϕw ($N/m^2K * 10^{-6}$)
BaF_2	4.83	82	(111)	65.8	25.1	0.31	1.75
CaF_2	3.18	158	(111)	110	42.5	0.29	2.93
LiF	2.635	110	(100)	110	45	0.22	4.85
MgF_2	3.18	415	(010),				
			(110)	137	53.9	0.26	1.74/2.53
NaF	2.588	60	(100)	76	30.7	0.24	3.35
SrF_2	4.24	130	(111)	89	34.6	0.29	2.27
SiO_2							
(α-Quarz)	2.65	741	None	95	44	0.08	1.28/0.71

12.7.5 Chemical properties

Type of crystal	Solubility in water (g/100 g water)	Melting point, T (K)
BaF_2	1620×10^{-4}	296
CaF_2	17×10^{-4}	293
LiF	2700×10^{-4}	291
MgF_2	$<2 \times 10^{-4}$	293
NaF	4.22	291
SrF_2	120×10^{-4}	296
SiO_2 (α-Quarz)	Insoluble	na

na, not available.

12.7.6 Thermal properties

Type of crystal	CTE (10^{-6}/K)	dn/dT (10^{-6}/K) (o: ordinary ray, e: extraordinary ray) λ (nm)		T (K)	T_S (K)	Thermal capacity, C_p (J/g * K)	Thermal conductivity, λ (W/m * K)	T (K)	Stress optical coefficient, K (10^{-6} mm²/N)
BaF_2	18.4	632.8	−16.3	313	1550	0.4474	12.0	300	5.5957
CaF_2	18.9	632.8	−11.5	293	1630	0.9113	9.7	300	3.3454
LiF	34.4	632.8	−17.0	313	1115	1.62	14	300	3.2716
MgF_2	$9.4\|\vec{a}$, $13.6\|\vec{c}$	632.8	1.12, 0.58e	293	1536	1.0236	$30\|\vec{a}$, $21\|\vec{c}$	300	9.2490, 6.4177
NaF	33.5	632.8	−13	313	1266	1.1239	22	250	7.5764
SrF_2	18.1	632.8	−12.5	313	1710	0.62	8.3	300	3.1447
SiO_2 (α-Quarz)	$12.38\|\vec{a}$, $6.88\|\vec{c}$	546	−6.2o, −7.0e	RT	845p*	0.74	$6.2\|\vec{a}$, $10.4\|\vec{c}$	300	3.1711, 5.3758

*p, phase change.

12.7.7 Form of delivery

Type of crystal	Manufacturer	Dimensions
BaF_2	Korth	Ø: 150 and 210 mm
CaF_2	Korth	Ø: 210 mm
LiF	Korth	Ø: 210 mm
MgF_2	Korth	Ø: 150 and 210 mm
NaF	Korth	Ø: 150 mm
SrF_2	Korth	Ø: 210 mm
SiO_2 (α-Quarz)		$120 \times 90 \times 30\,mm^3$

12.7.8 Applications and limitations

Current applications: UV / VUV spectroscopy, UV objectives, microlithography, excimerlaser components, PMT windows, detectors, space research, and military applications.
Limitations: Color center formation at high energy levels, mechanical and thermal damage of the surface, and more sophisticated finishing methods than glass.

12.7.9 Research and development

Research and development will focus on suppression of color center formation, production of material with very high transmission and low stress birefringence, and steady improvement of polished surfaces.

12.8 Crystals for IR Optics

J. Korth

12.8.1 Intended purpose of the crystals

High IR-transmittive optical elements and color-corrected optics.

12.8.2 Types of crystals

Type of crystal	Manufacturer	Crystal family/system
AgCl	Bicron, Korth, Russia	Rocksalt/cubic
BaF_2	Korth, Ohyo Koken	Fluorite/cubic
CaF_2	Korth and many others	Fluorite/cubic
CsI	Korth, Russia	CsCl-Typ/cubic
Ge	Umicore, Photonic Sense	Ccp*/cubic
KBr	Korth and others	Rocksalt/cubic
KCl	Korth	Rocksalt/cubic
Tl(Br, I) KRS-5	Korth, Russia	CsCl-Typ/cubic
NaCl	Korth and others	Rocksalt/cubic
Si	Wacker	ccp/cubic
ZnS	II-IV, Röhm & Haas, Vitron	Sphalerite/cubic
ZnSe	II-IV, Röhm & Haas, Umicore	Sphalerite/cubic

*Ccp: cubic close packed.

12.8.3 Optical properties

Type of crystal	n_d (0.58956 μm)	ν_d	Cutoff UV/VIS, λ (μm)	Internal transmission* (10 nm) λ_1 (3 μm)	λ_2 (10 μm)	λ_3 (20 μm)	Cutoff IR, λ (μm)	Stress optical coefficient, K (10^{-6} mm^2/N)
AgCl	2.066	20.8	0.42	1	1	1	28	na
BaF_2	1.472	68.2	0.14	1	1	0	13	na
CaF_2	1.434	94.9	0.12	1	0.093	0	10	na
CsI	1.654	30.4	0.25	0.973	1	1	62	na
Ge**	material is not transmissive at stated wave length	na	1.8	1	1	0.129	15	na
KBr	1.560	33.6	0.20	1	1	1	30.6	na
KCl	1.490	44.3	0.18	1	1	0.634	25	na
Tl(Br, I) KRS-5	material is not transmissive at stated wave length		0.58	1	1	1	42	na
NaCl	1.544	42.8	0.17	1	1	0.073	18	na
Si	material is not transmissive at stated wave length		1.1	1	1	0.295	6.5	na
ZnS	2.367	15.5	0.4	1	1	0	12.5	na
ZnSe**	2.622	8.2	0.5	0.823	0.799	0	20	na

*$T = (1 - R)^2 \times e^{-k \times d} \times e^{-A_{\text{surface}}} = T_{\text{theor}} \times T_{\text{interm}} \times T_{\text{surface}}$, with reflection loss $R = [(n-1)/(n+1)]^2$ (n-refractive index), volume absorption $k \times d$ (d-optical path, k-absorption constant), and surface absorption A_{surface}.

**Transmission without AR-coating.

na, not available.

12.8.4 Mechanical properties

Type of crystal	Density (g/cm^3)	Knoop hardness (kg/mm^2)	Cleavage	Young's modulus, E (10^8 N/m^2)	G-module (GN/m^2)	Poisson's ratio, μ	Thermal stress, ϕ_w ($N/m^2K * 10^{-6}$)
AgCl	5.56	9.5	None	22.9	8.1	0.41	1.26
BaF_2	4.83	82	(111)	65.8	25.1	0.31	1.75
CaF_2	3.18	158	(111)	110	42.5	0.29	2.93
CsI	4.51	1–2 Mohs*	None	18	7.3	0.26	1.18
Ge	5.35	800	(111)	132	54.8	0.20	0.94
KBr	2.75	7	(100)	18	7.2	0.30	0.99
KCl	1.984	9.3	(100)	22	8.5	0.29	1.13
Tl(Br, I) KRS-5	7.371	40	None	19.6	7.3	0.34	1.72
NaCl	2.165	18	(100)	37	14.5	0.26	2.06
Si	2.33	1150	(111)	162	66.2	0.22	0.54
ZnS (CVD)	4.04	178	None	74.5	nd	0.29	0.71
ZnSe (CVD)	5.42	137	None	70.3	nd	0.28	0.69

*Mohs is an empirical unit of hardness.
nd, not determined.

12.8.5 Physical and chemical properties

Type of crystal	Solubility in water (g/100 g water)	T (K)	(g/100 g water)	Melting point, T (K)	Remarks
AgCl	0.89×10^{-4}	283	21×10^{-4}	373	Sensitive to UV and VIS light
BaF_2	1620×10^{-4}	296			
CaF_2	17×10^{-4}	293			
CsI	44	273	160	334	
Ge	Insoluble				
KBr	53.4	273	102	373	
KCl	34.7	293	56.7	373	
Tl(Br, I) KRS-5*	0.05	293	0.25	341	
NaCl	35.7	273	39.1	373	
Si	Insoluble				
ZnS	0.65×10^{-4}	291			Poisonous H_2S gas with acids
ZnSe	Insoluble				Poisonous H_2Se gas with acids

* Data for TlBr.

12.8.6 Thermal properties

Type of crystal	CTE (10^{-6}/K)	λ (nm)	dn/dT (10^{-6}/K)	T(K)	Melting point, T_S (K)	Thermal capacity, C_p (J/g $*$ K)	Thermal conductivity, λ (W/m $*$ K)	T(K)	κ ($\lambda/C_p * \rho$) (10^{-6} m^2/s)
AgCl	32.4	633	−61	RT	728	0.3544	1.12	295	0.5691
BaF_2	18.4	632.8	−16.3	313	1550	0.4474	12.0	300	5.5957
CaF_2	18.9	632.8	−11.5	293	1630	0.9113	9.7	300	3.3454
CsI	48.6	633	−99.3	RT	898	0.2032	1.05	300	1.1429
Ge	5.7	2500	426	RT	1211	0.3230	59.9	300	37.3831
KBr	38.5	1150	−41.9	293	1007	0.4400	4.8	300	3.9526
KCl	36.5	1150	−36.2	293	1049	0.6936	6.7	300	4.8465
Tl(Br, I) KRS-5	58	1014	−240	298	687	0.16	0.32	300	0.2713
NaCl	41.1	1150	−36.4	293	1074	0.8699	6.5	300	3.4470
Si	2.62	2500	166	RT	1680	0.7139	140	300	8.5837
ZnS	6.8	1150	49.8	RT	1293p*	0.4732	16.7	300	8.8402
ZnSe	7.1	1150	59.7	RT	1790	0.339	13	300	7.0962

*p, phase change; RT, room temperature.

12.8.7 Form of delivery

Type of crystal	Manufacturer	Dimensions
AgCl	Korth	Ø: 70 mm
BaF_2	Korth	Ø: 150 and 210 mm
CaF_2	Korth	Ø: 210 mm
CsI	Korth	Ø: 70 and 100 mm
Ge	Umicore	Ø ≤300 mm
KBr	Korth	Ø: 180 mm
KCl	Korth	Ø: 180 mm
Tl(Br, I) KRS-5	Korth	Ø: 70 and 100 mm
NaCl	Korth	Ø: 180 mm
Si	Wacker	Ø ≤ 300 mm
ZnS		
ZnSe		

12.8.8 Applications and limitations

Current applications: IR spectroscopy and components for OEMs, IR laser optics, astronomical objectives, military equipment, Raman spectroscopy, and thermal imaging.
Limitations: Chemical and mechanical properties of the material.

12.8.9 Research and development

Research and development will focus on cheap production by automatic control.

12.9 Glass Ceramics

B. Schreder

12.9.1 Intended purpose of the glass ceramics

Reflective optics.

12.9.2 Types of glass ceramics

Type of glass ceramics	Manufacturer	Ceramics family/system
ZERODUR®	Schott	Lithium aluminum silicate
CLEARCERAM®-Z	Ohara	Lithium aluminum silicate
Astrositall®	LZOS (Lytkarino, Rus)	Lithium aluminum silicate
ZPF	Taiheiyo Cement	SiC/Si_3N_4/LAS-Ceramic

12.9.3 Optical properties

Type of glass ceramics	Refractive index, n_d	Abbe number, ν_d	Partial dispersion, $\Delta P_{g,F}$	Internal transmission λ_1 (600 nm)	λ_2 (2000 nm)	λ_3 (500 nm)	λ_4 (980 nm)	Stress optical coefficient, K (10^{-6} mm^2/N)
ZERODUR®	1.5424	56.1	na	>90% ($d = 5$ mm)	>90%	na	na	3.0
CLEAR-CERAM®-Z	1.546	55.5	na	na	na	>80% ($d = 10$ mm)	>97% ($d = 10$ mm)	na
Astrositall®	1.536		na	na	na	na	na	na
ZPF	na*	na*	na*	na*	na*	na*	na*	na

*ZPF nontransparent in Vis range.

12.9.4 Mechanical properties

Type of glass ceramics	Density, ρ	Knoop hardness, $HK_{0.1/20}$	Young's modulus, E (10^8 N/m^2)	Poisson's ratio, μ
ZERODUR®	2.53	620 MPa	90.3 GPa	0.243
CLEARCERAM®-Z	2.55	680 MPa	90	0.25
Astrosital®	2.46	na	92	0.28
ZPF	2.54	na	150	na

12.9.5 Chemical properties

Type of glass ceramics	Alkali resistance class (AR) (acc. ISO10629)	Acid resistance class (SR) (acc. ISO 8424)	Acid resistance class (SR) (Powder)	Climatic resistance class (CR) (acc. ISO/WD13384)	Hydrolytic resistance (WR) (acc. JOGIS)
ZERODUR®	1.0 A2 (ISO 695)	1.0 4 W (DIN 12116)	na	1	HGB 1 (ISO 719)
CLEAR-CERAM®-Z	A2 (ISO 695)	4 W (DIN 12116)	na	na	HGB 2 (ISO 719)
Astrositall®	na	na	na	na	na
ZPF	na	na	na	na	na

Link to http://www.schott.com/optics_devices/english/download/ (SR ISO).
Link to http://www.hoyaoptics.com/pdf/OpticalGlass.pdf (SR Powder).
Note, methods of determination of chemical stability are different for different manufacturers.
na, not available.

12.9.6 Thermal properties

Type of glass ceramics	Thermal expansion, CTE (0°C/50°C) (ppm/K)	Thermal expansion, CTE (−60°C/60°C) (ppm/K)	Thermal expansion, CTE (20°C/300°C) (ppm/K)	Temperature coefficients of refractive index, (dn/dt) (632.8 nm $* 10^{-6}$ + 20/+40°C)	Transformation temperature, T_g (°C)	Softening point, T_s (°C)	Thermal capacity, C_p (J/g * K)	Thermal conductivity, λ (W/m * K)	κ $(\lambda/C_p{*}\rho)$ $(10^{-6}\,m^2/N)$
ZERODUR®	0.013 (±20 ppb/K)	na	na	na	~670	na	0.8	1.46	0.72
CLEARCERAM®-Z	0.089 (±80 ppb/K)	na	na	na	na	na	na	1.52	na
Astrositall®	−0.043	±150 ppb/K	na	na	na	na	0.92	1.18	na
ZPF	0.02	na	na	na	na	na	0.8	na	na

na, not available.

12.9.7 Form of delivery

Type of glass ceramics	Manufacturer	Disks, blanks diameter, d (m)	Blocks, prisms, rods, cut pieces length, l (m)
ZERODUR®	Schott	2.5 reg. 4 spec. up to 8	2.5
CLEARCERAM®-Z	Ohara	0.7 reg. 1.7 spec.	3.2
Astrositall®	LZOS (Lytkarino, Rus)	2	3
ZPF	Taiheiyo Cement	<1	na

na, not available.

12.9.8 Applications and limitations

Current applications Zero expansion materials for lithography and astro applications (mirror blanks), high-precision measuring technology, wafer stages/prisms, and optical stages.
Potential application and outlook: EUV-lithography (mirrors for EUV objectives), molds.
Limitations: Mechanical and chemical stability, thermal conductivity, homogeneity (striae, bubbles), and surface roughness.

12.9.9 Links (company information)

- http://www.schott.com/optics_devices/english
- http://www.lzos.ru/en/glass_sitall.htm
- http://www.taiheiyo-cement.co.jp/english
- http://www.ohara-gmbh.com/d/produkte/Ohara_Clearceram-Z.pdf

12.9.10 Links (research and development)

- http://www.schott.com/ft/english

12.10 Opto-Ceramics

U. Peuchert

12.10.1 Types of opto-ceramics

Type of opto-ceramics	Manufacturer	Ceramics family/system
YAG	Konoshima	Yttria aluminia garnet
AlON	Surmet	Aluminium oxy-nitride
Lumicera	Murata	Ba-Ta-perovskite
Spinel	TA&T	Magnesium aluminum oxide
Y_2O_3	Konoshima	Sesquioxides

12.10.2 Optical properties

Type of opto-ceramics	Refractive index, n_d	Abbe number, ν_d	Partial dispersion, $\Delta P_{g,F}$	Internal transmission λ_1 (400 nm; 4.5 mm)	λ_2 (500 nm, 4.5 nm)	λ_3 (750 nm; 4.5 mm)	Stress optical coefficient, K (10^{-6} mm^2/N)
YAG	1.833; 1.81 (@1064 nm)	–	–	>80% (300–700 nm)	–	–	na
AlON	~1.795	–	–	~85% (400, 3000 nm; 2 mm)	~60% (5000 nm; 2 mm)	~20% (5400 nm; 2 mm)	na
Lumicera™	2.095	30.4	–	~51% (300 nm; 0.6 mm)	~76% (400 nm; 0.6 mm)	~79% (700–900 nm; 0.6 mm)	na
Spinel	~1.71–1.72	60.7	–	>80% (300 nm; 2.7 mm)	–	–	na
Y_2O_3	1.914	36	0.572 $\Delta P_{g,F} = -0.011$	0.83	0.93	>50%	na

na, not available.

12.10.3 Mechanical properties

Type of opto-ceramics	Density, ρ	Knoop hardness, $HK_{0.1/20}$	Young's modulus, E (10^8 N/m^2)	Poisson's ratio, μ
YAG	4.55	1400 kg/mm^2 (Hv)	–	–
AlON	~3.69	1850 kg/mm^2 (200 g load)	324	0.24
Lumicera™	7.5	800 (Hv)	–	–
Spinel	3.58	1645 kg/mm^2	275	0.26
Y_2O_3	5.01	760	–	–

12.10.4 Thermal properties

Type of opto-ceramics	Thermal expansion, CTE	Thermal expansion, CTE (100°C/300°C)	Thermal expansion, CTE (20°C/300°C)	Temperature coefficients of refractive index, dn/dt (632.8 nm $* 10^{-6}$ + 20/+40°C)	Transformation temperature, T_g	Softening point, T_s	Thermal capacity, C_p	Thermal conductivity, λ
YAG	8.0	–	–	–	–	–	–	11.7 W/m ∗ K
AlON	5.65 (30/200)	–	–	–	–	2150°C (MP)	–	0.023 @ 75°C (cal/cm-s-°C) = 12.6 W/m ∗ K
Lumicera™	–	–	–	–	–	–	–	3.3 W/m ∗ K
Spinel	8.0 (30/900)	–	–	–	–	–	–	14.6 W/m ∗ K
Y_2O_3	6–7	–	–	–	–	–	–	10–15 W/m ∗ K

Note, methods of determination of chemical stability are different for different manufacturers.

12.10.5 Form of delivery

Type of opto-ceramics	Manufacturer	Remarks
YAG	Konoshina	
Yttria	Raytheon	
AlON	Surmet	Windows, domes, plates, rods, tubes
Lumicera™	Murata	Cylinders, rods
Spinel	TA & T	Windows, domes, lenses
Y_2O_3	Konoshima	Slabs (no market presence)
Y_2O_3	Raytheon	Domes, windows (no optical quality)

12.10.6 Applications and limitations

Current applications: Windows, discharge burners, lasers, scintillator hosts, and armor.
Potential application and outlook: Improvement in the design of lenses.
Limitations: Price, scattering, and large sizes hard to achieve with suitable optical quality.

12.10.7 Links

- http://www.surmet.com/docs/Product_sheet_ALON.pdf (AlON)
- http://www.murata.com/ninfo/nr0572e.html
- http://www.murata.com/catalog/k99/elumi cer.pdf (LUMICERA)
- http://www.techassess.com/tech/spinel/spinel_prop.htm (SPINEL)
- http://industrialtechnologies.net/spinel_material.html (SPINEL)
- http://physics.nist.gov/Divisions/Div842/Gp3/DUVMatChar/PDF/ML5754-57.pdf (SPINEL) Handbook of Optical Materials (Spinel, YAG)
- http://www.konoshima.co.jp/en/yag.html (YAG)
- http://www.baikowski.com/components/pdf/ceramic_yag/YAG_DATA_SHEET.pdf (YAG)
- http://www.vloc.com/PDFs/YAGBrochure.pdf (YAG)
- http://baikowski.com/fr/technical_markets/tm_ceramicYAG.shtml
- http://www.murata.com/opt/index.html
- http://www.konoshima.co.jp/en/yag.html

12.11 Glasses for IR Optics

K. Seneschal

12.11.1 Intended purpose of the glass

Optical elements with high transmission in the IR range.

12.11.2 IR glass types

Glass type	Manufacturer	Glass family
IRG 1	Schott	IRG
IRG 2	Schott	IRG
IRG 3	Schott	IRG
IRG N6	Schott	IRG
IRG 7	Schott	IRG
IRG 9	Schott	IRG
IRG 11	Schott	IRG
IRG 15	Schott	IRG
IRG 100	Schott	IRG
IG 2	Vitron	IG
IG 3	Vitron	IG
IG 4	Vitron	IG
IG 5	Vitron	IG
IG 6	Vitron	IG
GASIR-1	Umicore	GASIR
GASIR-2	Umicore	GASIR
AMTIR-1	Amorphous materials	AMTIR
AMTIR-3 (1173)	Amorphous materials	AMTIR
As_2S_3	Amorphous materials	As_2S_3

12.11.3 Optical properties

Glass type	Refractive index, $n_{1.529.5\,\text{nm}}$	Abbe number, $\nu_{1.5} = (n_{1.5296} - 1)/n_{1.0140} - n_{2.3254}$	Refractive index, $n_{10\,\mu\text{m}}$	Abbe number, $\nu_{10} = (n_{10} - 1)/n_8 - n_{12}$	Transmission $\lambda_{(2\,\mu\text{m})}$ (thickness/τ%)	Transmission $\lambda_{3\ (4\,\mu\text{m})}$ (thickness/τ%)	Transmission $\lambda_{4\ (10\,\mu\text{m})}$ (thickness/τ%)	Stress optical coefficient, K (10^{-6} mm^2/N)
IRG 1	2.4380	26.3	na	na	5 mm/68%	5 mm/72%	5 mm/46%	na
IRG 2	1.8526	39.5	na	na	5 mm/83%	5 mm/79%	5 mm/0%	na
IRG 3	1.8089	35.2	na	na	5 mm/85%	5 mm/60%	5 mm/0%	na
IRG N6	1.5716	36.4	na	na	5 mm/90%	5 mm/78%	5 mm/0%	na
IRG 7	1.5442	32.4	na	na	5 mm/90%	5 mm/48%	5 mm/0%	na
IRG 9	1.4755	47.1	na	na	5 mm/92%	5 mm/65%	5 mm/0%	na
IRG 11	1.6581	41.4	na	na	5 mm/90%	5 mm/80%	5 mm/0%	na
IRG 15	1.5179	34.3	na	na	5 mm/92%	5 mm/48%	5 mm/0%	na
IRG 100	2.6577	na	2.6007	108	2 mm/63%	2 mm/64%	2 mm/65%	na
IG 2	na	na	2.4967	108	2 mm/76%	2 mm/78%	2 mm/70%	na
IG 3	na	na	2.7870	164	2 mm/55%	2 mm/62%	2 mm/65%	na
IG 4	na	na	2.6084	176	2 mm/63%	2 mm/65%	2 mm/68%	na
IG 5	na	na	2.6038	102	2 mm/63%	2 mm/65%	2 mm/69%	na
IG 6	na	na	2.7775	161	2 mm/60%	2 mm/65%	2 mm/68%	na
GASIR-1	na	na	2.4944	na	2 mm/68%	2 mm/68%	2 mm/68%	na
GASIR-2	na	na	2.6	na	na	na	na	na
AMTIR-1	2.5469	na	2.4977	na	10 mm/68%	na	6 mm/68%	na
AMTIR-3 (1173)	na	na	2.6023	na	na	na	na	na
As2S3	2.4380	na	na	na	2 mm/71%	2 mm/70%	2 mm/60%	na

Link to definition of optical properties: Schott Technical Information at http://www.schott.com/optics_devices/english/download/ and http://www.oharacorp.com/swf/og.html.
na, not available.

12.11.4 Mechanical properties

Glass type	Density $\delta\rho$ [g/cm^3]	Knoop hardness (H.K.)	Young's-modulus, E (10^3 N/m^2)	Poisson ratio, μ	Thermal stress, ϕ_w (N/mm^2)
IRG 1	3.19	107	17	0.274	na
IRG 2	5.00	481	96	0.282	na
IRG 3	4.47	541	100	0.287	na
IRG N6	2.81	623	103	0.276	na
IRG 7	3.06	379	60	0.216	na
IRG 9	3.63	346	77	0.288	na
IRG 11	3.12	608	107	0.284	na
IRG 15	2.80	430	70	0.237	na
IRG 100	4.67	150	21	0.261	na
IG 2	4.41	143	21.5	na	na
IG 3	4.84	138	22.0	na	na
IG 4	4.47	114	na	na	na
IG 5	4.66	115	22.1	na	na
IG 6	4.63	106	18.3	na	na
GASIR-1	4.40	na	18	0.28	na
GASIR-2	4.70	na	19	0.27	na
AMTIR-1	4.40	170	22	0.27	na
AMTIR-3 (1173)	4.67	150	21	0.26	na
As2S3	3.2	109	15	0.24	na

Link to definition of mechanical properties: Schott Technical Information at http://www.schott.com/optics_devices/english/download/ and http://www.oharacorp.com/swf/og.html.

na, not available.

12.11.5 Chemical properties

Glass type	Alkali resistance class (AR) (acc. ISO 10629)	Acid resistance class (SR) (acc. ISO 8424)	Climatic resistance class (CR) (acc. ISO WD 13384)	Phosphate resistance (PR) (acc. ISO 9689)
IRG 1	na	na	1	na
IRG 2	na	na	1	na
IRG 3	na	na	1	na
IRG N6	na	na	1	na
IRG 7	na	na	1	na
IRG 9	na	na	2	na
IRG 11	na	na	4	na
IRG 15	na	na	1	na
IRG 100	na	na	na	na
IG 2	na	na	na	na
IG 3	na	na	na	na
IG 4	na	na	na	na
IG 5	na	na	na	na
IG 6	na	na	na	na
GASIR-1	na	na	na	na
GASIR-2	na	na	na	na
AMTIR-1	na	na	na	na
AMTIR-3 (1173)	na	na	na	na
As2S3	na	na	na	na

Link to definition of chemical properties: Schott Technical Information at http://www.schott.com/optics_devices/ english/download/ and http://www.oharacorp.com/swf/og.html.
na, not available.

12.11.6 Thermal properties

Glass type	Thermal expansion, CTE (20°C/200°C) [$10^{-6} \times 1/K$]	Temperature coefficients of refractive index, $dn_{3.39\,\mu m}/dT$ (+20; +30°C) [$10^{-6}/K$]	Temperature coefficients of refractive index, $dn_{10\,\mu m}/dT$ (+20; +30°C) [$10^{-6}/K$]	Transformation temperature, T_g (°C)	Softening point, T_s (°C)	Thermal capacity, C_p (J/gK)	Thermal conductivity, λ (W/mK)	κ, $\lambda/C_p \cdot \rho$ (10^{-6} m^2/s)
IRG 1	22.9	na	na	183	na	na	na	na
IRG 2	8.8	na	na	700	na	0.452	0.91	0.40
IRG 3	8.1	na	na	787	na	0.481	0.87	0.40
IRG N6	6.3	na	na	713	na	0.808	1.36	0.60
IRG 7	9.6	na	na	413	na	0.632	0.73	0.38
IRG 9	16.1	na	na	421	na	0.695	0.88	0.35
IRG 11	8.2	na	na	800	na	0.749	1.13	0.48
IRG 15	9.3	na	na	522	na	0.746	0.96	0.46
IRG 100	15.0	na	na	280	351	na	0.30	na
IG 2	12.7	71	60	368	445	0.33	0.24	0.16
IG 3	13.6	130	145	275	360	0.32	0.22	0.14
IG 4	22.4	20	36	225	310	0.37	0.18	0.11
IG 5	14.3	76	91	285	348	0.33	0.25	0.16
IG 6	21.8	35	41	185	236	0.36	0.24	0.14
GASIR-1	16.9	na	52	292	na	0.36	0.28	0.17
GASIR-2	16.0	na	50	264	na	0.34	0.23	0.14
AMTIR-1	12.0	77	72	362	na	0.29	0.25	0.19
AMTIR-3 (1173)	13.5	58	56	278	na	0.27	0.22	0.17
As_2S_3	21.4	na	na	180	na	0.45	0.17	0.12

Link to definition of thermal properties: Schott Technical Information at http://www.schott.com/optics_devices/english/download/ and http://www.oharacorp.com/swf/og.html.

na, not available.

12.11.7 Form of delivery

Glass type	Manufacturer	Precision gobs	Spheres	Strips
IRG 1	Schott	na	na	na
IRG 2	Schott	na	na	na
IRG 3	Schott	na	na	na
IRG N6	Schott	na	na	na
IRG 7	Schott	na	na	na
IRG 9	Schott	na	na	na
IRG 11	Schott	na	na	na
IRG 15	Schott	na	na	na
IRG 100	Schott	na	na	na
IG 2	Vitron	na	na	Diameter max 200 mm, thickness max 50 mm
IG 3	Vitron	na	na	Diameter max 200 mm, thickness max 50 mm
IG 4	Vitron	na	na	Diameter max 200 mm, thickness max 50 mm
IG 5	Vitron	na	na	Diameter max 200 mm, thickness max 50 mm
IG 6	Vitron	na	na	Diameter max 200 mm, thickness max 50 mm
IG 6	Vitron	na	na	Diameter max 200 mm, thickness max 50 mm
GASIR-1	Umicore	na	na	Diameter max 150 mm, thickness max 80 mm
GASIR-2	Umicore	na	na	Diameter max 150 mm, thickness max 80 mm
AMTIR-1	Amorphous materials	na	na	na
AMTIR-3 (1173)	Amorphous materials	na	na	na
As_2S_3	Amorphous materials	na	na	na

na, not available.

12.11.8 Applications and limitations

Current applications: Fire fighting, weapon sights, predictive maintenance, process control, security, surveillance, medicine, search and rescue, night vision, driver's vision enhancers, smart sensing, and aerospace.
Potential application and outlook: Free-form lenses via precision glass molding.
Limitations: Temporary exemption from RoHS compliance (e.g., materials containing As, Se, Ge); contamination during the fabrication process, for example, with oxygen, hydrogen, and carbon, causing undesirable absorption in the transmission range of chalcogenide glasses.

12.11.9 Further reading

J.A. Savage, *Infrared Optical Materials and Their Antireflection Coatings*, Adam Hilger Ltd. (1985).

12.11.10 Links

- http://www.schott.com/english/index.html
- http://www.vitron.de/english/optical_glasses.html
- http://optics.umicore.com/umicore_optics/index.asp
- http://amorphousmaterials.com/home.htm

12.11.11 Research and development

Research and development will focus on new chalcogenide glasses and glass ceramics.

Chapter 13

Processing Technologies

13.1 Zonal Grinding Process

H. Buchenauer, M. Haag-Pichl

13.1.1 Basic assessment of the technology

Today, there are different manufacturing methods available for the fabrication of aspherical surfaces. Among these methods, machining with CNC has as its biggest advantage a very high flexibility regarding geometry and material. No special form tools are necessary, and nearly all brittle materials are machinable. Additionally, the highest form accuracy of the lens surface can be reached with the subsequent ultraprecision finishing processes.

The typical process chain for the machining of aspherical surfaces with CNC consists of three steps:

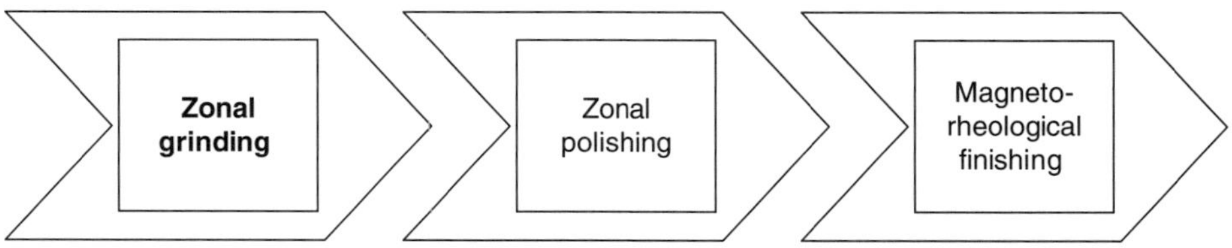

The zonal grinding process, which is described in this chapter, has the role of generating a highly precise but still microscopic rough surface, which achieves its optical quality in a subsequent polishing step.

13.1.2 Intended purpose of the technology

The traditional loose abrasive grinding process of spherical and flat precision optics surfaces had several drawbacks regarding the flexibility and repeatability of the process. Special form tools for each specific radius of curvature and a lot of manual work, with iterative steps, were necessary. To improve this situation, the advantages of modern CNC technology were transferred to optical fabrication. The traditional lap process with form tools and loose diamond grain has been replaced by universal cup tools with bonded diamonds.

Early zonal grinding was carried out, for example, using the copy grinding technique, but surface form accuracy has been limited by an insufficient stiffness of the machines and inaccuracies through the manufacturing and scanning of the template [1].

In 1993, the transition from CNC-controlled spherical fabrication (line contact) to aspherical fabrication (point contact) was realized through the first commercially available CNC center with integrated asphere technology. The patented Schneider Aspheroline [2] (with grinding, polishing, and metrology) made it possible to machine precise aspheres in series production.

13.1.3 The technology's typical features

The basic idea behind the introduction of spherical CNC machining with cup tools and fixed abrasives was to obtain a flexible and reproducible fabrication with standard tools. Different radii of curvature are producible by variation of the cup tool angle.

Based on the possibility of a precisely controllable positioning of the tool and a transition from a line contact (during spherical grinding) to a point contact, a reproducible and deterministic generation of rotational symmetric aspherical surfaces (in the following named 'aspheres') has been achieved. This led to zonal grinding without any form tools, and it was possible to fabricate aspheres of nearly any specification just by defining the form in the machine software and generating the exact contour using high-precision CNC axes.

An alternative to using the common cup tools for zonal grinding is the use of disc grinding tools. With these tool types, high stock removal and contour accuracy can be achieved in combination with a very good process stability. The processing of very steep concave surfaces is limited by the disc diameter. In such cases, cup tools are preferable.

13.1.4 Description of process

13.1.4.1 Operating resources (Fig. 13.1)

Pre-, fine-, and ultrafine-grinding require

- A CNC machine with at least two contour-controlled axes;
- A high-precision, speed-controlled workpiece and tool spindle;
- Disc and cup tools with bonded diamond grain;
- Separate tool spindles for pre-/fine- and ultrafine-grinding;

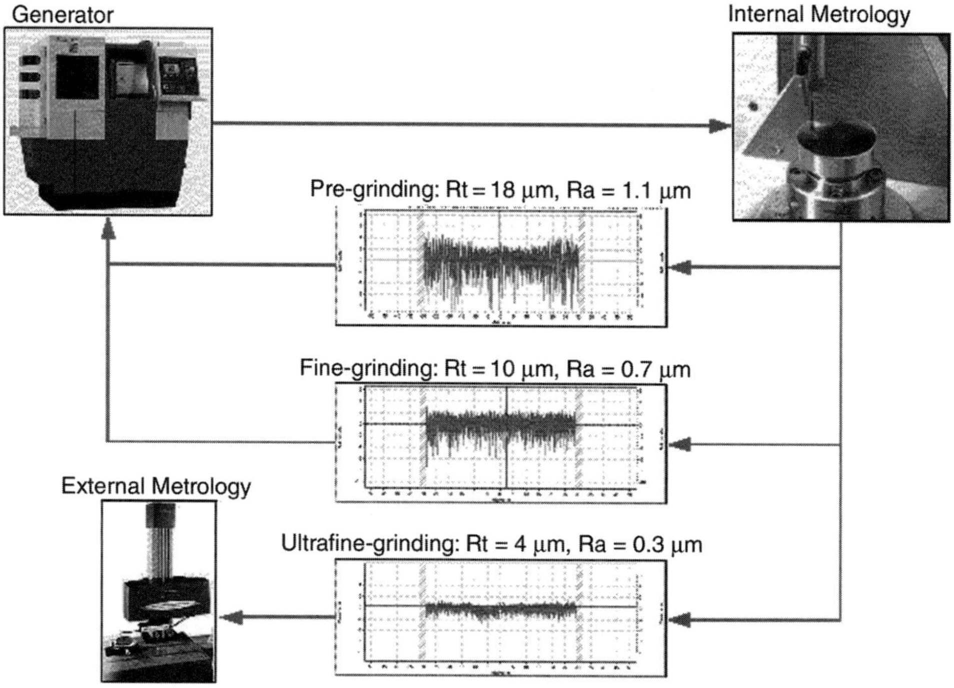

Figure 13.1 Integration of grinding process and metrology.

- A multistage process with successive smaller diamond grain from pre- to ultrafine-grinding for the minimization of roughness and subsurface damage;
- An integrated automated tool changer for the pre- and fine-grinding processes;
- Precise and reproducible clamping technology for the tools and workpiece;
- Internal metrology for
  - tool (contour);
  - workpiece (center thickness and form);
- CNC embedded correction with different smoothing algorithms;
- Interface to external metrology.

External metrology should include

- Tactile measurement of contours for ground surfaces with a profilometer or 3D coordinate measuring machine;
- An interface for data transmission to the generator.

13.1.4.2 Mode of operation

Cup wheel grinding

The cup-shaped grinding tool (Fig. 13.2) is tilted with a constant angle to the surface normal. With this angle, a point contact between tool and lens is achieved and the

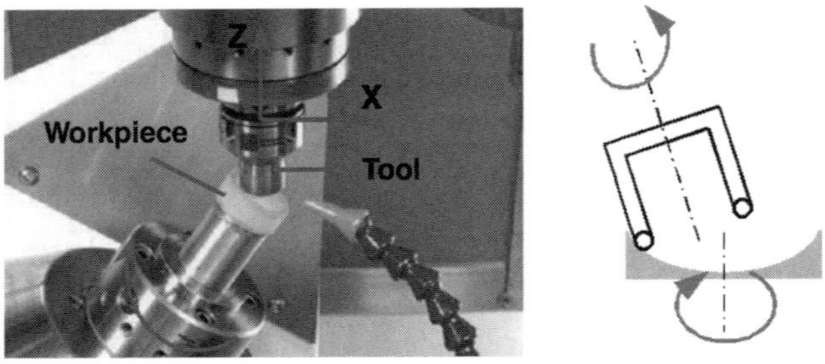

Figure 13.2 Cup wheel grinding.

aspherical surface is generated using precise contour-controlled axes (one swivel and two linear).

Disc wheel grinding

The disc-shaped grinding tool (Fig. 13.3) is moved with precise contour-controlled axes (two linear) to generate the aspherical surface.

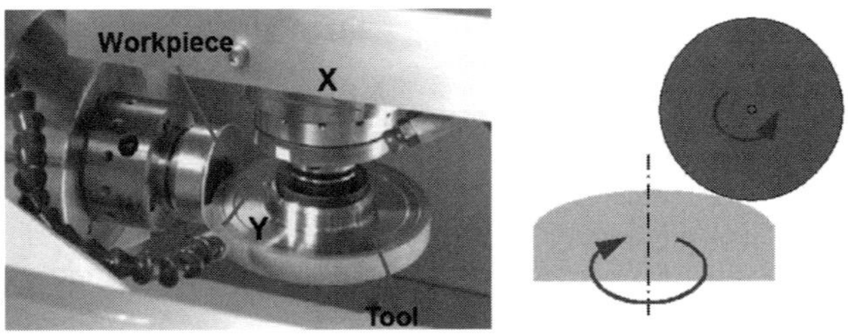

Figure 13.3 Disc wheel grinding.

Figure 13.4 A representative grinding center for generating spheres and aspheres.

13.1.5 Versions (state of the art)

Asphere generators are available in different configurations from several machine manufacturers. Differences can be found in the general construction of the machines, software, and level of automation. Figure 13.4 shows a representative grinding center from Schneider Opticmachines in Germany, with which spheres and aspheres can be generated up to a diameter of 125 mm.

13.1.6 Data for the zonal grinding process (Tables 13.1–13.6)

Table 13.1 Typical specifications.

Size	Asphericity	Surface deviation	Surface roughness
Some mm up to >1000 mm	Free	pv < 1 μm	R_a < 200 nm

Table 13.2 Typical operation parameters.

Parameter	Minimum	Maximum
Workpiece spindle	25 rpm	2500 rpm
Tool spindle	2000 rpm	15000 rpm

Table 13.3 Machinable dimensions.

Geometrical dimension	Minimum	Maximum
Diameter	Some mm	Around 1000 mm
Radius	Some mm	∞

Table 13.4 Process measurement technique.

Measured variable	Method of measurement
In-line	
Center thickness	Tactile
Form 2D	Tactile
Off-line	
Center thickness	Tactile
Form 2D or 3D	Tactile

Table 13.5 Cost drivers.

Cost factor	Amount
Machine investment	Medium, 100T€–500T€ depending on size
Setup time	Small because of CNC-driven technology and universal tools
Process time	1–30 min per piece depending on size and material
Operator	Skilled operator (optician)
Tools and supplies	Low, very long lifetime of diamond tools and coolant

Table 13.6 Limits of the technology.

Parameter	Limiting value	Reason for limitation
Concave radius	Half diameter of grinding wheel	Part radius < tool radius

13.1.7 Conclusions

The zonal grinding process with CNC machines dramatically changed the fabrication of apherical surfaces. Nearly any asphere can be generated within a very short setup time, as universal diamond tools are used, and the accuracy of the surface is determined by precise contour-controlled machine axes.

With its high flexibility, zonal grinding is predestined for the fast fabrication of prototypes and for the low- and mid-sized series production of any brittle material. Roughness and subsurface damage, which is created in the generating process, needs to be removed by a subsequent polishing step.

13.1.8 Further reading

The motivation for the use of aspherical optics and a walkabout through the history of manufacturing of aspherical lenses can be found in [3]. Reference [1] delivers a deep understanding of the mechanisms of the zonal grinding process. A methodological analysis of abrasive optical fabrication techniques is given in [4].

1. E.N. Koch, "Technologie zum Schleifen asphärischer optischer Linsen, Thesis RWTH" *Aachen* (1991).
2. G. Schneider and H. Buchenauer, "Procedure of and device for fabricating aspheric lens surfaces," EP 0 685 298 B2 (1994).
3. E. Heynacher, "Ashperic optics—How they are made and why they are needed," *Physics in Technology*, Vol. 10, pp. 124–131 (1979).
4. O.W. Fähnle and H. van Brug, "Novel Approaches to Generate Aspherical Optical Surfaces," in *Optical Manufacturing and Testing III*, Ph. H. Stahl (ed.), SPIE, Vol. 3782, pp. 170–180 (1999).

13.1.9 Links

Plant engineering and construction:

- Schneider GmbH & Co. KG; http://www.schneider-om.com
- Optotech GmbH; http://www.optotech.de
- Satisloh GmbH; http://www.satisloh.com
- Moore Nanotechnology Systems; http://www.nanotechsys.com

Toolmakers and suppliers of auxiliary materials:

- Schneider GmbH & Co. KG; http://www.schneider-om.com
- Saint-Gobain Diamantwerkzeuge GmbH & Co. KG; http://www.winter-diamantwerkz-saint-gobain.de
- Dr. W. Müller, Diamantmetall; http://www.muedia.de
- Günter Effgen GmbH; http://www.effgen.de

Research and development:

- Fraunhofer-Institute für Production Technology IPT; http://www.ipt.fraunhofer.de
- IWT Foundation Institute for Materials Science; Division Manufacturing Techn.; http://www.iwt-bremen.de/ft
- University Aalen; Center for Optical Technologies; http://www.htw-aalen.de/ extern/zot
- Center for Optics Manufacturing COM/University of Rochester; http://www.opticsexcellence.org

13.2 Zonal Polishing Process

H. Buchenauer, M. Haag-Pichl

13.2.1 Basic assessment of the technology

In the common process chain for the machining of aspheres, the main stock removal and the shaping of the aspherical surface to accuracies in the sub micrometer range is done by zonal grinding. After that process, the surface is still microscopic rough and therefore does not have optical quality.

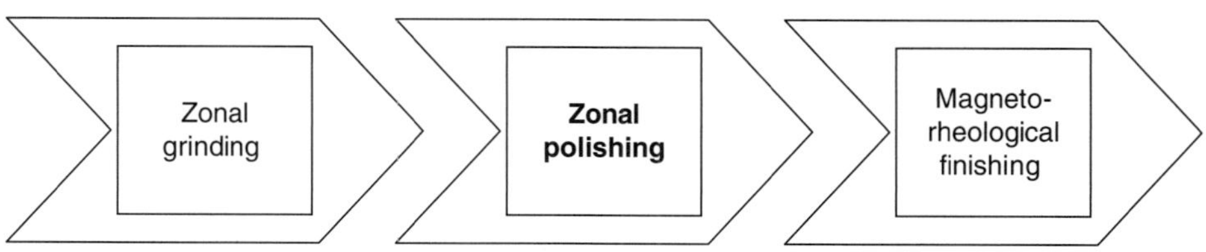

The role of the zonal polishing process described in this chapter is to remove roughness, subsurface damage, and mid-frequency surface errors after the grinding process under preservation and corrective improvement of the form accuracy.

13.2.2 Intended purpose of the technology

Conventional pitch polishing of spherical and flat precision optics surfaces had several draw-backs regarding the flexibility and repeatability of the process. Due to a complex correlation between the many technology parameters, several empirically based iterative steps are necessary, with a requirement for highly skilled opticians as operators.

To improve this situation, the advantages of modern CNC technology have been transferred to optical fabrication. For spherical production, traditional batch production with polishing pitch has been replaced by a single-part production on CNC machines with full aperture form tools and a polyurethane polishing foil ("synchrospeed kinematics"; [1, 2]).

For the fabrication of aspheres, a controlled zonal polishing method is necessary instead of the area-averaging traditional form tool polishing. Many manufacturers of optical elements had their own (predominantly manual and empirical) solutions for zonal polishing. With the availability of computer technology, it became possible to automate the process. Early work on computer-assisted zonal polishing methods was published in 1972 by Aspden et al. [3]. The described workpieces are aspheres with large diameters (>1 m). However, the zonal polishing process can also be scaled down to smaller parts used in devices such as taking, enlarging, and projection lenses.

The first commercially available CNC machines for zonal polishing of aspheres in the diameter range <100 mm entered the market in the mid 1990s.

13.2.3 The technology's typical features

The primary goal of the polishing process is the quick removal of roughness, subsurface damage, and mid-frequency errors. In the spherical polishing process, this is done with full aperture form tools that are much larger than the workpiece and that have exactly the negative radius of the lens.

Because of the local curvature variation of aspherical surfaces, it is not possible to use full aperture polishing tools. Instead, adaptive subaperture tools must be used, which are adapted to the surface, for example, by air pressure or an elastic cushion behind the polishing pad. The contour is mostly controlled by a variation of the dwell time. This benefits from the fact that the local material removal is proportional to the dwell time (Preston equation; [4]). Algorithms calculate a dwell time profile for the tool to maintain the surface accuracy as good as possible or to correct deviations of the surface form. Modern contour-controlled precision CNC machines are used in conjunction with feedback from metrology.

13.2.4 Description of process

13.2.4.1 Operating resources

For polishing, these include

- A CNC machine with at least three contour-controlled axes;
- A speed-controlled workpiece and tool spindle;
- A flexible and pressure-controlled subaperture tool for zonal polishing;
- Software for calculation of the dwell time profile.

For external metrology, these include

- Tactile contour measurement for 2D profiles;
- Interferometric 3D measurement with a computer-generated hologram for polished surfaces;
- An interface for data transmission to the polishing machine.

Figure 13.5 shows a typical workflow for a corrective zonal polishing process. The tool removal function and the surface form deviation are inputs into a computer program that calculates a dwell time profile based on Preston's equation [4]. The CNC machine moves the subaperture tool over the surface using the dwell time profile. A significant decrease of surface form deviation is the result. This process may be carried out in several iterative steps.

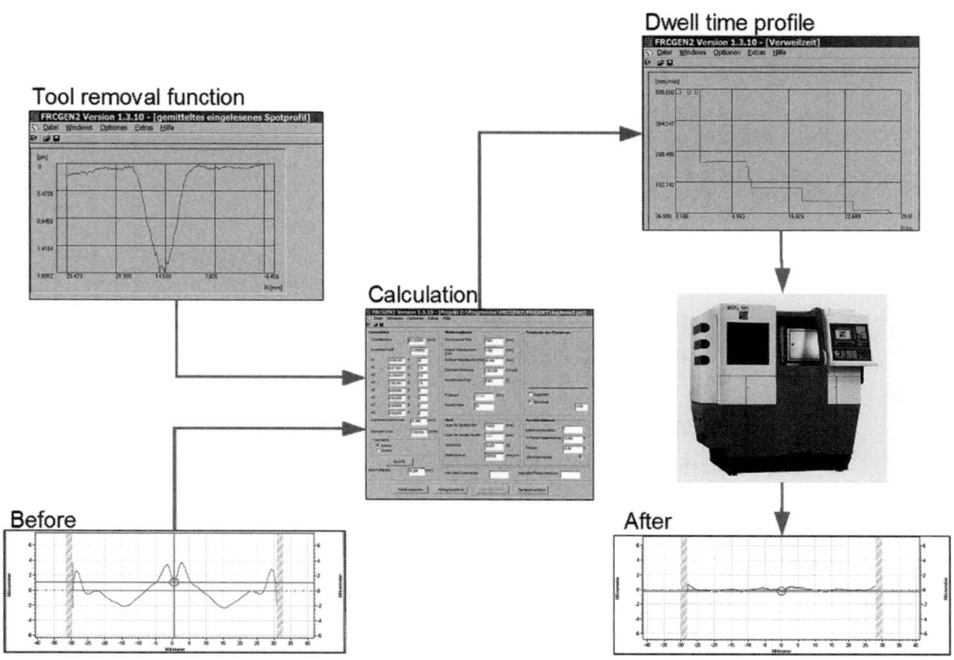

Figure 13.5 Typical workflow for a corrective zonal polishing process.

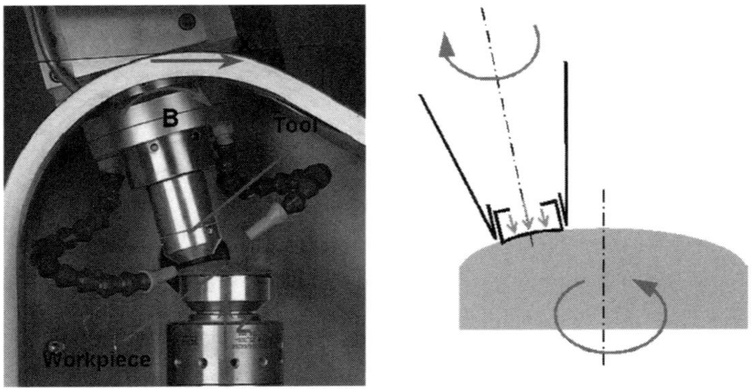

Figure 13.6 Elastic pad zonal polishing.

13.2.4.2 Mode of operation

Elastic pad zonal polishing

The subaperture tool (Fig. 13.6) is adapted to the surface shape of the lens by an applied pressure or an elastic cushion behind the polishing pad, which is usually a polyurethane material. The polishing slurry used is similar to that in the conventional spherical process. The path of the tool is controlled by the CNC axes of the polishing machine and kept perpendicular to the aspherical form. In most cases, the dwell time of the tool on a local part of the lens is used to control the amount of removal, but, as can be seen from Preston's equation, pressure or relative velocity can also be used as control parameters.

An alternative tool concept is the use of a polishing wheel instead of an elastic pad.

13.2.5 Versions (state of the art)

Asphere polishers are available in different configurations from several machine manufacturers. Differences can be found in the general construction of the machine, software, and level of automation. A representative polishing center, from Schneider Opticmachines in Germany, is shown in Fig. 13.4 with which spheres and aspheres can be polished up to a diameter of 125 mm.

13.2.6 Data for the zonal polishing process (Tables 13.7–13.12)

Table 13.7 Typical specifications.

Size	Asphericity	Maximum surface deviation	Surface roughness
Some mm up to >1000 mm	Free	pv < 1 μm	$R_a < 1$ nm

Table 13.8 Typical operation parameters.

Parameter	Minimum	Maximum
Workpiece spindle	500	2500
Tool spindle	500	2500

Table 13.9 Machinable dimensions.

Geometrical dimension	Minimum	Maximum
Diameter	Some mm	Around 1000 mm
Radius	Some mm	∞

Table 13.10 Process measurement technique.

Measured variable	Method of measurement
In-line	
na	na
Off-line	
Form 3D	Tactile or interferometric with CGH
Roughness	AFM

na, not available.

Table 13.11 Cost drivers.

Cost factor	Amount
Machine investment	Medium, 100T€–500T€ depending on size
Setup time	Small because of CNC-driven technology and universal tools
Process time	2–60 min depending on size, material, and prework conditions
Operator	Skilled operator (optician)
Tools and supplies	Long lifetime of tools and slurry

Table 13.12 Limits of the technology.

Parameter	Limiting value	Reason for limitation
Diameter	Polishing tool size	No zonal polishing possible

13.2.7 Conclusions

The zonal polishing process with CNC machines does significantly reduce the vast number of iterations that have been necessary to obtain an aspherical surface within its specifications by traditional manual corrections. With a precise contour-controlled path of the subaperture tool over the workpiece and a computer-based calculation of a dwell time profile, the results of a polishing run are much more predictable than in the past. A small number of iterations is still necessary because of slight variations in the tool removal function due to wear and the necessary high stock removal due to subsurface damage after the grinding process.

13.2.8 Further reading

A comprehensive selection of publications describing the early work in zonal polishing from 1972 to 1991 is available in [3]. The fundamental Preston equation, which is the basis for many removal algorithms, is derived in [4].

1. E. Brück, "Optische Linsen herstellen," *Industrieanzeiger*, Vol. 99 (1989).
2. S. Hambücker, "Technologie der Politur sphärischer Optiken mit Hilfe der Synchrospeed-Kinematik," *Thesis, RWTH Aachen* (2001).
3. R.A. Jones, "Selected papers on computer-controlled optical surfacing," *SPIE Milestone Series* **MS40** (1991).
4. F.W. Preston, "The theory and design of plate glass polishing machines," *J. Soc. Glass Technol.*, Vol. 11 (1927).

13.2.9 Links

Plant engineering and construction:

- Schneider GmbH & Co. KG; http://www.schneider.om.com
- Optotech GmbH; http://www.optotech.de
- Satisloh GmbH; http://www.satisloh.com
- Precitech, Inc.; http://www.precitech.com

Toolmakers and suppliers of auxiliary materials

- Schneider GmbH & Co. KG; http://www.schneider.om.com
- Pieplow & Brandt GmbH; http://www.pieplow-brandt.de

Research and development:

- Fraunhofer-Institute für Production Technology IPT; http://www.ipt.fraunhofer.de
- Center for Optical Technologies/University Aalen; http://www.htw-aalen.de/extern/zot

13.3 Magnetorheological Finishing

H. Buchenauer, M. Haag-Pichl

13.3.1 Basic assessment of the technology

After the zonal polishing process, the surface roughness of the processed aspherical surface usually has an optical quality, but it is difficult and time-consuming to reach surface accuracies that are well below 1 μm. Therefore, an additional step is necessary for a deterministic correction of the surface deformation to the specified accuracy.

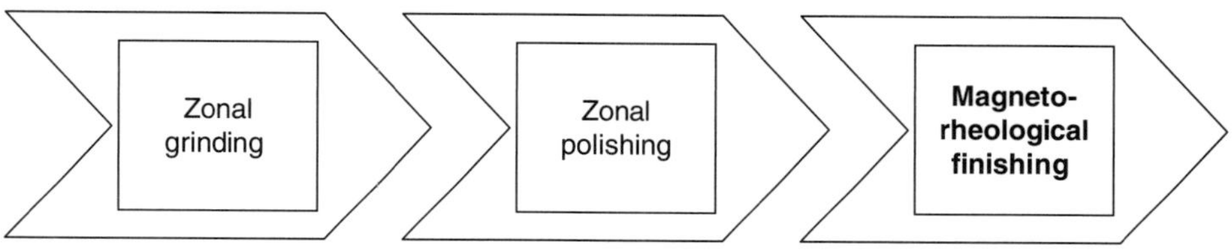

The magnetorheological finishing process described in this chapter has the role of achieving an ultraprecision finishing of the zonally polished surface.

13.3.2 Intended purpose of the technology

The traditional ultraprecision finishing of aspheres with zonal polishing is an extremely iterative, empirical process. On the basis of an interferometric measurement, an experienced operator applies a dwell time profile either manually or with simple algorithms. Because of variations of the tool function, the target specification is reached only after many iterations. This time-consuming procedure led to high production costs and limited the application of aspheres to individual solutions.

The patented method of magnetorheological finishing (MRF, [1]) was invented in the late 1980s in Minsk, Belarus. It was further developed by a team around W. Kordonski and the Center for Optics Manufacturing (COM) in Rochester, USA, for a deterministic fabrication of optical surfaces (in particular, aspheres). In 1998, QED Technologies [2] introduced the first commercial product using MRF. The complete process chain with MRF for the production of aspherical lenses was first presented in the Optatec Show 2000 in conjunction with Schneider Opticmachines.

With the use of the innovative magnetorheological (MR) fluid, it was possible to significantly decrease the variation of the tool removal function, the main problem in the traditional polishing process.

13.3.3 The technology's typical features

The achievable predictability of the polishing result of ultraprecise aspherical surfaces with the usual pad polishing tools is strongly limited because of the inherent variation of the tool removal function. This requires an iterative procedure with many measuring steps.

With the highly innovative application of a fluid that can be controlled in its viscosity by an externally applied magnetic field (magnetorheological fluid), the influence of tool wear can be eliminated. The basic idea behind this technology is to stabilize the tool removal function by a permanent regeneration of the contact zone between tool and workpiece and a stabilization of the fluid viscosity. With such an approach, a deterministic ultrafine finishing, down to an accuracy of some nanometers, is possible.

A local tool removal profile taken with static tool and workpiece is interferometrically measured and serves as a basis for the computer simulation of the complete material removal. The measured surface form deviation of a polished asphere is predictably corrected by a dwell time control calculated in a computer simulation.

13.3.4 Description of process

Magnetorheological finishing (MRF) requires what is detailed in Fig. 13.7.

- A CNC machine with a minimum of four contour-controlled axes;
- A speed-controlled tool and workpiece spindle;

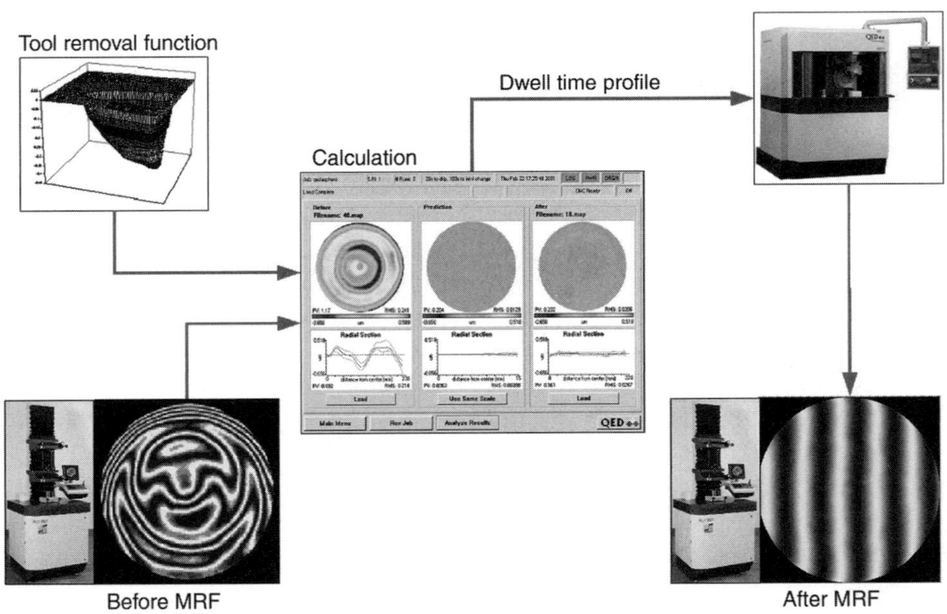

Figure 13.7 Typical workflow for magnetorheological finishing.

- Exact stabilization of the magnetorheological fluid;
- Integrated software for the calculation of the dwell time profile.

External metrology includes

- Full aperture interferometry with computer-generated hologram for polished surfaces;
- An interface for data transmission from metrology to the machine.

The tool removal function ("spot") and the surface form deviation of the finished lens are inputs into a computer program. This program calculates, with the use of sophisticated algorithms, a dwell time profile to complete a deterministic corrective finishing of the surface to its specifications.

13.3.4.1 Mode of operation

The MR fluid is directed on a rotating wheel, which is encased by strong electromagnets. When a magnetic field is applied around the wheel, the fluid is stiffened and has an abrasive effect on the workpiece. Behind the wheel, the fluid is collected and circulated in a closed loop. The position of the lens is controlled by precise CNC axes, and the tool can be either moved on a spiral path or a raster path depending on the application and machine configuration (Fig. 13.8).

Two types of MR fluids are currently in industrial use. One composition consists of cerium oxide in an aqueous suspension of magnetic carbonyl iron powder and is appropriate for almost all optical glasses and low-expansion glass ceramics. The second composition uses nanodiamond powder as the polishing abrasive and is more suitable to finish calcium fluoride, IR glasses, hard single crystals like silicon and sapphire, and very hard polycrystalline ceramics.

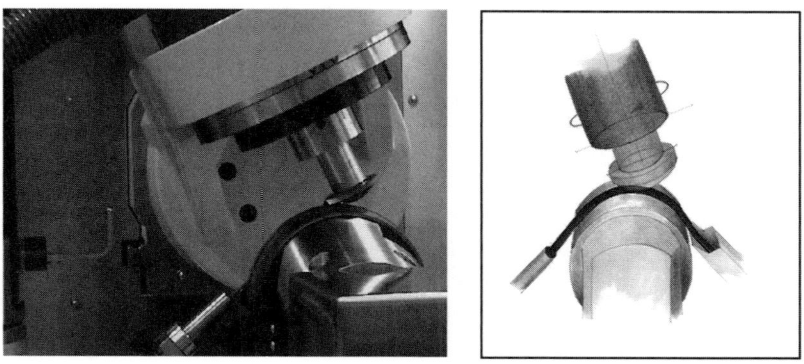

Figure 13.8 Magnetorheological finishing.

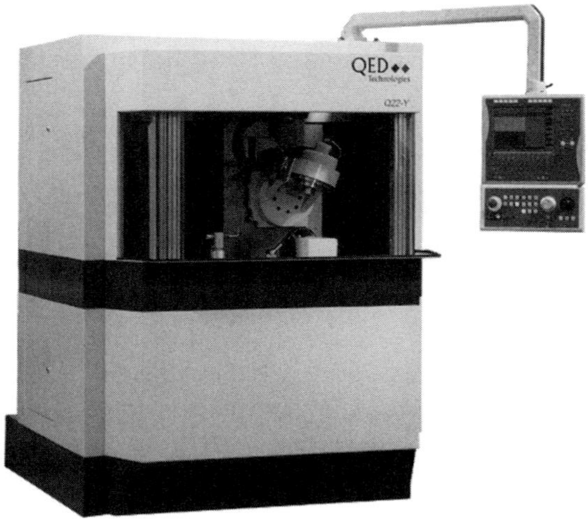

Figure 13.9 MRF equipment.

13.3.5 Versions (state of the art)

Machines with the technology of MRF are exclusively provided by the U.S. company QED Technologies in Rochester. Different sizes and axis configurations are available to cover a large range of geometries. A typical version is shown in Fig. 13.9, which can finish workpieces up to a diameter of 200 mm.

13.3.6 Data for magnetorheological finishing (Tables 13.13–13.18)

Table 13.13 Typical specifications.

Size	Asphericity	Maximum surface deviation	Surface roughness
Some mm up to >1000 mm	Free	pv < 10 nm	$R_a < 0.5$ nm

Table 13.14 Typical operation parameters.

Parameter	Minimum	Maximum
Workpiece spindle	0	550

Table 13.15 Machinable dimensions.

Geometrical dimension	Minimum	Maximum
Diameter	Some mm	Around 1000 mm
Radius	convex hemisphere/concave 15 mm	∞

Table 13.16 Process measurement technique.

Measured variable	Method of measurement
In-line	
Center thickness	Tactile
Off-line	
Form 3D	Interferometric with computer-generated hologram

Table 13.17 Cost drivers.

Cost factor	Amount
Machine investment	Medium–High, >250T€
Setup time	Small because of CNC technology and no form tools
Process time	2–60 min, depending on size, material, and prework conditions
Operator	Skilled operator (optician)
Tools and supplies	Lifetime of MR fluid around 2 weeks; no other tools and supplies needed

Table 13.18 Limits of the technology.

Parameter	Limiting value	Reason for limitation
Concave radius	Half diameter of wheel	Part radius < tool radius

13.3.7 Conclusions

MRF revolutionized the fabrication of aspherical surfaces. With a totally different approach, it uses an innovative way to achieve a permanent regeneration of the polishing tool in the contact zone with the workpiece. As a consequence, it is a method for achieving a predictable, deterministic, ultraprecise finishing of nearly any surface, in particular aspheres. Because no special tools are needed, setup time is minimal and the flexibility is high, therefore this fabrication technology is ideally suited for the fast fabrication of prototypes and for the low- and mid-sized series production of aspheres.

13.3.8 Further reading

1. W. Kordonski, et al., "Magnetorheological polishing devices and methods," US pat. No. 5 449 313, 1995.
2. Homepage of QED Technologies: http://www.qedmrf.com.

3. S.D. Jacobs, S.R. Arrasmith, I.A. Kozhinova, L.L. Gregg, A.B. Shorey, H.J. Romanofsky, D. Golini, W.I. Kordonski, P. Dumas and S. Hogan, "An Overview of Magnetorheological Finishing (MRF) for Precision Optics," in *Finishing of Advanced Ceramics and Glasses*, Sabia, R., Greenhut V.A., and Pantano, C.G., *Ceramic Transactions*, Vol. 102, pp. 185–199 (1999).

13.3.9 Links

Plant engineering and construction:

- QED Technologies Inc.; http://www.qedmrf.com
- Schneider GmbH & Co.KG; http://www.schneider-om.com.

Toolmakers and suppliers of auxiliary materials:

- QED Technologies Inc.; http://www.qedmrf.com.

Research and development:

- QED Technologies Inc.; http://www.qedmrf.com.
- Center for Optics Manufacturing (COM)/University of Rochester; http://www.opticsexcellence.org.

13.4 Robotic Polishing

R. Börret

13.4.1 Basic assessment of the technology

CCP (computer control polishing) or CCOS (computer-contolled optical surfacing) is the corrective polishing step in the classical process chain (Fig. 13.10).

CCP involves material removal in localized areas of the optical element based on measurement performed in advance. The goal of the CCP process step is to improve the optical performance of the treated element by a corrective polishing step on the surface of the element.

Figure 13.10 Classical fabrication process for optics.

13.4.2 Intended purpose of the technology

Local, computer-controlled polishing was developed between 1968 and 1976 [1–3]. This is related to the availability of appropriate computers, necessary for the calculation of the algorithms.

As with other local corrective polishing steps (e.g., ion beam figuring), CCP was first applied to large astronomical mirrors used in telescopes or in space applications. These astronomical parts are designed with aspherical shapes that require subaperture tools to polish the surface, despite the local change of curvature. Before the development of CCP, material removal in localized areas was carried out manually by highly skilled opticians, for example, by manually applying pressure and circular movements. With the introduction of CCP, the entire optical fabrication process became substantially deterministic and more productive (Fig. 13.11).

13.4.3 The technology's typical features

As formulated by Preston [4] in 1927, removal of material by polishing is a function of tool pressure, relative velocity between the tool and optical element, and polishing time. Taking one of these parameters as a function $f_{\text{removal}}(x, y)$ of tool position

(a) (b)

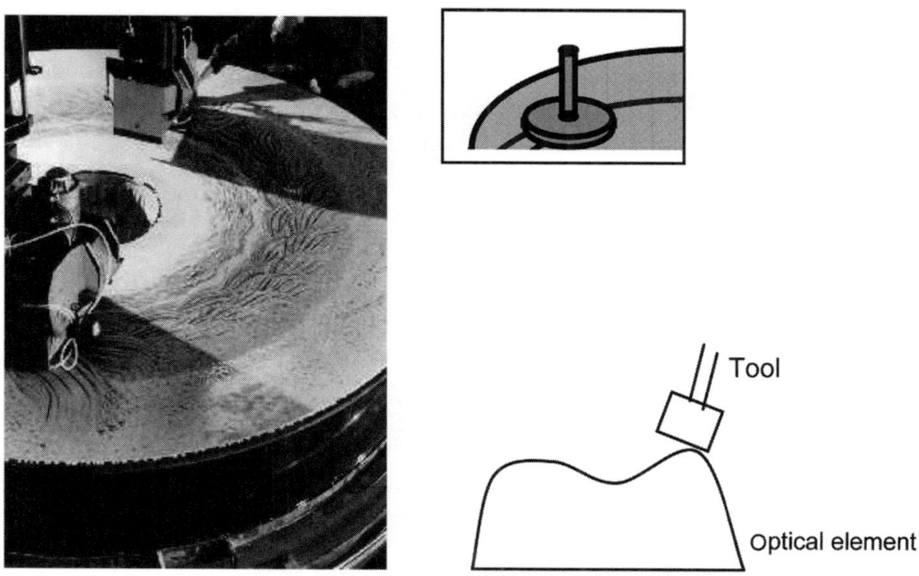

Figure 13.11 (a) Computer-controlled polishing process of the NTT ESO 3.5 m mirror. (b) Principle of the CCP process.

would allow in theory the modulation of local removal rates in a controlled manner. Keeping all the other parameters constant during the complete polishing run leads to a predictable change of the surface figure. Suitable measurement methods are necessary to characterize the starting surface and sophisticated algorithms to calculate $f_{\text{removal}}(x, y)$.

13.4.4 Description of process

The principle of CCP is shown in Fig. 13.11(b). A subaperture tool is moving relative to the polished surface. According the Preston equation, tool wear is modulated by varying one of the parameters mentioned above. Today, for most CCP processes, dwell time $t(x, y)$ is calculated as a function of position, pressure p, and relative velocity, which is kept constant. The relative velocity is generated by a spinning or precession tool. As for the other subaperture processes (MRF, IBF, fluid jet, and so on) the dwell time function $t(x, y)$ is calculated by a deconvolution of the desired removal profile and the wear function (indent function[3] or spot[4]) or by an optimization algorithm.

13.4.5 Versions (state of the art)

CCP belongs to the core know-how of optical fabrication in large and small optics companies. CCP is used in different configurations in the precision optics workshops of CCOS (computer control optical surfacing) at Tinsley (USA), in the computer-aided polishing technique at Sagem (France), or as small-tool CCP at Zeiss (Germany). Beyond that, there are commercially available CCP machines such as the Precessions CNC polishing machine IRP200 from Zeeko/Loh (Fig. 13.12).

Common to all varieties of CCP processes are a spinning or precession tool made of flexible material, some kind of CNC machine with up to seven axes, and a dwell time calculation algorithm.

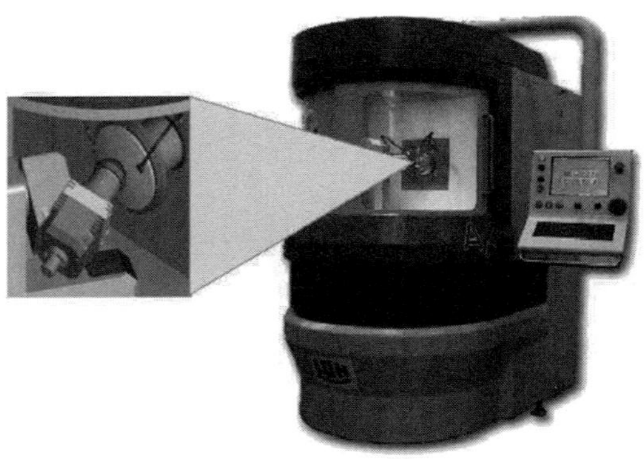

Figure 13.12 Zeeko precision CNC polishing machine (Zeeko/Satisloh).

The details of the machine, tool setup, and dedicated algorithm may vary. The polishing surface may be polyurethane, and a cerium oxide slurry is mostly used. The tool rasters the surface or makes diametric passes over the rotating workpiece, controlled by the calculated dwell time function. These kinds of motion are similar to the other subaperture processes.

Important parameters of the CCP process include

- Stability of the polishing process (slurry, tool wear),
- Spot (size, stability),
- Accuracy of position system, tool path,
- Algorithm, and
- Accuracy of metrology data.

13.4.6 Data for robotic polishing (Tables 13.19–13.24)

Table 13.19 Typical specifications.

Size	Surface roughness	Defect of form
Some mm up to 8 m	~0.5 nm rms	~1 nm rms

Table 13.20 Typical operation parameters.

Parameter	Minimum	Maximum
Workpiece rpm	5	1000
Tool rpm	100	1000

Table 13.21 Machinable dimensions.

Geometrical dimension	Minimum	Maximum
Diameter	Some mm	8 m
Radius	Some mm	∞

Table 13.22 Process measurement technique.

Measured variable	Method of measurement
In-line	
None	None
Off-line	
Shape 3D	Interferometric, e.g., with computer-generated hologram
Surface roughness	Micro-interferometer, atomic force microscope (AFM)

Table 13.23 Cost drivers.

Cost factor	Amount
Machine investment	Medium, 100 T€–500T€ depending on size
Setup time	Small, fast setup possible compared with, for example, MRF
Process time	Similar to other subaperture processes
Operator	Skilled operator; similar to other subaperture processes
Tools and supplies	Cheap compared to other subaperture processes

Table 13.24 Limits of the technology.

Parameter	Limiting value	Reason for limitation
Surface roughness	0.5 nm rms	Polishing surface, slurry
Final shape	1 nm rms	Deterministic behavior of process (around 80% deterministic for one run)

13.4.7 Conclusions

Of all the different kinds of subaperture processes, CCP is the most utilized process within the optics industry based on experience of over 30 years. A lot of companies that produce optics have developed their own proprietary CCP process. For nearly all types of optics with diameters from a few mm up to 8 m and specifications of $\lambda/2$ and below, a subaperture polishing process is applied, in most cases using CCP. Applying classical polishing pads and slurries makes use of the whole polishing experience concerning contamination, slurry–workpiece interaction, and so on, from the last 50 years. The limitations are tool wear, which results in changing removal rates over time, and stability of the chemomechanical material removal process.

13.4.8 Further reading

A detailed example of fabricating a nonrotational optical element by a CCP process is given in [5]. Details of a commercially available CCP process is reported in [6] and related papers.

1. R.A. Jones and P.L. Kadakia, *Appl. Opt.*, Vol. 7, pp. 1477 (1968).
2. R. Aspden, R. McDonald, F.R. Nitchie, "Computer assisted optical surfacing," *Appl. Opt.*, Vol. 12, pp. 2739–2747 (1970).
3. D. Bajuk, "Computer controlled generation of rotationally symmetric aspheric surfaces," *Optical Engineering*, Vol. 15, pp. 401–406 (1976).
4. F.W. Preston, "The theory and design of plate glass polishing machines," *J. Glass Tech.*, Vol. 11, pp. 124 (1927).

5. T. Hellmuth, A. Bich, R. Boerret, A. Holschbach and A. Kelm, "Variable phaseplates for focus invariant optical systems," *SPIE EOD2005*, Jena, Germany, 12–16 September (2005).
6. D. Walker, D. Brooks, A. King, R. Freemann, R. Morton, G. McCavana and S.-W. Kim, "The pressions tooling for polishing and figuring flat, spherical and aspheric surfaces," *Optics Express*, Vol. 11, issue 8, pp. 958–964 (2003).
7. W. Kordonski, A. Shorey and M. Tricard, "Precision finishing with magnetorheological (MR) jet technology," *Tech. Digest of SPIE*, Vol. TD03, pp. 1–3 (2005).

13.4.9 Links

Plant engineering and construction:

- http://www.zeeko.co.uk

Engineering and construction of commercially available tools are carried out by Zeeko and Satisloh, but there are proprietary machines available from different optics manufacturers.

Toolmakers and suppliers of auxiliary materials:

- http://www.zeeko.co.uk

Research and development: Research is performed within the different companies, for example, at the University of Rochester, IPT Aachen, and HTW Aalen.

- http://www.sagem-ds.com
- http://www.asphere.com
- http://www.zeiss.de/de/service/jen/home.nsf
- http://www.zeeko.co.uk/

13.5 Subaperture Robotic Polishing

A. Schwarzhans

13.5.1 Basic assessment of the technology

For some optical applications, conventional polishing is the preferred manufacturing method due to its excellent smoothness and microroughness. Combining decades of experience in chemomechanical polishing and state-of-the-art technology led to the development of subaperture robotic polishing. Designed to correct local errors of optical components on the nanometer scale, it offers an economical and interesting alternative to commercially available computer-controlled polishing (CCP) methods.

13.5.2 Intended purpose of the technology

Subaperture robotic polishing dates back to the beginning of CCP [1]. It marks the milestone when a manipulator equipped with a polishing head was used to achieve accurate material removal in localized areas instead of a highly skilled optician. Developed in the 1970s and 1980s, these methods were focused on the manufacturing of large aspherical astronomical mirrors, which required subaperture tools due to their local change of curvature. Since then, robotic polishing has been pushed and adapted by some leading optical manufacturers. It is used for various applications, from prepolishing to fine correction, for planar to optical free-form shapes. However, little about the potential of such systems is generally known in public. In this article, one representative of this technology, called the Advanced Lens Polishing System (ALPS [2]), is discussed. It is built on a standard industrial manipulator, achieves precision by statistics and probability, and overcomes the unpredictable face of classical polishing.

13.5.3 The technology's typical features

A lot of an optical workshop's capabilities still depend on the extensive know-how of mechanical-chemical polishing and the knowledge of which slurries and pads work best for a specific substrate material. With this knowledge and fairly simple technical equipment, experienced pitch polishers are also able to manufacture to the tightest of specifications. The problem is that this kind of work is not always very predictable. The reason for that is often seen in the mechanical chemical removal process itself, which is fairly complex and difficult to control. Therefore, more stable material removal processes have been developed, which generally achieve good surface correction capabilities, but suffer from rather wavy surface structures and bad surface roughness for different substrate materials.

This is different to mechanical chemical polishing, where these characteristics are the other way around. Subaperture robotic polishing takes advantage of the strength of the mechanical chemical polishing process and tackles its weakness. Analyzing the conventional polishing process shows that strengths like smooth surfaces and excellent microroughness are directly related to statistics. It can be shown that the amount of statistics applied has a direct impact on the surface structure. To demonstrate this idea, experimental results of a polishing tool, turning its cylindrical axis and polishing at a fixed location on a substrate material, are shown in Fig. 13.13. Tests were performed with the polishing area of the tool lying perpendicular and parallel to its rotation axis. If all other parameters of the two tests are kept the same, the results for the microroughness differ by a factor of eight. Differences like the size of the polishing area of the tool and its cutting directions (constant, different for any point) have a significant influence on the amount of statistics introduced into the polishing process and explain the different outcomes.

These test results also explain that deterministic removal methods taking place in one preferred direction are generally not suited to improve the surface roughness of a substrate material. On the other hand, very primitive pitch machines can achieve

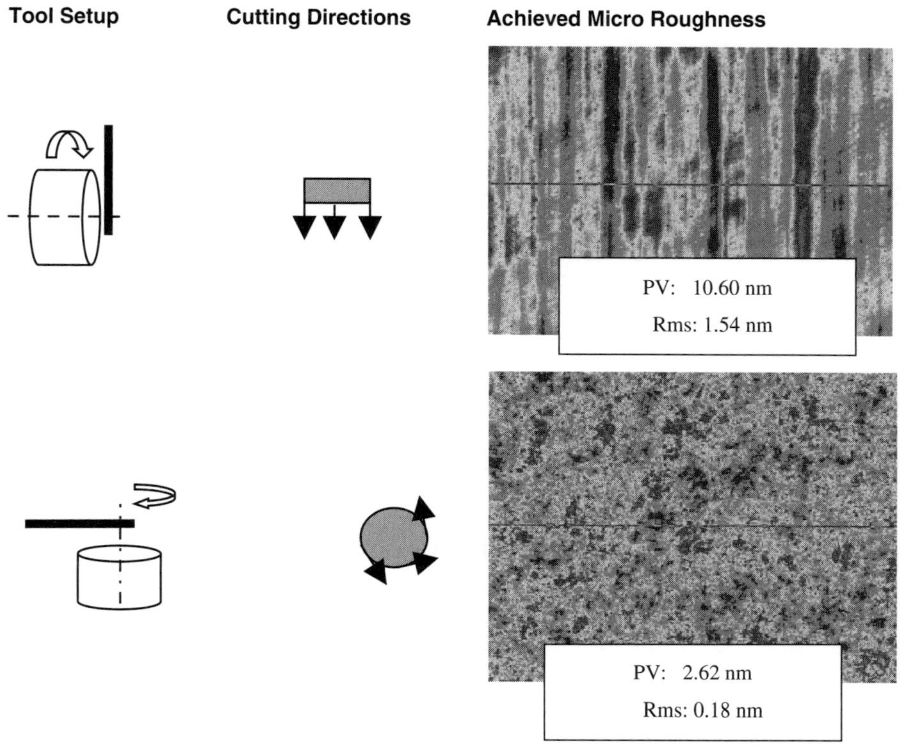

Figure 13.13 Surface microroughness achieved by different effective polishing areas of a polishing tool.

smooth surfaces and excellent microroughness. In this case, statistics is the elementary key to precision. So this raises the question: Is it also possible to apply statistics to improve the convergence of the mechanical chemical surface figuring process?

Starting from the basic principles of statistics, the mean value of a measurable outcome of a multiple repeated event with random errors is the best guess for its true value. Moreover, the influence of the errors becomes less important for a stable mean the more often the single action is repeated. However, this is only true if the single actions are free of systematic errors. In the application of mechanical chemical polishing, this means that the systematic errors have to be reduced. It turns out that with a fairly moderate use of precision, the main systematic errors can be eliminated. This includes, for example, nonperfect areal contact and working pressure between tool and substrate and polishing slurry properties, like concentration and temperature. To further reduce the influence of remaining systematic errors, soft polishing pads acting as dampers and polishing pads with chaotic patterns introducing additional randomness are used. Furthermore, the polishing tool is larger than the distance between two neighboring tool paths and the material removal at a single point on the substrate is the superposition of many neighboring tool paths. In order to avoid constructive superposition of the remaining systematic errors, a certain amount of randomness is added to the positioning of the polishing tool

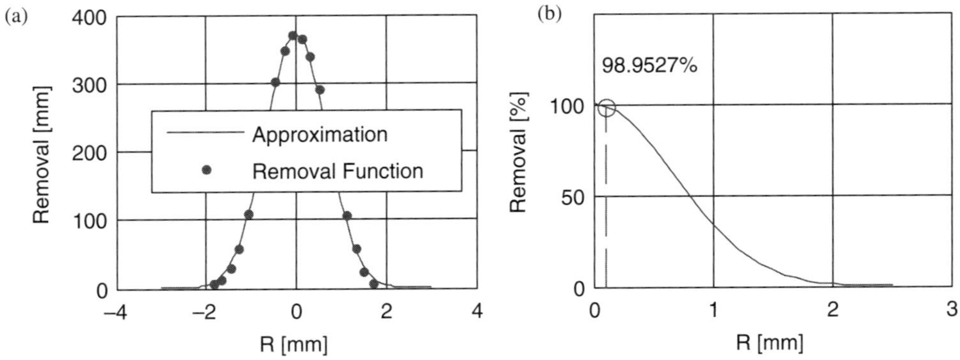

Figure 13.14 (a) Approximation of actual removal function by a Gaussian; (b) Influence of positioning errors on correction capability (removal function displayed relative to its center maximum).

across the area of possible error superposition. This causes a random distribution of the systematic errors and leads to very smooth substrate surfaces.

The idea of systematic error elimination and artificially introduced randomness is implemented by a subaperture polishing tool achieving a small local removal within one turn.

By means of a large number of necessary turns to reach the desired local material removal, the mean value of the removal rate could be stabilized for ALPS within 1%.

Additionally, the shape of the removal function has an important impact. On the one hand, it should be smooth to generate smooth surfaces; on the other hand, it should be narrow banded to achieve excellent local correction characteristics. It turns out that a Gaussian-like removal function is ideal. It is smooth and generates most of its removal within a small center region. Moreover, it limits the necessary positioning accuracy of the polishing tool, which allows more statistics to smoothen out the surface. For the actual removal function, a Gaussian removal function turns out to be the best approximation (Fig. 13.14a). It can be shown that a positioning error of 0.1 mm still generates a removal rate that represents 98.9% of the expected value in the center of the removal function (Fig. 13.14b). In other words, also

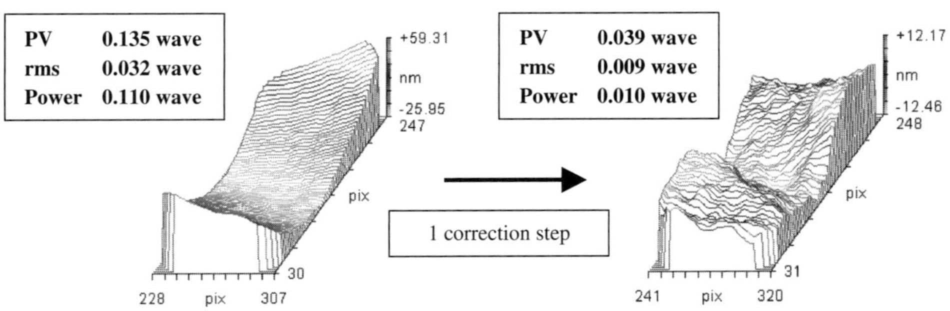

Figure 13.15 Correction of a prism surface (25 × 68 mm, free optical aperture 95%) by ALPS.

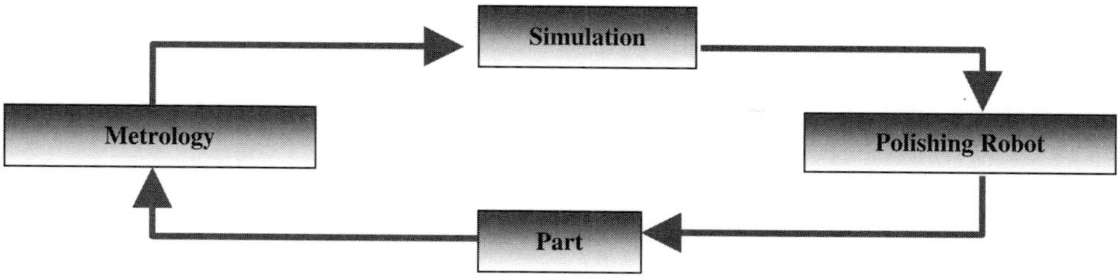

Figure 13.16 Principle structure of ALPS.

with a fairly high positioning error of 0.1 mm, the correction has a convergence of 98.9%. In practice, ±0.1 mm of positioning errors can be actively used to avoid orange structures by introducing statistics. A serial kinematics turned out to be the best choice for this job, which is the reason for applying an industrial robot as a polishing machine.

The result of applying statistics and probability to a mechanical chemical polishing technique turned out to be a very deterministic polishing process. Figure 13.15 shows a practical example of a prism surface that had to be corrected below a PV value of 0.04 waves.

13.5.4 Description of process

The process itself is very similar to other CCP approaches. It is a closed loop of metrology, simulation, and polishing (Fig. 13.16). The polishing movements are determined by a simulation software taking surface metrology data and the material removal function of the polishing tool as an input [3, 4]. The polishing strategy itself is executed by a robot, moving a small mechanical tool with feed control across the surface deviations.

ALPS is designed to be used on the work floor level and is kept as simple as possible for the operator. This is achieved by extensive software integration and automation. As there is little interaction with the operator, an unskilled worker without any experience in optics or robotics would be able to operate the system within one day of training.

13.5.5 Data for subaperture robotic polishing (Tables 13.25–13.27)

Table 13.25 Typical operation parameters.

Shape accuracy	λ/20–λ/40, dependent on metrology used
Surface roughness	Depending on substrate materials, e.g., fused silica 1.2 Å rms
Gradients	Similar to conventional polishing
Machinable dimensions	10 mm–1 m
Radius	From planar to hemisphere (CV), from planar to 30 mm (CC)

Table 13.26 Cost drivers.

Investment	~100,000€
Process time	Similar to other subaperture processes
Operator	Unskilled worker with one day of training
Tools and supply	Similar to conventional polishing

Table 13.27 Limits of the technology.

Metrology	Only what can be measured can be corrected
Geometry	Concave shapes with curvature smaller than the tool ($r < 30$ mm)

13.5.6 Conclusions

As a representative for subaperture robotic polishing, ALPS proves that a mechanical chemical polishing process in combination with a standard industrial robot is applicable for accurate CCP polishing. It combines the advantages of conventional polishing with a deterministic correction process based on the experiences of the past. The strength lies in its flexibility, mainly achieved by using a standard six-degree-of-freedom manipulator as a polishing machine.

These systems can generally be used for planar surfaces to optical free-form shapes, from small diameters to large, and for a large variety of different substrate materials. The concept proves that there are ways to achieve precision by statistics and probability and not only by precision itself. Due to its excellent repeatability, SwissOptic is already extensively using subaperture robotic polishing for the correction of optical components.

13.5.7 Status

From a technical point of view, a sophisticated subaperture robotic polishing system strongly depends on the quality and possibilities of metrology. Whatever errors

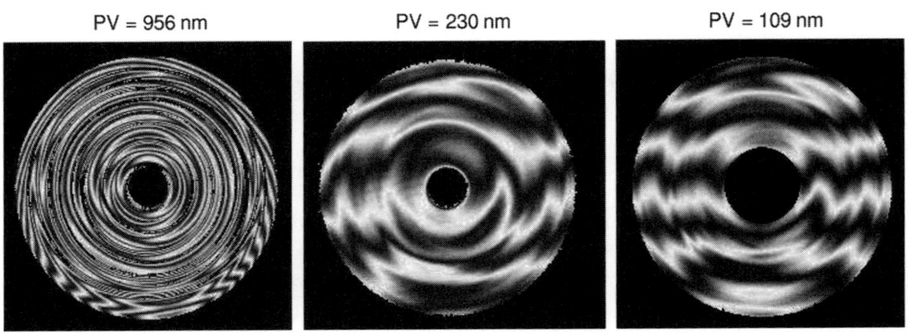

Figure 13.17 Fringe maps of correction steps performed by ALPS (parabola, $r = 301$ mm, $\varnothing = 300$ mm).

that can be measured and are related to the surfaces of an optical component can be corrected for. Advances in metrology will enable the manufacturing of off-axis aspheres and optical free-forms as well as the correction of whole systems. Figure 13.17 shows, for example, consecutive correction steps for a large aspherical surface, performed with ALPS.

From an economical point of view, the setup of a robotic polishing system opens new perspectives. By adding different grippers and tools, much more than just fine correction can be done, including the handling or even the prepolishing of parts.

13.5.8 Further reading

1. R.A. Jones and P.L. Kadakia, "An automated interferogram analysis technique," *Appl. Opt.*, Vol. 7, pp. 1477 (1968).
2. http://www.swissoptic.com/news.
3. D.D. Walker, D. Brooks, A. King, and R. Freeman, "The precessions tooling for polishing and figuring flat, spherical and aspheric surfaces," *Optics Express*, Vol. 11, No. 8, OSA (2003).
4. T. A. Porsching and C.A. Hall, "Approximation methods and the computer numerically controlled fabrication of optical surfaces," in *Advanced Optical Manufacturing and Testing II*; Proc. SPIE, Vol. 1531, pp. 205–215 (2001).
5. A. Schwarzhans, "Freeform fabrication," *SPIE E-magazine*, Vol. 3, No. 5, pp. 27–28 (2003).

13.6 Robot-Assisted Fluid Jet Polishing (FJP)

O. Fähnle

13.6.1 Basic assessment of the technology

This is a new polishing technique under development, capable of finishing flat, spherical, and aspherical surfaces. A commercially available subaperture polishing machine based on this technology is currently being constructed.

13.6.2 Intended purpose of the technology

Robot-assisted fluid jet polishing has two applications:

- Shape correction of a previously polished optical surface, and
- Polishing starting from a fine ground starting surface.

13.6.3 The technology's typical features

Because FJP causes no tool wear, and the surface is being cooled within the polishing spot, it is a stable abrasive polishing technique enabling a deterministic finishing.

The footprint geometry can be altered in size and shape in order to optimize the influence function for shape corrections.

13.6.4 Description of process

FJP represents a kinetic finishing technique where the shape of the applied footprint is generated by process parameters rather than by a given tool geometry. It is a subaperture fabrication technique that applies an abrasive slurry jet for polishing of optical surfaces in brittle materials such as glass. A premixed slurry is guided through a nozzle to the surface of the workpiece at pressures ranging from 3 to 20 bar. The footprint is determined by the shape of the nozzle as well as by the orientation of the slurry beam with respect to the local surface normal. No tool wear occurs, and the footprint remains constant during the manufacturing process, allowing shape corrections in a deterministic way (Fig. 13.18a).

13.6.5 Versions (state of the art)

A variation of the FJP technique (Fig. 13.18a) is arbitrarily named Jules Vernes (JV) (Fig. 13.18b). Similar to PACE (plasma-assisted chemical etching), a small cup-wheel-like nozzle is positioned close to the glass surface to be processed. FJP slurry is fed into the nozzle along its axis of symmetry, and, because of the circular gap between tool and workpiece, the suspension is radially accelerated to high velocities depending on the working pressure and the stand-off between the nozzle and the glass surface. Thus, an FJP-like material removal is

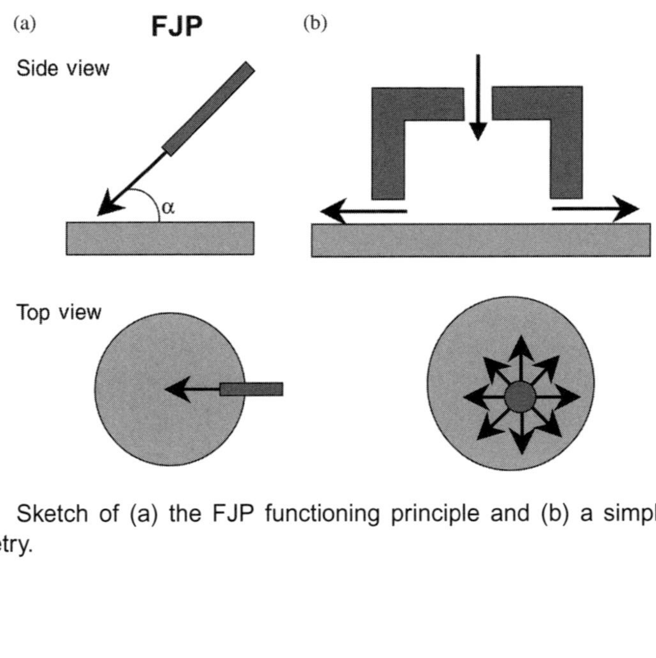

Figure 13.18 Sketch of (a) the FJP functioning principle and (b) a simple Jules Verne nozzle geometry.

achieved along the edge of the rotating nozzle. Whereas, FJP typically applies a nozzle stand-off distance of millimeters to centimeters, JV uses a stand-off distance down to 50 μm. The objective is to generate a nondirectional fluid flow parallel to the surface, which is specifically suited to reducing the surface roughness.

13.6.6 Performance and applications

A machine based on FJP is currently under development and will allow the following features:

- FJP does not need pH 7 polishing liquid, so all known polishing suspensions can be used.
- It is possible, by variation of parameters, to achieve an ablation rate larger than needed for prepolishing. This allows prepolishing and form correction to be performed at the same machine, using the same polishing slurry.
- Form correction of a prepolished spherical or aspherical glass surface with an accuracy of $<\lambda/20$.
- Fine polishing of a fine-ground spherical or aspherical glass surface (the process chain consists of a milling and an FJP polishing tool, with no need of prepolishing).
- The generation and form correction of mini-aspherical elements (sub-mm footprints) is possible.

Figure 13.19 shows a sketch of the prototype layout. A CNC optics fabrication machine has been altered to enable an FJP and JV polishing of rotary symmetric optical surfaces. Typical operation parameters are shown in Table 13.28.

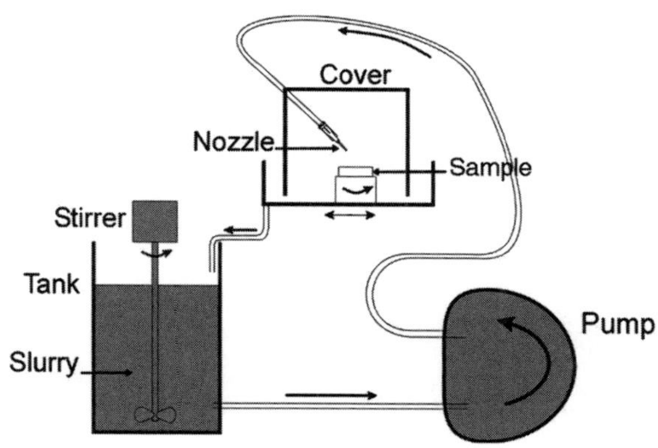

Figure 13.19 Sketch of the layout of a CNC machine capable of applying FJP and JV.

13.6.7 Data for robot-assisted fluid jet polishing (Tables 13.28–13.32)

Table 13.28 Typical operation parameters.

Parameter	Minimum	Maximum
Workpiece rotary speed	0	Up to 500 rpm
Tool rotary speed		
(not needed) Slurries	All types of slurries and polishing grains applicable—similar to pitch polishing on a traditional spindle machine	
Stand-off distance	0.05 mm	20 mm
Pumping pressures	5 bar	20 bar
Nozzle diameters	Sub-millimeter	4 mm (depending on pumping power)
Footprint geometry	Adjustable (depending on α and the cross section of the applied nozzle)	
Water rate	As in traditional polishing	
Wear mechanism	Kinetic finishing (two body process)	
Footprint stability	$<\lambda/20$	
Footprint size	Sub-millimeter	Depending on pumping system, currently up to 10 mm

Table 13.29 Machinable dimensions.

Geometrical dimension	Minimum	Maximum
Diameter	5 mm	Limited by machine dimensions; currently 240 mm

Table 13.30 Process measurement technique.

Measurement variable	Method of measurement
In-line	
Shape	Interferometer or profilometer
Off-line	
Shape	Interferometer or profilometer

Table 13.31 Cost drivers.

Cost factors	Comparable with
Machine	Because there is no rigid contact between tool and workpiece, there are lower requirements for stiffness of the machine. Although the position relative to the local surface normal has to be maintained (accuracy comparable with standard CNC machines), the positioning accuracy along the axis of the nozzle is less stringent.
Tool costs	No tool wear: standard nozzles from the water-cutting industry are applicable.
Energy consumption	Standard CNC machine.
Slurry	Standard slurries applicable. FJP applies a closed-loop slurry system. Costs are comparable to a continuous spiindle polishing machine
Operator	Skilled operator (optician)

Table 13.32 Limits of the technology.

Parameter	Limitation	Reason
Footprint minimum size		Jet stability
Footprint maximum size		Capacity of pumping system
Footprint stability	Currently $\lambda/20$	Depends on pumping system layout
Smallest diameter of lens	Currently 5 mm	Depends on nozzle design

13.6.8 Status

A feasability study has been finished, and a machine is currently under development. An in-house prototype is being used at Fisba Optik AG, St. Gallen, Switzerland (http://www.fisba.com).

13.6.9 Further reading

1. O.W. Fähnle, H. van Brug, and H.J. Frankena, "Fluid jet polishing of optical surfaces," *Applied Optics*, Vol. 37, No. 28, pp. 6771–6773 (1998).
2. S.M. Booij, O. Fähnle, *et al.*, "Jules Verne—A New Polishing Technique Related to FJP," in *Proceedings of SPIE Annual Meeting San Diego*, CA, USA, Optical Manufacturing and Testing IV, H. Philip Stahl (2003).
3. C.B. Zarowin, "Comparison of the smoothing and shaping of optics by plasma-assisted chemical etching and ion milling using the surface evolution theory," *Applied Optics*, Vol. 32, No. 16, pp. 2984–2991 (1993).

13.6.10 Links

- http://www.fisba.com
  Plant engineering and construction Prototype currently by
  - Fisba Optik AG, Rorschacher Strasse 268, CH-9016 St. Gallen, Switzerland; http://www.fisba.com
  - Sartisloh, Wilhelm-Loh Str. 2–4, D-35578 Wetzlar, Germany
  - R&D Fisba Optik AG

13.7 Ion Beam Polishing

R. Börret

13.7.1 Basic assessment of the technology

IBF (ion beam figuring), or ion beam milling, is a special kind of computer-controlled polishing (CCP), the corrective figuring step in the classical process chain (Fig. 13.20). IBF involves material removal in localized areas of the optical element based on measurements performed in advance. The goal of the IBF process step is to improve the optical performance of the treated element by a corrective figuring step on the surface of the element. Instead of a polishing pad for CCP, fluid jet, or an MR fluid (MRF), a beam of charged particles (ions) is used as a polishing tool. The accelerated ions act like sand blasting on an atomic scale.

13.7.2 Intended purpose of the technology

Early work on ion figuring of optical components was performed by Gale [1]. This work was carried on by Wilson [2] at the University of New Mexico. The first ion figuring system in an industrial environment was developed at Eastman Kodak by Allen and colleagues [3]. The Kodak ion figuring system is capable of processing optical components up to 2.4 m in diameter. It was developed in 1988 for space optics and became operational in 1990. The Kodak ion figuring system was a result of the Strategic Defense Initiative (SDI), a system proposed by U.S. President Reagan to use space-based systems to protect the United States from attacks by strategic nuclear missiles.

IBF is a local corrective polishing step and an excellent complement to conventional figuring. The optics is first polished (ground) conventionally, and the final figure (Fig. 13.21a) is milled by IBF, which works on a molecular level. The optics

Figure 13.20 Classical fabrication process for optics.

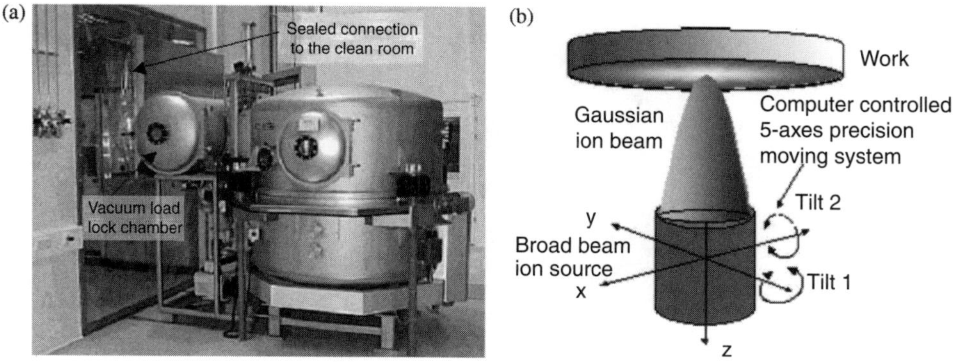

Figure 13.21 (a) Outside view of an IBF system. (b) Principle of IBF (Photo NTGL).

is inserted into a high-vacuum chamber facing down. The IBF then directs a beam of argon ions upward to the glass. The glass is removed on a molecular level. The beam itself is translated across the optical surface through a computer-controlled mask or a dwell time controlled path, removing figuring errors and surface roughness left-over from conventional polishing.

13.7.3 The technology's typical features

IBF removes material on a molecular level, resulting in a highly predictable and stable removal rate. Compared to CCP, where chemical interaction (hydrated layer) and mechanical interaction (scratching) remove the glass material, IBF is like sand blasting with argon ions. The sputter rate of the ion beam can be calculated very accurately by the physical laws of elastic and inelastic scattering. The absence of chemical interaction is one of the reasons for the stability of the process. The Gaussian beam profile (Fig. 13.21b) allows a simple and stable version for the well time calculation algorithm, which is performed in a similar manner to CCP.

Removing glass material, atomic layer by atomic layer, conserves the microroughness of the original surface. That means that very smooth surfaces do not change surface roughness during IBF, but rough surfaces are also not improved.

The main advantages of the IBF process:

- It is a noncontact technique with (a) no force loading on the workpiece, (b) no edge roll-off, (c) minimized support structure print through means that the bars from the lightweight structure of the back side are visible on the front optical surface;
- No tool wear;
- Stable wear function, which can be analytically described (Gaussian profile);
- Removal of material on the molecular level.

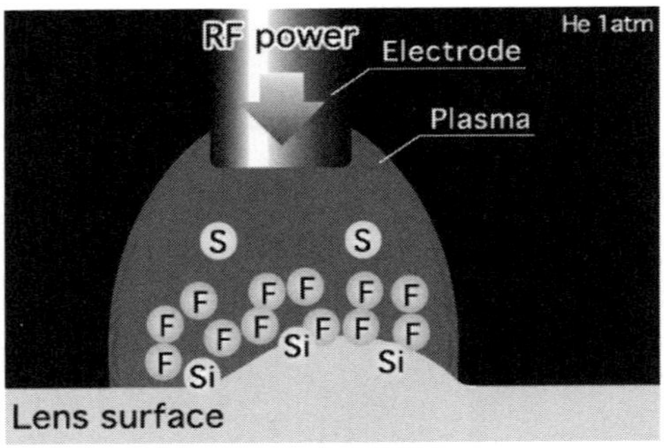

Figure 13.22 Principle of plasma CVM as further development of IBF.

The material removal rate is between one and two glass atoms per incoming argon ion. This allows very fine correction of surface deformations but also has the disadvantage that large deviations from the desired shape result in exorbitant manufacturing times.

Further development results therefore in a chemical-assisted IBF, called PACE [6] (plasma assisted chemical etching) or Plasma CVM (http://www.nikon.co.jp) (Chemical Vaporization Machining). The principle is shown in Fig. 13.22. SF_6 is used as the basis to generate Fluor radicals, which react chemically with the Si atoms in the glass. Although the IBF process requires ultrahigh vacuum conditions, PACE has also been tested under a rare gas atmosphere. The removal rates of PACE are a factor 10 to 20 higher compared to IBF.

13.7.4 Description of process

The principle of the IBF process is shown in Fig. 13.23(a). The beam profile moves relative to the polished surface. The dwell time function $t(x, y)$ is calculated by a deconvolution of the desired removal profile and the beam profile or by an optimization algorithm.

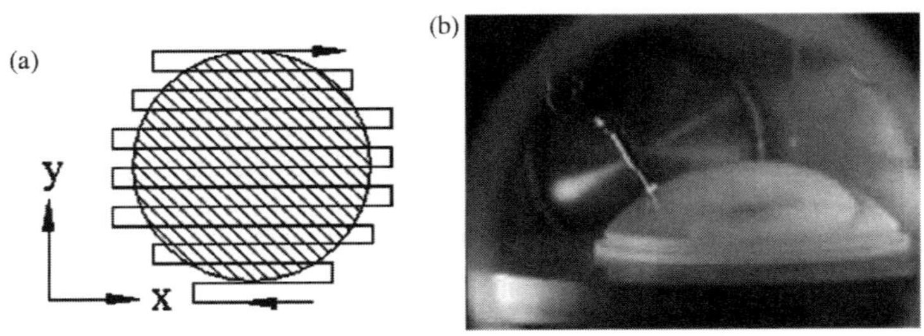

Figure 13.23 (a) Principle of material removal with IBF, line by line. (b) Beam moving across the surface (Photo NTGL and Nikon).

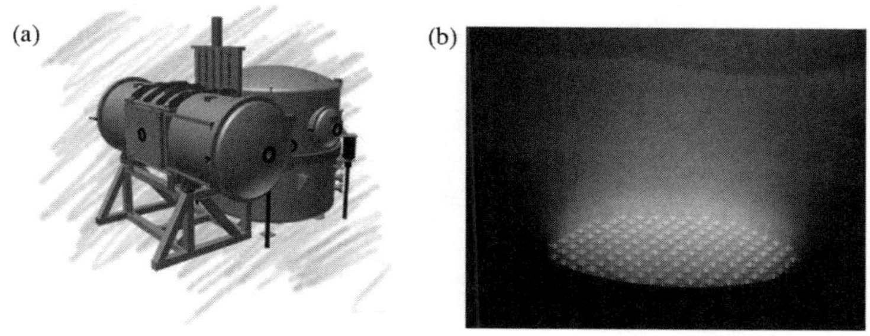

Figure 13.24 (a) Commercially available IBF system from NTGL (Germany). (b) Ion process from RC optical system (USA).

13.7.5 Versions (state of the art)

IBF is in most cases, the final figuring process within the optical fabrication of high-end optics. The ion milling process is used in different situations in the precision optics workshops at available Eastman Kodak, Tinsley (USA), Sagem (France), or Zeiss (Germany). Beyond that, there are commercially available IBF systems such as the UPFA from NTGL (Germany) (Fig. 13.24).

The detailed engineering of machine, ion source, and dedicated algorithm are different at each company. As for CCP, the tool rasters the surface controlled by the calculated dwell time function. Similar to CCP, the important parameters of the IBF process are

- Stability of the ion beam,
- Accuracy of position system and tool path,
- Algorithm,
- Accuracy of the metrology data.

13.7.6 Data for ion beam polishing (Tables 13.33–13.38)

Table 13.33 Typical specifications.

Size	Surface roughness	Defect of form
Some mm up to 2.4 m	~0.2 nm rms	~0.2 nm rms

Table 13.34 Typical operation parameters.

Parameter	Minimum	Maximum
Gaussian beam shape (FWHM)	Some mm	200 mm
Ion beam current	Some mA	Several 100 mA
Removal rate	About 1 nm/min	Several 100 nm/min

Table 13.35 Machinable dimensions.

Geometrical dimension	Minimum	Maximum
Diameter	Some mm	2.4 m
Radius	Some mm	∞

Table 13.36 Process measurement technique.

Measured variable	Method of measurement
In-line	
None	None
Off-line	
Shape 3D	Interferometric; e.g., with computer-generated hologram
Surface roughness	Micro-interferometer, atomic force microscope (AFM)

Table 13.37 Cost drivers.

Cost factor	
Machine investment	High, >500T€ depending on size and vacuum conditions
Setup time	Small, fast setup possible compared with CCP
Process time	Large for IBF, fast for PACE
Operator	Skilled operator; similar to other subaperture processes Basic know-how on vacuum necessary
Tools and supplies	Medium, but high cost on vacuum spare parts

Table 13.38 Limits of the technology.

Parameter	Limiting value	Reason for limitation
Surface roughness	0.2 nm rms	Limited by roughness of prepolished surface and fine structure of material
Final shape	0.1 nm rms	Metrology data used as input for figuring process

13.7.7 Conclusions

Without question, ion beam figuring is the biggest breakthrough ever made in optical manufacturing. Instead of conventional grinding (scratching) of a mirror, ion milling removes glass on a molecular level, resulting in a very accurate surface with minimum deviation from nominal shape. This extremely high surface figure and smoothness reduces aberrations and scatter, and results in the highest contrast possible.

The limitations of the process are the high investments required in vacuum technology, the low removal rate for IBF, and the fact that PACE works only on a few materials, such as silica or fused silica.

13.7.8 Further reading

A detailed description of the process and related algorithm is found in [5]. The actual limitation about final surface quality is reported in paper [4]. Details of a commercially available IBF system are available at the NTGL homepage [http://www.ntgl.de].

1. A.J. Gale, "Ion machining of optical components," *Optical Society of America Annual Meeting Conference Proceedings*, Nov. (1978).
2. S.R. Wilson and J.R. McNeil, "Neutral ion beam figuring of large optical surfaces," in *Current Developments in Optical Engineering*, Bellingham, WA, SPIE, pp. 320–323 (1987).
3. L.N. Allen and R.E. Keim, "An ion figuring system for large optics fabrication," in *Current Developments in Optical Engineering*, Bellingham, WA, SPIE, pp. 1168 (1989).
4. H. Handschuh, J. Fröschke, M. Jülich, M. Mayer, M. Weiser and G. Seitz, "Extreme ultra-violet lithography at Carl Zeiss: Manufacturing and metrology of aspheric surfaces with angstrom accuracy," *Journal of Vacuum Science & Technology B: Microelectronics and Nanometer Structures*, Vol. 17, pp. 2975–2977 (1999).
5. T.W. Drueding, T.G. Bifano, and S.C. Fawcett, "Contouring algorithm for ion figuring," *Precision Engineering*, Vol. 17, pp. 10–21 (1995).
6. G. Boehm, W. Frank, A. Schindler, A. Nickel, H.-J. Thomas, F. Bigl, and M. Weiser, "Plasma jet chemical etching—a tool for the figuring of optical precision aspheres," *Precision Science and Technology for Perfect Surfaces*, eds. Y. Furukawa, Y. Mori and T. Kataoka, The Japan Society for Precision Engineering, Tokyo (1999) pp. 231–236 (Proc. of the 9th ICPE, Osaka/Japan) (1999).

13.7.9 Links

Plant engineering and construction:

- http://www.ntgl.de

Engineering and construction of commercially available tools are carried out by NTGL, but there are also proprietary machines available at different optics manufacturers.

Toolmakers and suppliers of auxiliary materials:

- http://www.ntgl.de (commercial tools)

Research and development (performed within different companies and at the IOM Leipzig):

- http://www.sagem-ds.com/eng/bds_optro_01_00.htm
- http://www.ornl.gov/info/ornlreview/rev30-12/text/techtran.htm
- http://www.ntgl.de/
- http://www.rcopticalsystems.com/ionmill.html

- http://www.nikon.co.jp/main/eng/portfolio/about/technology/nikon_technology/plasma_cvm_e/index.htm
- http://www.iom-leipzig.de/index_d.cfm

13.8 Precision Glass Molding

C. Klein, R. Müller

13.8.1 Basic assessment of the technology

Precision glass molding (PGM) is a technology suitable for cost-efficient, large-scale production of imaging precision and lighting components from special glasses, by hot forming without cold post-processing of the optical surfaces.

13.8.2 Intended purpose of the technology

Optical components with flat and spherical surfaces can be fabricated cost-efficiently and at a satisfactory quality by grinding and polishing. However, aspherical lenses, mirrors, and optical components with optical free-forms can be fabricated with classical grinding and polishing techniques, only at high costs. Optical components with diffractive structures cannot be fabricated by grinding and polishing.

PGM works particularly well for the large-scale fabrication of optical components from special glasses that cannot be fabricated cost-efficiently with the established fabrication techniques. Since the 1990s, especially in Asia, optically imaging aspheres (with a diameter between 0.2 and 20 mm) and lighting components (condenser and relay lenses, integrator boards, and so on) for mass products such as mobile phones, digital still cameras, D-SLR, telecommunication, optical storage, optical sensorics, and so on, have been fabricated in large delivery quantities (>200 million pieces per year). Typical production batches for mass products, such as mobile phones, and cameras, are 500,000–1,000,000 per month, or for DSC 50,000–100,000 per month.

13.8.3 The technology's typical features

PGM differs essentially from other hot-forming techniques by the fact that the plastic deformation takes place at distinctly higher viscosities (see Sec. 13.8.4) than with the classical hot glass forming techniques. Classically, glass is deformed at viscosities of 10^4 dPas forming viscosity of standard hot forming. With PGM, the tight fit of the glass with the pressing die is established by plastic deformation of a semifinished product under high pressure. During the following downcooling controlled phase nearly to the transformation temperature (low cooling rates), the tight fit of the glass has to be maintained in order to replicate the contour of the pressing die. As the material shrinks during cooling of the glass, the tight fit only can be guaranteed if the glass sticks to one half of the die (usually the aspherical die) (single-sided PGM) or if, during the cooling process, the glass is still deformed by pressing (double-sided PGM). The optical components fabricated in this way can normally be used without any further processing of the surfaces. There may be one exception for the post-processing of the edges, if this is necessary for reasons of design (centering bevel, D-cut, and so on, especially if after pressing during the PGM process the edge has been overpressed).

13.8.4 Description of process

The operating resources (with the powertrain control module supporting technologies) as shown in Fig. 13.25.

13.8.4.1 Operating resources

Glasses used

Special glasses are used with special optical characteristics, according to the specifications for the optical design, and with special mechanical and thermal characteristics (so-called "low-T_g glasses"). By the use of these glasses, the pressing temperature is reduced to 350–650°C, according to the glass system necessary

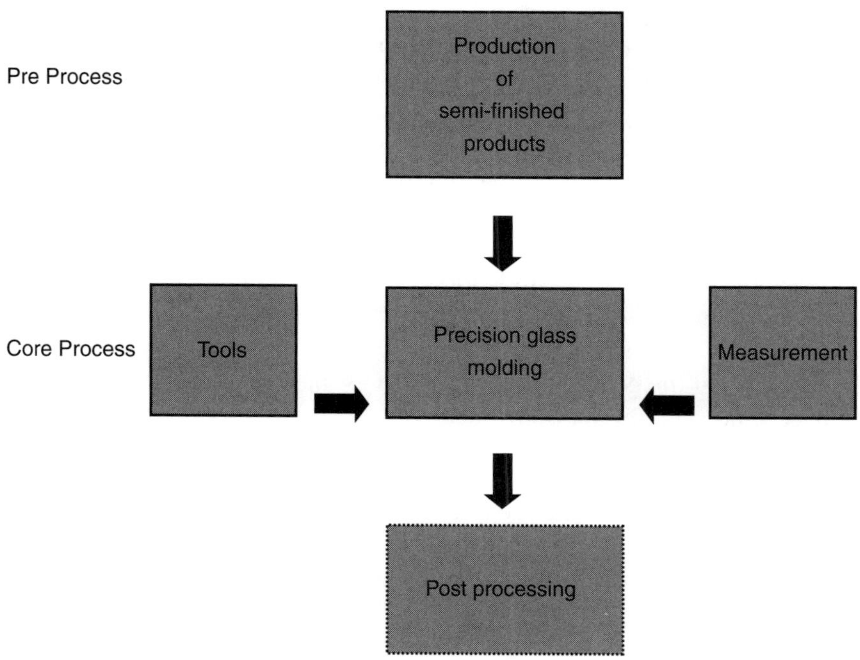

Figure 13.25 Operating resources, including supporting technologies.

premise for the long production life of the molds, and the process times are reduced (short cooling time because of a steep viscosity–temperature curve). During reheating of the semifinished products and during pressing, the glasses must not devitrify. Volatile glass elements (oxides) must not outgas during PGM, as the condensation sublimates on the surface of the mold, which leads to surface defects and to short production lives of the interface (separating layer glass/mold) in particular.

Semifinished products used

For PGM, various semifinished products are used, dependent on the use of the pressings. For the fabrication of components for lighting optics, usually nonportioned semifinished products are used; for the fabrication of components for optically imaging precision optics, portioned semifinished products from special glasses are used.

Nonportioned semifinished products (rods)

These are bars ("rods") of ~1 m in length, with a round profile (diameter between 20 and 40 mm) and with a polished surface from the manufacturing process. The founding of the glasses and the fabrication of the rods is carried out with current techniques and equipment used for the fabrication of optical glasses.

Portioned semifinished products (preforms)

These are glass batches with an order-specific weight and with a blank surface ("preforms"). The geometry of the semifinished products has to be chosen in such a way that no air can normally be enclosed between the die and the semifinished product, to avoid overbending during pressing. If required, the geometry of the semifinished products can be similar to the components to be pressed, so that during pressing only a little material has to be deformed. The semifinished products must have a surface (roughness, cleanliness, surface defects) in optical quality, as during blank pressing only the global contour is formed, but the quality of the surface cannot be improved. Fabrication of the preforms from low-T_g glasses is carried out with current techniques and equipment used for the fabrication of optical glasses.

Depending on the geometry of the aspheres and the quantity to be produced, different semifinished products are used:

- *Precisions balls*: Mostly used for small aspheres with a maximum diameter of 5 mm, for example, mobile phones and large-scale production.
- *Precision gobs*: Cost-efficient preforms for large quantities. Disadvantage: Cannot be fabricated from every glass and are not suitable for every aspheric geometry.
- *Disks*: Cost-efficient preforms, because only the flat abutting faces have to be polished. However, disks are not suitable for every aspheric geometry.

- *Near shape preforms*: Spheres, which will be pressed into aspheres. Advantage: The conversion is relatively small. Disadvantage: Significantly more expensive than precision balls and precision gobs.

Precision glass molding

- *Ambient conditions*: The blank pressing machines are normally used in a clean room of category 10,000 to avoid contamination of the molds, preforms, and aspheres being contaminated. The measure of the air quality of a clean room is described in Federal Standard 209. Clean rooms are rated as "Class 10,000," where there exists no more than 10,000 particles larger than 0.5 microns in any given cubic foot of air; "Class 1000," where there exists no more than 1000 particles; and "Class 100," where there exists no more than 100 particles. Hard disk drive fabrication requires a Class 100 clean room. In places where the surfaces of the preforms, or of the molds, are exposed to the environment, a clean room of category 100 is locally targeted to avoid dust particles from depositing on the preform, or on the mold, which might damage the mold.
- *Pressing dies*: The material used for the dies has outstanding heat resistance, good grinding and polishing properties, and, by a uniform fine grain, guarantees a minimum of roughness. Among other materials, tungsten carbide or SiC are used. A disadvantage of these materials is the time-consuming processing required by grinding and polishing. The contour modifications of the pressings, because of shrinking during downcooling to ambient temperature, have to be taken into account in the geometry of the die. As a rule, multiple correction cycles have to be run through to optimize the contour of the die, to guarantee the necessary exactness of the contours of the pressings.

  The surface of the dies usually gets a coating (interface), which has to fulfill the following functions:
  - Avoiding the glasses from sticking to the die surface during pressing;
  - Avoiding chemical reactions in the interfacial area glass/forming die;
  - Avoiding oxidation of the die surface;
  - Maximizing the operational life of the dies;
  - Protecting the die surface from defects (e.g., scratches).

Hot forming (Fig. 13.26)

- *PGM from rods (usually for components of lighting optics)*: The rods are heated outside the press, positioned between the two halves of the press, and portioned by "overpressing". The glass sticks to one half of the die (usually to the die half with the aspherical contour) in such a way that the tight fit is maintained during downcooling until nearly ambient temperature. The contour of the die is replicated, and in this way an optical surface is created by blank pressing (single-side pressing). The material retraction (shrink) takes place at the opposite side. The second optical surface (usually a spherical one) is

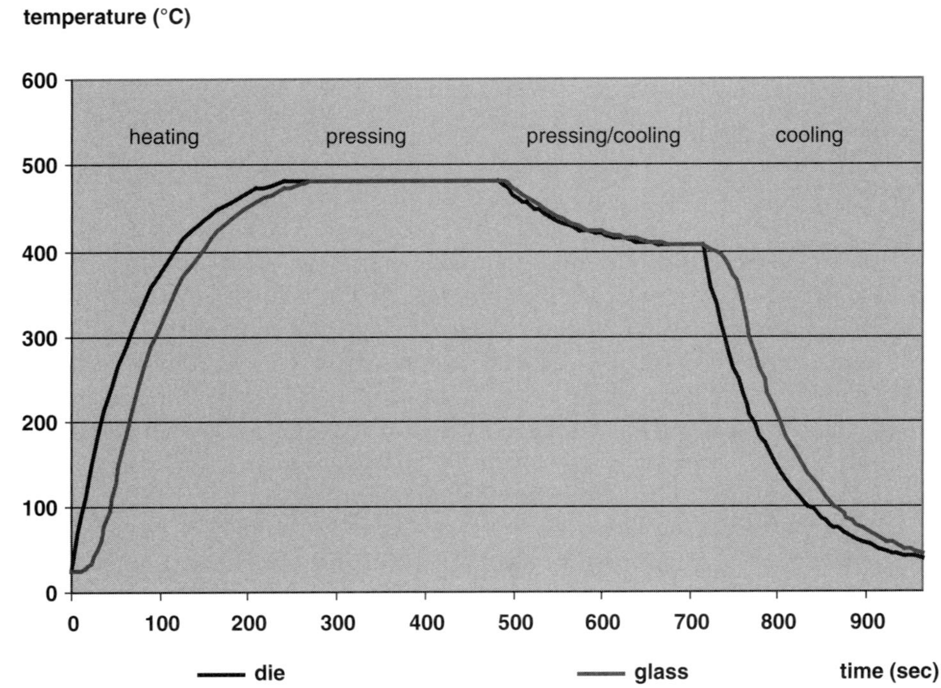

Figure 13.26 Isothermal pressure molding.

subsequently created by grinding and polishing. For the cooling process, the pressings are removed from the die and are usually cooled in a continuous annealing furnace.

- *PGM from preforms (usually for precision components of imaging optics)*: The cold preforms are placed in the press, between the top and the bottom die. Die and glass are heated up to the desired pressing temperature in such a way that both glass and die attain the same temperature at the end of the heating period. Usually the heating process determines the total processing time.

Between the top and the bottom die, at a viscosity between 10^8–10^{10} dPas, the preforms are plastically deformed, with a defined press load, until a tight fit is achieved with the optical surfaces (isothermal pressure molding). When the load values are set, it has to be taken into account that the contact surface between glass and forming die changes during the transformation process. The pressure exercised on the glass surface is thus not only dependent on the applied press load but also on the actual pressing path. Best pressing results are obtained if the course of pressing is selected in such a way that the pressing speed is relatively constant and both optical surfaces are pressed (usually without overpressing). During downcooling until nearly the transformation temperature, the press load is maintained to preserve the tight fit by viscoelastic flow of the glass, notwithstanding the material shrinkage. Thus, both optical surfaces are created by blank pressing (double-side pressing).

The pressings are usually cooled in the die. Only in special cases are they submitted to a further thermal post-processing.

13.8.4.2 Mode of operation

Process differences in hot forming (state of the art)

To reduce production costs and to maximize performance, PGM has different process variants:

- For the standard process, all process steps run in a stationary die, chronologically one after another. The process steps run serially. The performance (number of pressings per time) of the equipment is low.
- For the improved process, the process steps (heating, pressing, cooling) run in different locations. Various dies therefore run through the press, and the process steps run in parallel in time. The performance of the equipment is significantly improved.
- For small sizes, usually several pressings are pressed into one die (multiple pressings).
- Often the edge is overpressed for technological or logistical reasons. As a consequence, post-processing of the edge is necessary.

13.8.5 Data for precision glass molding (Tables 13.39–13.43)

13.8.5.1 Typical specifications

Optically imaging aspheres

Tables 13.39 and 13.40 show, for different size categories of optically imaging aspheres, the quality characteristics PGM can achieve. The values in the table are meant to be benchmarks only. With the available technology, optically imaging aspheres with a maximum diameter of up to ~40 mm can be fabricated with the quality characteristics in these tables.

A typical example of a blank pressed asphere for optical applications is shown in Figure 13.27.

Table 13.39 Typical specifications.

Size category		Contour exactness	Surfaces roughness
Diameter (mm)	Weight (g)	PV (μm)	PV (nm)
>30	>20	4*	<3
30–20	20–4	2**	<3
20–10	4–0.4	1	<3
10–5	0.4–0.1	0.5	<3
5–2	0.1–0.01	0.2	<3
<2	<0.01	0.1	<3

Ad* : In the center 2 μm.
Ad**: In the center 1 μm.

Table 13.40 More specifications.

Angular misalignment	<1.5 minute
Offset of axes	20–30 μm
Surface defects: 3 × 0.16–3 × 0.4 (max. number × max. length (pcs × mm))	3 × 0.16–3 × 0.4

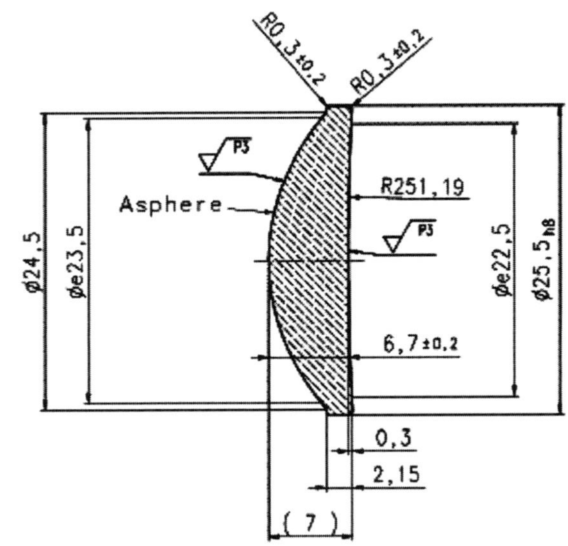

Figure 13.27 Example of a blank pressed asphere for optical applications.

Optical lighting components

The requirements for contour exactness are usually lower than for the optically imaging components by a factor of 5 to 100.

For specific applications, aspheres, pressed from the rod, can also be used without post-processing the second surface.

13.8.5.2 Typical operation parameters

Table 13.41 shows typical operation parameters.

13.8.5.3 Process measurement technique

The PGM process is mainly controlled by temperature press load, and processing time (Table 13.42).

Table 13.41 Typical operation parameters.

Parameter	Minimum	Maximum
Pressing temperature	325°C	650°C
Press load		
Cycle time	15 min	20 min
Number of cavities	1	~30

Table 13.42 Process measurement technique (The PGM process is mainly controlled by temperature, press load, and processing time).

Measured variable	Method of measurement
In-line	
Temperature	Thermocouples
Force	Force sensor
Off-line	
na	na

na, not applicable.

13.8.5.4 Cost drivers

Table 13.43 shows the strengths and weaknesses of the various process variants for favored products and markets regarding the following criteria:

- Start-up costs (development and investment costs standardized on yearly performance);

Table 13.43 Cost drivers.

	Process variables					
	Pressing from rods with one station press (large to medium components of lighting optics)		Pressing from preforms with one station press (medium and small components for imaging optics)		Pressing from preforms with multiple station presses (medium and small component for imaging optics)	
Markets	Industry	Consumer	Industry	Consumer	Industry	Consumer
Start-up costs	**	***	**	***	**	*
Pressing costs	**	***	**	**	*	*
Mould costs	**	**	**	**	***	**

***, high; **, medium; *, low.

- Production costs (without start-up and mold costs);
- Mold costs (measured by the prices to be achieved).

13.8.6 Conclusions

PGM is suited to the fabrication of optical lighting components and for the fabrication of precision optical components from special glasses. The quality characteristics, contour, and roughness of the optical surfaces are comparable with those achieved by grinding and polishing.

The process is particularly suitable for the large-scale production of components with aspherical surfaces or free-forms that cannot be produced cost-efficiently with established processes such as grinding and polishing.

The various process variants are particularly suitable for the following business fields:

- *Single-sided PGM from rods* (*one station press with stationary die*): For large to medium components of lighting optics in small to medium delivery quantities (business fields "Industry" and "Consumer"); in particular, also for optically imaging components.
- *Double-sided blank pressing from preforms (one station press with stationary die)*: For medium and small components of imaging optics with small and medium delivery quantities (business fields "Industry" and "Consumer").
- *Double-sided blank pressing from preforms (multiple station press with non-stationary dies)*: For medium and small components of imaging optics, with medium and large delivery quantities (business fields "Industry" and "Consumer").

13.8.7 Status

Additional developments are necessary to further reduce mold costs. A further challenge is the adaptation of the interface layer to newly developed glasses for PGM, such as highly diffractive glass with a refractive index $n_d > 1.90$.

13.9 Tools for Precision Glass Molding

T. Bergs, B. Bresseler

13.9.1 Basic assessment of the technology

Precision glass molding (PGM) is an economic alternative to traditional manufacturing technologies like grinding and polishing, especially for

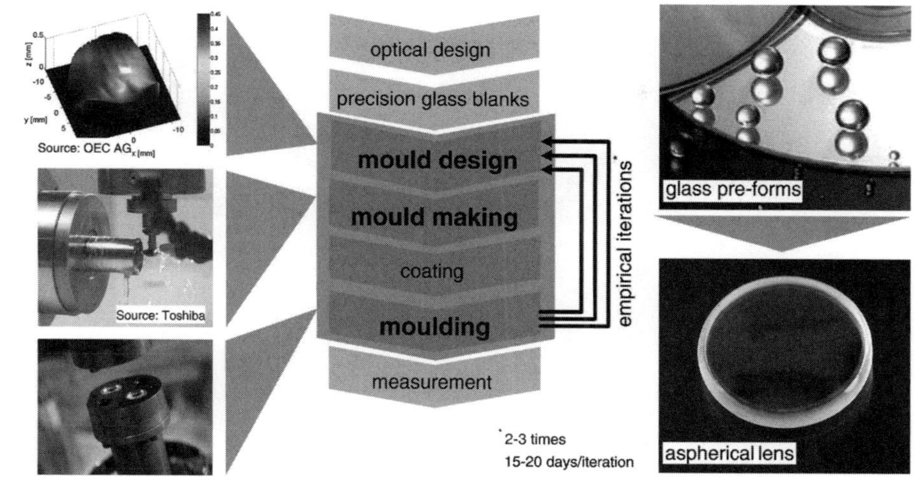

Figure 13.28 Process steps.

complex-shaped optics (Fig. 13.28). Being originally developed for high-volume production, it can also be utilized for medium or even small batch production, particularly if traditional technologies are not feasible. In comparison to traditional molding, PGM does not require any subsequent finishing operation for the optical functional surfaces. Great geometrical freedom, high form accuracy, and excellent surface qualities are the main advantages of this technology.

Asian manufacturers have been using precision molding for years to produce lenses, for example, for digital cameras. The technology has so far not been able to establish itself in Europe; the necessary tools have only been available in Asia, making tool manufacture, maintenance, and repair so costly that production would not have been worthwhile in Europe. Due to the replicative character of PGM, the quality requirements for the moulds in terms of form accuracy and surface finish are at least as high as for the optical elements themselves. In addition, the molds have to stand high-temperature cycles over a long period and have to be oxidation-resistant so as not to react with the hot glass surface. The rising need for complex optical geometries, like aspheres or micro-arrays in combination with the high quality requirements mentioned and usually low lot sizes of molds, lead to the application of highly flexible ultraprecise cutting technologies.

13.9.2 Intended purpose of the technology

Ultraprecision manufacturing technologies (Fig. 13.29) have been developed to machine special materials at extreme qualities, which usually means mirror surfaces with roughness values in the few nanometers range. Ultraprecision manufacturing technologies are characterized by the application of highly precise machine tools. The accuracy of motion has to be better than 1 μm, which often requires the use of linear motors. Every vibrational influence has to be eliminated so as not to damage the surface quality. Therefore, mainly aero- or hydrostatic spindle and

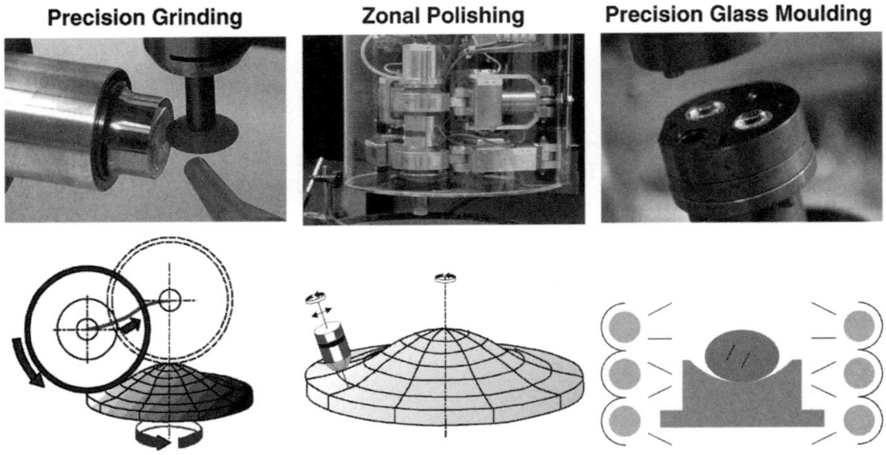

Figure 13.29 Ultraprecision manufacturing technologies.

guide systems are applied. Cutting forces have to be decreased to a few newtons to be able to achieve high form accuracies. Ultraprecision manufacturing technologies are highly flexible, which qualifies these processes, especially for low-volume or single-item production.

13.9.3 The technology's typical features

PGM of optical components leads to strong requirements for the molds applied. For most optics, the base material requirements are set by high molding temperatures (>450°C) (Fig. 13.30). Due to this high temperature and the relatively long contact time between glass and mold, the reactivity between both is high. This leads to the requirement for high corrosion and oxidation resistance to minimize the effect of adhesion. For avoiding sticking of the glass over a long process period, additional coating systems are applied on the mold surface. Thermal expansion similar to the glass material as well as good thermal conductivity is required for

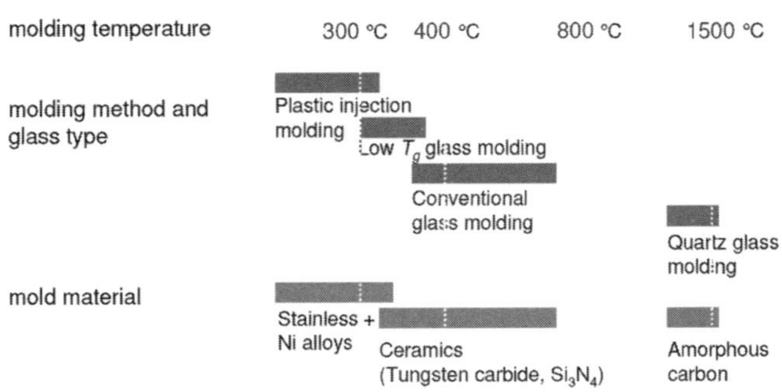

Figure 13.30 Temperature range for molding processes.

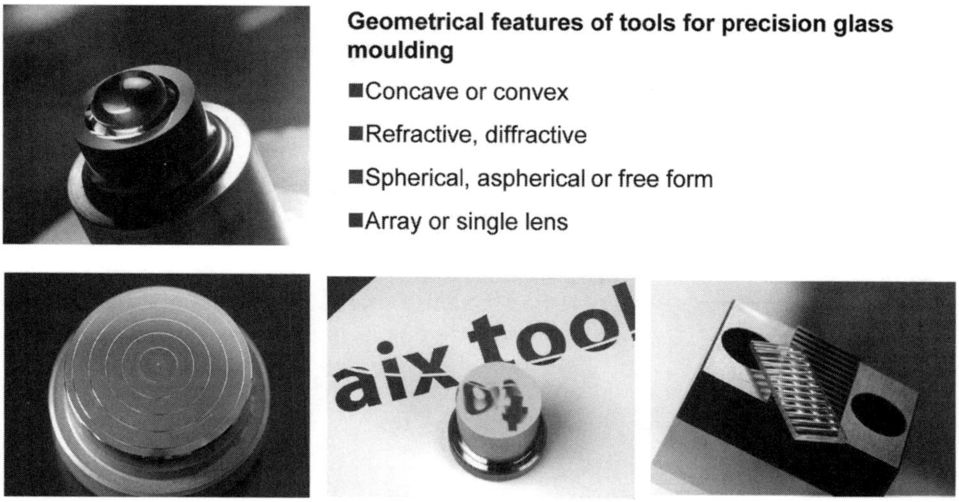

Figure 13.31 Geometrical features of tools for PGM.

good molding results. The base material has to be hard and wear-resistant and has to be machinable in optical quality. Different materials have been investigated and qualified, and include, in particular, advanced cemented carbides and ceramics. These materials are characterized by hard and brittle material properties and have to be machined by processes with a nondefined cutting edge. Only the latest development of low-T_g glasses with transformation temperatures below 300°C allows the application of nickel-coated steel molds, which are machinable by ultraprecision cutting with single-crystalline diamond tools and offer wide ranges of geometrical flexibility. However, the variety of low-T_g glasses available is small, to date.

Lens design determines the requirements in terms of mold geometry, form accuracy, and surface finish. For current optical applications, requirements are

- Convex or concave shapes,
- Spherical, aspherical, or free-form surfaces,
- Rotational-symmetric or cylindrical configurations,
- Single or array lenses,
- Lenses with a diameter ≤ 2 mm up to large lenses with a diameter >50 mm, and with deep cavities and steep sides,
- Form accuracies between PV 100 nm and 2 μm, and
- Surface roughness between R_a 1 and 10 nm.

13.9.4 Description of process (Figs 13.32 and 13.33)

The manufacturing of precision glass molds typically comprises the following steps:

1. Mold design and mold material selection;

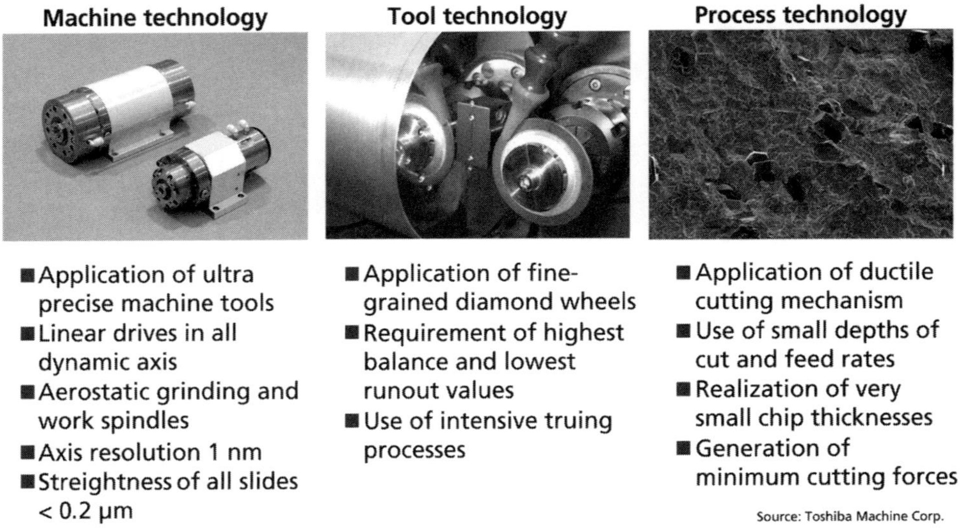

Figure 13.32 Process layout.

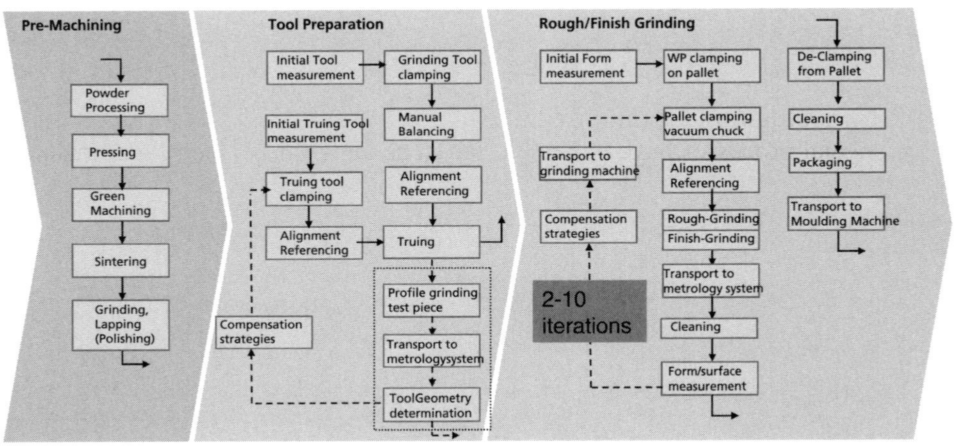

Figure 13.33 Mode of operation.

2. Rough, fine machining of the mold and components;
3. Ultraprecision machining of the mold cavities;
4. Measurement of mold properties (esp. form and roughness);
5. Form correction (step 3, 4).

This is followed by further steps:

6. Try-out molding tests;
7. Coating;
8. Further form correction.

13.9.4.1 Operating resources

Mold design and mold material selection

First, mold design has to meet the requirements of the lens design. However, more complex geometries result in difficult processing, problems in accuracy, and consequently in high costs. Thus, steep geometries, deep cavities (of the mold), small radii between faces, and very small or large diameters should be avoided. Finally, the ability to process all important faces (in particular the pressing face) within one clamping increases form accuracy after processing. Typical materials use are tungsten carbide, ceramics, super alloys, or (for pressing temperature $<400°C$) NiP-coatings. Mold material strongly depends on pressing temperature and required surface roughness. Moreover, thermomechanical characteristics (like high thermal hardness, thermal conductivity, or thermal expansion of the glass to be pressed and other mold components) have to be considered.

Rough, fine, and finish machining of the mold

Processes have to be selected depending on geometry and material. However, the main criteria for selection are costs. Possible processes are grinding, turning, EDM, but also preforming of the raw material (e.g., hot isostatically pressed (HIPed) powder). Special processes have to be selected depending on geometry and material. Typically, the required accuracies result in highly accurate machine tools for standard processes, such as grinding (in particular, ductile grinding), ultraprecision turning or single-point diamond turning (SPDT), but also special processes like magnetorheological finishing (MRF) or electrolytic in-process dressing (ELID) grinding.

To reach the final roughness of the surface, polishing processes are used. In the case of aspherical geometries, polishing is done manually; however, computer-controlled polishing (CCP) by zonal controlled polishing is growing in importance. Polishing typically worsens form accuracy.

Measurement of mold properties

Apart from the standard geometrical quantities that are measured, special quantities to be assessed are

- Form accuracy: tactile or interferometer assessed;
- Centricity: tactile or interferometer assessed;
- Roughness: tactile, atomic force microscope (AFM), or interferometer assessed;
- Layout; and
- Mode of operation.

13.9.5 Data for tools for precision glass molding (Tables 13.44–13.47)

Table 13.44 Typical specifications.

Topic	Quantity	Unit	Comments
Diameter range	0.5–70	mm	Application-specific
Surface roughness (R_a)	<3	μm	Depending on diameter
Form error (PV)	<1	μm	Depending on diameter
Flank angle	<60	deg	Limited metrology
Diffractive structures	–	–	Infrared applications

Table 13.45 Process measurement technique.

Measured variable	Method of measurement
Off-line	
Form accuracy	Interferometer, tactile, optical CMS
Surface roughness/topography	Tactile, atomic force microscope (AFM), white light interferometry, optical microscope, stereo loupe
Overall form error	CMS

CMS, control measuring system.

Table 13.46 Cost drivers.

Parameter	Typical expense	Comments
Ultraprecision machine	300–800T€	Depending on substrate material
Measurement systems	100–600T€	Depending on mold complexity
Try-out molding machine	350T€	Depending on molding technology

Table 13.47 Limits of the technology.

Parameter	Reason for limitation
Use of advanced mold substrates	Thermal expansion, surface finish
Use of special coatings	Adhesion of glass, long cycle times
Comprehensive iteration steps	Nondefined machining operation

13.9.6 Conclusions

The manufacturing of glass molds mainly defines the feasibility as well as the efficiency of the actual PGM process. It requires comprehensive know-how in tool layout, ultraprecision machining, as well as in the actual molding process. Each step in the process chain is usually associated with extreme technological challenges and thus may lead to high costs. However, high-value precision molds enable economical molding of advanced optics in glass, especially in those cases where traditional technologies fail.

13.9.7 Further reading

1. F. Klocke and G. Pongs, "Precision glass moulding of optical components," *WGP Annals* (2004).
2. F. Klocke and G. Pongs, Abformung präzisionsoptischer Komponenten aus Glas mit keramischen Werkzeugen, wt Werkstattstechnik online, Ausgabe 3 (2004).
3. F. Klocke, G. Pongs, O. Dambon, and M. Heselhaus, "Innovative Ultra-Precision Finish-Machining Processes for Optical Components," in *Proceedings of the 2nd Glass Colloquium Glass Processing*, Aachen (2002).
4. F. Klocke, G. Pongs, B. Bresseler, and M. Heselhaus, "Precison Technologies for Micro-Structuring of Glass," in *Proceedings of CIRP Seminar on Micro and Nano Technology 2003*, Copenhagen, Denmark (2003).

13.9.8 Links

Tool supplier

- http://www.aixtooling.de; Aixtooling GmbH, Steinbachstraße 17, D-52074 Aachen, Germany.

Research and development

- http://www.ipt.fraunhofer.de; Fraunhofer-Institut für Produktionstechnologie IPT, Steinbachstr. 17, D-52074 Aachen, Germany.

13.10 Injection Molding of High-Precision Polymer Optics

S. Gold, R. Mayer

13.10.1 Basic assessment of the technology

Commonly well known conventional injection molding is not sufficient for optical parts production. Hence, by optimization of process parameters, a high-precision optical molding process is developed.

High-precision injection molding (using high-grade optical polymers) is a technology that helps to satisfy growing market demands due to its unique features, which are

- Versatility,
- Ready-to-use production, and no finishing required (except coating),

- High shot-to-shot repeatability,
- Low-cost process,
- Low-cost parts,
- Suitability for complex aspherical and free-form surfaces,
- Integrated optical/mechanical functionality,
- Environmentally safe,
- Energy effective.

The development of high-precision molding processes and suitable polymer materials is far from being complete. In the range of high-precision optical components, it is essential to consider optical design, mechanical design, mold-process development, and mold-machine development as parts of an integrated process with very strong interactions. You cannot do any one without the others!

13.10.2 Intended purpose of the technology

Precision injection molding was developed in order to enable an economic mass production of precise spherical, aspherical, and free-form plastic lenses and mirrors with high accuracy and optical surface finish. Examples of injection-molded parts are presented in Figs. 13.34–13.36.

13.10.3 The technology's typical features

When considering injection-molded parts, there are several features considered unique to injection-molded optics:

- *Low weight*: Optical polymers have approximately half the density of glass. Hence, low-weight designs are possible.

Figure 13.34 Automotive rain sensor optics.

Figure 13.35 Aspherical lenses for money-control systems.

Figure 13.36 Camera focusing screen.

- *Low material costs*: Optical polymers cost within a range of 5–30€/kg. Compared to optical glass, this constitutes a noticeable difference.
- *Suitable for mass production at competitive costs*: Injection-molded lenses are finished in one step to optical quality, without the need for additional finishing steps like polishing or cleaning. Compared to glass, cycle times are very low, which makes injection molding very suitable for mass production.
- *High degrees of freedom in optical designs*: Using injection molding, nearly every surface shape a designer can think of is feasible. Hence, this process is very well suited for mass production of aspherical optical elements at lower costs than glass aspherics. Furthermore, some surface geometry is nearly impossible to reproduce in quantities in glass. Because of this, glass optics and polymer optics are sometimes used side by side (e.g., as a field lens in an all-glass CCD-lens) to improve imaging quality at reasonable costs.
- *Good automation possibilities*: Modern injection-molding machines are fully automated and computer controlled in every parameter. Together with an autonomous handling system and advanced process control, it is easy to set up flexible manufacturing cells. These cells are capable of running whole process chains like molding, testing, coating, and packaging.

- *Integration of mechanical functionality*: Not only will the optical design benefit from the possibilities of injection molding, but injection molding enables the designer to incorporate mechanical mounts, like lens mounts, snappers, and other fixature elements, together with optical functionality into one part, which reduces the number of elements or may increase alignment accuracy of optical components.

13.10.4 Description of process

The process layout, from design to part, is presented in Fig. 13.37.

Variation of material properties during a manufacturing period, like changes of the refractive index from batch to batch, can be taken into account. Using sophisticated metrology, such as a high-resolution interferometer, it is possible to detect those variations in the molded part (not in a probe!), which are compensated by slight modifications of the optical surface in the mold insert.

13.10.4.1 Operating resources

In order to manufacture high-precision optical components from polymers in an injection-molding process, it is essential to utilize very advanced technologies and molding machines. Step-by-step quality control during the tool-making process is essential too.

Looking at the process layout one can realize four different departments that should work hand in hand: the Optical and CAD design departments, the Tool shop, the Metrology department, and the production facility.

The necessary resources for the Optical and CAD design department include

- In optics and mold design, skilled and experienced design engineers;
- 3D-CAD systems with special mold-designing features;
- Optical design software (which helps to understand the parts function and necessary tolerances);
- Mold flow analysis software; and
- Finite-element analysis software and experienced operators.

The necessary resources for the Tool shop include

- A CAD/CAM interface;
- In tool setting and tweaking, experienced toolmakers;
- Precise high-speed cutting machines;
- Precise (wire and sunk) EDM equipment;
- Grinding machines for rounds and flats;
- Ultra-high-precision diamond turning and milling machines (e.g., Moore, Precitech, and others); and
- Polishing equipment and experienced personnel.

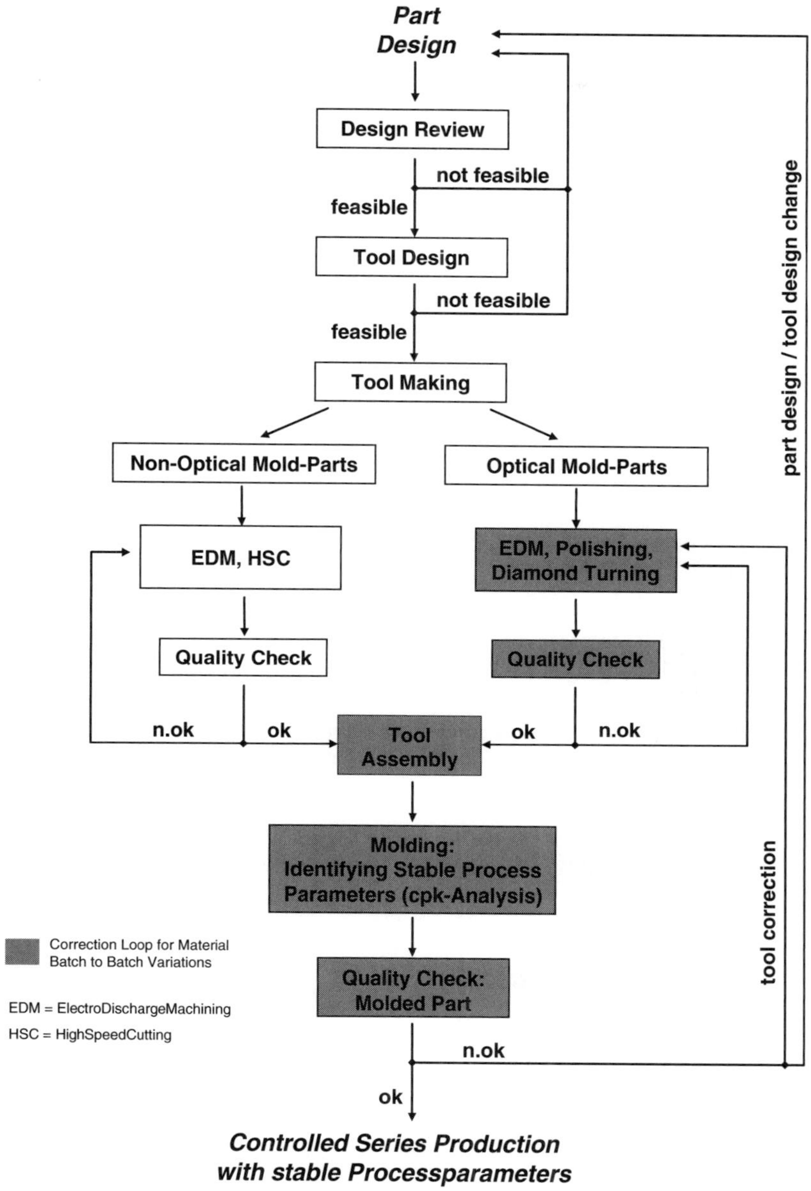

Figure 13.37 Controlled series production with stable process parameters.

The necessary resources for the Metrology department include

- In optics metrology, experienced personnel;
- High-resolution digital interferometers with single- and double-pass capability and holographic wavefront correction for checking aspheres;
- Wavefront sensors;
- Focal length test equipment;
- Foucault analysis;

- Test equipment for internal stress and birefringence;
- Test equipment for refractive index and Abbe number;
- MTF test equipment;
- Highly precise mechanical measurement equipment, such as tactile surface probes, nontactile/tactile 3D-metrology, and video test equipment; and
- An atomic force microscope for roughness measurement.

The necessary resources for the Production facility include

- Experienced and well-trained machine operators/controllers;
- Clean and air-conditioned premises;
- To the highest standard, cleaned and dried process-pressure air;
- The most advanced injection-mold machinery, preferably electric-driven and position-regulated, with high motion accuracy and the ability for combined injection and embossing processes;
- A highly precise process control, e.g., motion/position control of moving axles in the micrometer range;
- Very precise and stable temperature control units for controlling mold temperature within a range of ± 0.1°C within 24 h;
- Reliable and effective dryers and feeders for polymer granulate;
- Automated part-handling systems, for part handling during and after the injection process and for packing parts as well;
- Special devices to cut sprue without distortion in the part;
- Production metrology (modified, simple-to-use interferometers for part control during the manufacturing process);
- Semi-automatic geometry test equipment;
- Ideally, a closed manufacturing cell without the need for human interaction during manufacturing.

13.10.4.2 Mode of operation

Because of the wide variety of processes and machinery involved during the production of aspherical optical elements, only a small selection of important tools and processes are shown here.

Optical design

Access to in-house optical design software enables you to get an idea of optical tolerances and their impact on real-world measurement equipment such as interferometers. This is very time saving in finding the correct setup and starting point for the metrology equipment. Furthermore, understanding the optical design enables and helps understanding of error mechanisms during production (Fig. 13.38).

Single-point diamond turning on recent machines enables the manufacturing of spherical, aspherical, and even free-form surfaces in optical quality, with very low tolerances to 0.1 μm and less, and surface roughness figures of 8 nm R_a and lower.

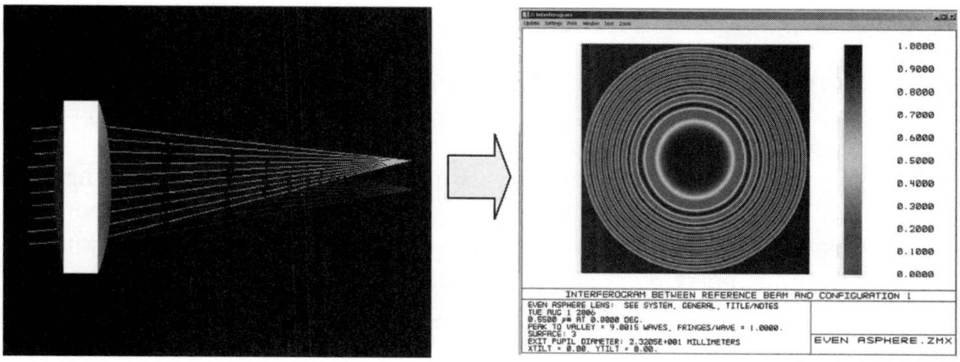

Figure 13.38 Simulation of an aspherical lens in an optical design software: optical surfaces are transformed to expected interferogram from the second surface for process control.

Furthermore, diamond turning is an essential tool in establishing a correction loop by "precorrecting" the mold insert surfaces once a stable and repeatable injection mold process has been established that unfortunately does not yield the right geometry. The loop is based on the measured molded part deviation, which is the negative shape of the diamond turned part (with some math). Hence, a wide variety of symmetric and asymmetric surface deformations can be removed from the mold part. To run such loops, sufficiently accurate metrology is necessary.

Another important diamond-machining process is grinding with diamond and CBN grinding wheels directly into hardened tool steel (despite the heavy tool wear during grinding), which enables direct machining of the mold-insert surfaces without the need for nickel-plating, which is necessary in single-point diamond turning.

Figure 13.39 Diamond turning equipment.

Injection molding

One may ask what the difference is in process parameters between precision injection molding and conventional injection molding? Table 13.48 is a brief attempt to differentiate between the processes.

Injection molding delivers ready-to-use products. Hence, further finishing, such as polishing or cleaning before coating (except where there is unwanted contamination), is neither required nor yields any positive influence on part performance. A schematic of an injection molding machine is shown in Fig. 13.40.

There are basically two different injection mold processes: standard injection molding and advanced injection molding (injection molding + embossing stroke).

Table 13.48 Features of precision injection molding and conventional molding.

	Precision injection molding	Conventional molding
Process features		
Critical phase	Post-filling	Filling
Mold temperature	High	Low
Polymer temperature	High	Low
Cycle time	Long	Short
Packing pressure	High	Medium
Injection velocity	Low	High
Major difficulties	Sink marks, warpage, shrinkage	Short shots, flash
Molding of thin-walled parts	Easy	Difficult
Molding of thick-walled parts	Difficult	Easy
Material features		
Glass transition temperature	High	Medium
Water uptake	Very low	na
Compliance	Low	na
Stiffness	High	na
Melt viscosity	Low	Low

na, not applicable.

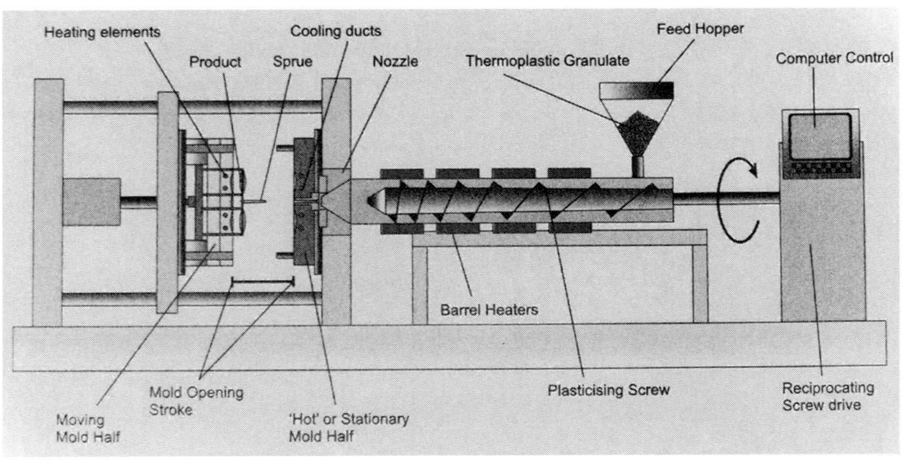

Figure 13.40 Schematic of an injection mold machine.

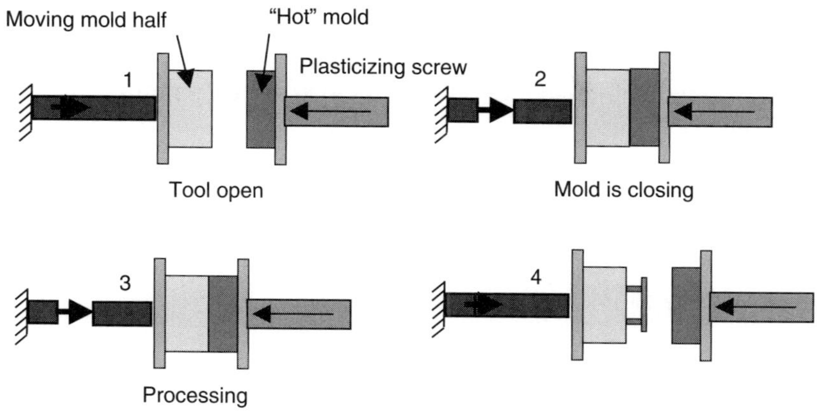

Figure 13.41 Standard injection mold process.

Brief schematics of both processes are shown in Figs. 13.41 and 13.42. Both processes may be optimized for speed (conventional injection molding) or for accuracy (precision injection molding).

The combined injection/embossing mold process enables the manufacture of polymer optics to the highest standards (for plastic parts). Furthermore, there is a very positive result for birefringence in the molded parts.

To utilize the full potential of injection molding processes, it is favorable to use electric-driven machines, which provide the widest variety of processing parameters and guarantee the highest degree of process accuracy and stability. Considering the injection molding process itself, the following issues are most important for parts quality:

- Process parameters of stability and repeatability,
- Precise temperature control in all parts of the machine and tool,

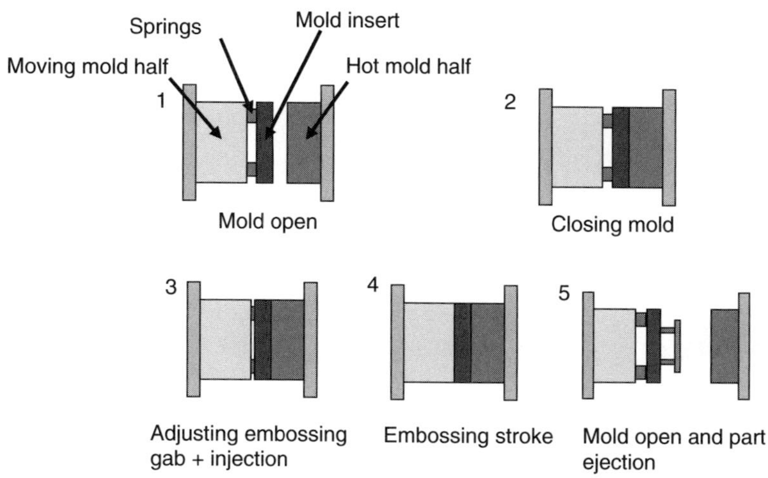

Figure 13.42 Advanced injection mold process with embossing stroke.

- Mold cavity sensor technology: pressure, temperature, and so on,
- Material pretreatment,
- Manufacturing of mold inserts by diamond turning/milling or controlled polishing,
- Process study and productibility study,
- Process control,
- Metrology for mold inserts and molded parts, and
- Stable environmental conditions.

Summarizing injection molding process parameters, the most important are

- Mold temperature: 90–170°C,
- Compound temperature: 180–330°C,
- Cycle time: 30 s to several minutes,
- Packing pressure: part-, material-specific,
- Injection velocity: mold-, part-, material-specific.

The mold design itself also has a similar impact on part properties.

Mold processes may also be simulated with numerical methods. However, the experience of the operators, the toolmakers, and the tool designers still counts for a lot when it comes to top quality molded products.

13.10.5 Data for injection molding of high-precision polymer optics (Tables 13.49–13.53)

Typical tolerances for injection-molded lenses (up to 20 mm diameter) are listed in Table 13.49. Looking at this table, one should never forget that tolerances are very

Table 13.49 Typical specifications.

	Low-cost quality	Standard quality	High-end quality
Focal length error	±3–5%	±2–3%	±0.5–1%
Curvature error	±3–5%	±2–3%	±0.5–1%
Irregularities (for = max. 25 mm diameter)	6–10 Fringes	2–6 Fringes	0.5–2 Fringes
Geometry error (for free-form surfaces)	20–50 μm	5–20 μm	1–5 μm
Surface finish (scratch/dig)	80/50	60/40	40/20
Roughness R_a	10–15 nm	5–10 nm	2–5 nm
Centering error	±3 min	±2 min	±1 min
Center thickness	±0.1 mm	±0.05 mm	±0.01 mm
Diameter	±0.1 mm	±0.05 mm	±0.01 mm
Shot-to-shot repeatability	1–2%	0.5–1%	0.3–0.5%

Table 13.50 Machinable dimensions.

Geometrical dimension	Minimum	Maximum
Lens diameter	Ø 1 mm	Ø 100 mm
Thickness	1 mm	30 mm
Diameter–thickness ratio	1:1	5:1
Optical used area	1 mm^2	50,000 mm^2

Table 13.51 Process measurement technique.

Measured variable	Method of measurement
In-line	
Surface shape, Zernike, Seidel	Interferometer, double and single pass
Wavefront aberrations, Zernike, Seidel	Wavefront sensor
MTF	MTF probe
Paraxial focal length	Focus probe
Surface shape, backfocal length	Foucault analysis
Birefringence	Polarizing microscope
Transmission	Spectrometer/photometer
Off-line	
Surface shape	Contour probe
Common part geometry	3D-coordinate measuring machine
Surface roughness	Atomic force microscope
Material impurities	Stray light analysis
Refractive index	Ellipsometer

Table 13.52 Cost drivers.

Cost factor	Volume
Injection molding machinery	70,000–300,000€
Part handling	30,000€
Metrology	70,000€ upwards, no limits
Mold tools	15,000–100,000€
UPM (e.g., diamond turning)	300,000€
Personnel in production/operator	1/3 machines
Engineering	2 designs/engineer

part-, geometry-, and material-specific. Hence, you cannot generalize tolerances; they have to be evaluated in process studies as they arise. A huge variety of process metrologies is available; hence, Table 13.51 represents only a small selection of commonly used methods. The main difference between in-line and off-line metrology classification is speed. In-line metrology has to work within the cycle time of the injection mold process, which can be a tough requirement for some precise metrology.

Table 13.53 Limits of the technology.

Parameter	Limitations	Reason for limitation
Service temperature	−40–150°C (depending on expected performance)	Thermal expansion, dimensional stability, temperature sensivity of refractive index
Index of refraction	1.5–1.7	Material limitations
Abbe number	28–58	Material limitations
Accuracy of molded parts	See Table 13.49	Shrinkage, warpage during processing, thermal and mechanical stress, water absorption, geometry, heat deflection
Birefringence	Down to intrinsic birefringence of material	Thermal stress, photoelasticity, stress optical law, material properties, geometry

13.10.6 Further reading (nonrepresentative)

1. S. Bäumer (ed.), *Handbook of Plastic Optics*, Wiley-VCH (2005).
2. J. Greener and R. Wimberger-Friedl, *Precision Injection Molding*, Hanser-Gardner (2006).
3. F. Johannaber, *Injection Molding Machines*, Hanser-Gardner (1994).
4. H. Rees, *Understanding Product Design for Injection Moulding*, Hanser-Gardner (1996).

13.10.7 Links (nonrepresentative)

- http://www.viaoptic.de (Viaoptic GmbH, Ludwig-Erk-Straße 7, D-35578 Wetzlar, Germany)
- http://www.dki-online.de; Deutsches Kunststoff Institut
- http://www.nanotechsys.com; Moore high-precision diamond machining equipment
- http://www.precitech.com; Precitech high-precision diamond machining equipment
- http://www.arburg.com; injection molding machines
- http://www.fanuc.co.jp; injection molding machines
- http://www.ticona.com; polymer manufacturer
- http://www.zeonchemicals.com; polymer manufacturer
- http://www.fisba.ch; metrology
- http://www.spotoptics.com; metrology
- http://www.bte.com; coating facility
- http://www.schott.com/coated_components/english/; coating facility

13.11 Aspherical Microlenses Manufactured by Wafer-Based Technology

M. Eisner, R. Völkel, K.J. Weible

13.11.1 Basic assessment of the technology

Microlenses and microlens arrays are used for fiber coupling, collimation, beam shaping, beam homogenizing, wavefront sensing, imaging, and other applications. Different manufacturing techniques, from ion diffusion (GRIN), injection molding, embossing, and grinding, to wafer-based technologies, are suitable and well developed. For the production of high-quality micro-optics, the wafer-based microlens technology is most promising [1–3]. It uses well-established processes from the semiconductor industry like resist coating, lithography, and reactive ion etching. This technology permits the manufacture of microlenses from 10 microns to 2 mm in diameter, and aspherical lens profiles with excellent uniformity and lateral position accuracy in the submicron range. SUSS MicroOptics has successfully established wafer-based manufacturing in 200 mm (8″) wafer technology, allowing a maximum array size of 180 mm in diameter (Fig. 13.43c).

Special care has to be taken for the testing of aspherical microlenses. Interferometrical and mechanical profilometers are available. However, due to the small size of the microlens apertures, the accuracy of the measurements is limited by the number of fringes and the accuracy of the sampling. Additionally, diffraction effects might influence the optical properties of refractive microlenses with small F-numbers.

13.11.2 Intended purpose of the technology

Wafer-based microlens manufacturing technology uses standard semiconductor processes and equipment. The semiconductor industry is a frontrunner, with incredible high annual investments in new high-end processing tools and equipment. The micro-optics industry commonly uses equipment that has been optimized by the semiconductor industry some years before. This allows manufacturing and packaging of high-quality micro-optics at reasonable costs. As all processes are based

(a) (b) (c)

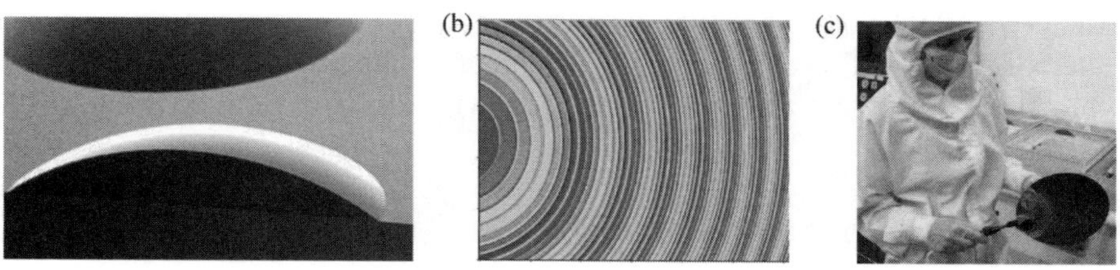

Figure 13.43 (a) Scheme of a refractive microlens (photograph courtesy of J.-C. Roulet); (b) Eight-level diffractive microlens; (c) 8″ microlens wafer.

on standard semiconductor technologies, the achieved quality is merely a question of budget and effort to optimize to the very limits.

In contrast to the well-established manufacturing technology, the instruments available for testing of microlenses are still at a low level of development. For interferometrical testing of small microlenses, an interferometer is combined with a microscope to allow sufficient magnification of the observed fringe pattern. For testing of large microlens arrays and a full-wafer mapping, an automated, fast, and precise multi-axis tool is required. Such instruments do not serve much purpose for other tasks than micro-optics testing, and it is not very attractive to develop special equipment for a very limited market. This lack in adequate metrology tools is a severe constraint for the micro-optics industry.

13.11.3 The technology's typical features

Wafer-based manufacturing technology is well suited for microlenses from 10 μm to 2 mm diameter. The microlenses are typically arranged in quadratic or hexagonal arrays. More sophisticated arrangements like Nipkow-discs or a random-like distribution of different microlenses are also possible. The lateral position accuracy depends on the precision of the photomask pattern and is usually better than ±100 nm. After manufacturing and testing, the 8″ wafers are diced to smaller units by using a wafer saw or laser dicing. Dicing marks on the wafer allow a precision for the outer dimensions of better than ±2 μm.

The lenses are planoconvex with aspherical lens profiles. The conic constant typically ranges from $k = +10$ to $k = -20$. A diffraction-limited optical performance and excellent array uniformity is achieved.

Wafer-based manufacturing includes a dry-etching step (see Chapter 15 for details), which limits the choice of wafer material for the microlens arrays. The described wafer technology works very well for fused silica, Pyrex (Corning 7740 or Schott BF33), and silicon wafers. Other dry-etchable materials, such as GaAs, GaP, and InP, are possible by using chlorine gases.

For the mounting of microlens arrays, the classical approaches using mounts and holders with springs, rings, balls, or other mechanical supports are well suited. For larger numbers of small microlens arrays, the wafer-based manufacturing technology allows implementation of packaging and alignment strategies from the semiconductor industry. This includes double-sided microlens arrays with a front-to-backside accuracy of ±1 μm, alignment marks, aperture or pupil arrays, structured color or dielectric filters, and advanced packaging techniques like wafer-level bonding and flip-chip bonding. For large arrays, special care is necessary to avoid bending of the planar substrates by gravity or stress from the holder.

13.11.4 Description of process

A photoresist layer, typically 1 μm to 100 μm thick, is spin coated on a wafer. Typical wafer material is fused silica, Pyrex, or silicon. A binary chromium-on-glass

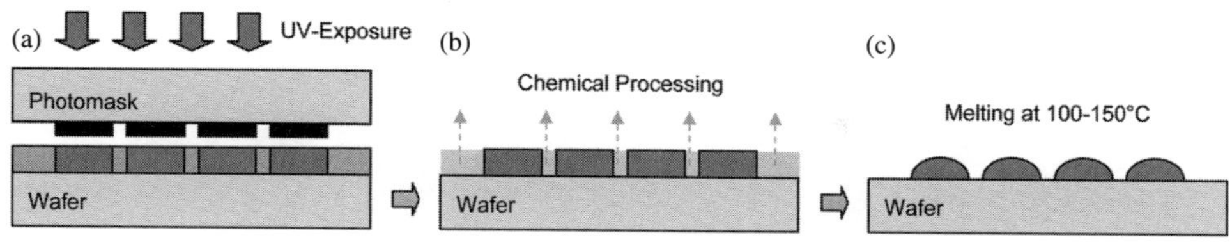

Figure 13.44 Scheme of the fabrication of refractive microlenses in photoresist by the reflow method. (a) Photolithography, (b) developing, and (c) melting of the photoresist structure.

photomask is contact-copied in a mask aligner as shown in Fig. 13.44. The exposed resist is resolved in a standard wet-chemical developing process. The resist structures are now molten at a temperature of 100–150°C on a hot plate or in an oven. Surface tension forms a lens-like drop, the shape of the drop is preserved after cooling and repolymerization of the resist. The remaining resist microlens is typically very close to a spherical lens profile.

Figure 13.45 shows the reactive ion etch (RIE) transfer process of resist microlenses in fused silica. Melted resist lenses are fabricated on a fused silica wafer. The resist shape is transferred in fused silica by RIE. Atoms from the resist surface and the silica are removed simultaneously by energetic ions until the lens shape is completely etched into the substrate. The etch rate of the photoresist and the silica depends strongly on the RIE parameters. A spherical microlens profile will be slightly deformed after the RIE transfer. Usually, the lenses are steeper at the rim and flatter at the vertex. Spherical aberrations are severely enhanced due to the profile change. A profile modification is achieved by changing the etch rate during the RIE step. The resist is etched more quickly at the beginning and more slowly at the end. This procedure also allows precise profile shaping of aspherical microlenses from conic constant $k = +10$ to -20.

Figure 13.46 shows the surface profile of an aspherical microlens with a lens diameter of 1.1 mm, a radius of curvature of 8.85 mm, and a conic constant of $k = -11.75$ measured in a Wyko NT3300 white light profilometer in PSI mode. The microlens surface is measured in reflection. A spherical reference surface is used for calibration. The radius of curvature (ROC) of the reference surface is well known and close to the aspherical microlens surface to be measured. The reference sphere serves to determine the measurement errors due to the interferometer. A file containing these errors is subtracted from each asphere measurement.

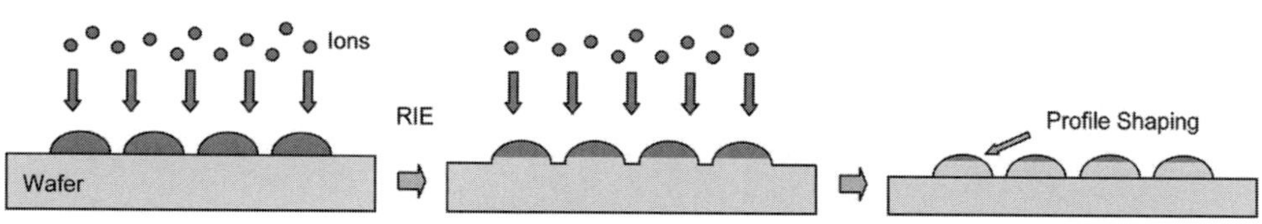

Figure 13.45 Scheme of the reactive ion etching (RIE) transfer process of resist microlenses in fused silica. A correction of the lens slope is obtained by changing the etch rate between the resist and fused silica during the etching process.

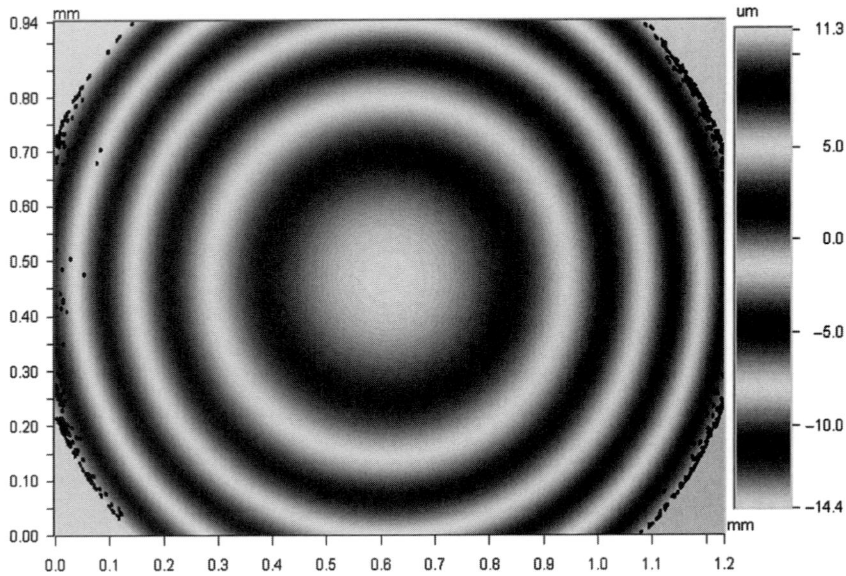

Figure 13.46 Aspherical microlens with lens diameter Ø $=$ 1.1 mm, ROC $=$ 8.85 mm, and conic constant $k = -11.75$ measured in a Wyko NT3300 white light profilometer in PSI mode.

The aspherical data with the error file subtracted are taken into a fitting routine, and the best-fitting aspherical surface is calculated using the rms deviation as an optimization constraint. Comparison of the measured microlens profile (gray dots)

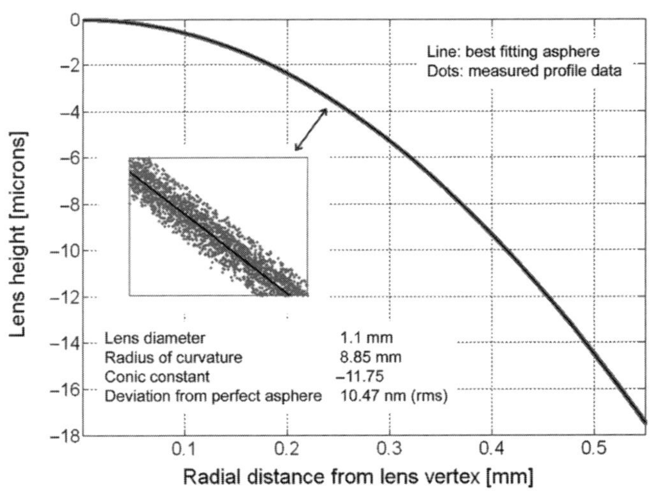

Figure 13.47 Comparison of the measured microlens profile (gray dots) with the best-fitting asphere (solid line). The gray dots represent the polar coordinates of all measured 3D surface data. A deviation from the perfect asphere of $\pm$10.47 nm (rms) is achieved.

with the best-fitting asphere (solid line). The gray dots represent the polar coordinates of all measured 3D surface data (Fig. 13.47). A deviation from the perfect asphere of ±10.47 nm (rms) is achieved.

13.11.5 Data for aspherical microlenses manufactured by wafer-based technology (Tables 13.54–13.56)

Table 13.54 Typical operation parameters.

Shape accuracy	<±20 nm (rms)
Surface roughness	1.1–1.4 nm (rms)
Gradients	
Machinable dimensions	190 mm diameter for 8″ wafer technology
Radius	5 μm up to 200 mm

Table 13.55 Cost drivers.

Investment	Cleanroom, process equipment, metrology
Process time	2–3 weeks
Operator	Less important for fully automatic equipment
Tools and supply	Wafer cleaning, resist-coating, photomasks, lithography, RIE, metrology, interferometers, microscope

Table 13.56 Limits of the technology.

Metrology	Diffraction effects and sampling limit the accuracy of profile measurements for small microlens diameters
Geometry	Lens diameters are typically 10 μm to 2 mm Lens sag typically from 100 nm to 100 μm

13.11.6 Conclusions

Aspherical microlens and microlens arrays are key components for medical applications, telecom networks, metrology, laser, and illumination systems. Even though the micro-optics is only a tiny passive component within such complex systems, these components are often very decisive for the overall performance of the system.

The described wafer-based manufacturing technology is well established and controlled. Tools for simulation of microlens arrays within optical design programs and adequate metrology to test and characterize microlens arrays exist, but they are not yet fully developed.

Although the basic principles of aspherical microlenses and microlens arrays are well understood, proper definition and integration in complex optical systems is often very difficult. For a large asphere, the lens material, the ROC, and the aspherical coefficient k are sufficient to predict optical performance. This is not the case for aspherical microlenses and microlens arrays:

- For microlenses with a low Fresnel number $\text{FN} = (\text{Ø}^2/4\lambda f) \leq 1$, diffraction effects significantly shorten the observed focal length [4].
- Microlenses have a small space-bandwidth product and do not transmit many pixels in an imaging system [5, 6].
- Illuminating a microlens array with a periodic lens pitch with coherent laser light might lead to interference effects in the far-field.

Miniaturization of optics leads to an increase of diffraction effects [7], and array optics suffers from interference. From the first draft of a system to the final acceptance tests, an optical engineer has to understand how the microlens or the microlens array will improve the performance of his optical system and where potential risks lie. If the engineer is not an expert in micro-optics, he needs qualified partners and suppliers who can educate him and provide him with all of the information required to make the right decisions. Only customer education will lead to customer satisfaction [8]. It will be a joint effort of the micro-optics manufacturers, research institutes, and universities [9] to better educate and train the next generations of optical engineers in micro-optics.

13.11.7 Status

The front-running semiconductor industry and the recent progress in microtechnology research and development constantly improve the manufacturing equipment and processing also usable for wafer-based microlens technology. Therefore, further improvement of aspherical microlenses and microlens arrays is merely a financial question. Despite the high initial costs, wafer-based microlens technology is perfectly suited for mass production. No industry has been more successful in cost reduction of highly sophisticated products than the semiconductor industry. So far, micro-optics is still a niche product with small batches of wafers—and is therefore still expensive.

13.11.8 Further reading

1. Z.D. Popovic, R.A. Sprague, and G.A. Neville-Connell, "Technique for monolithic fabrication of microlens arrays," *Appl. Opt.*, Vol. 27, pp. 1281 (1988).
2. D. Daly, R.F. Stevens, M.C. Hutley, and N. Davies, "The manufacture of microlenses by melting photoresist," *J. Meas. Sci. Techn.*, Vol. 1, pp. 759–766 (1990).
3. R. Völkel, H.-P. Herzig, P. Nussbaum, and R. Dändliker, "Microlens array imaging system for photolithography," *Optical Engineering*, Vol. 45(11), pp. 3323–3330 (1996).
4. P. Ruffieux, T. Scharf, H.P. Herzig, R. Völkel, and K.J. Weible, "On the chromatic aberration of microlenses," *Opt. Express*, Vol. 14, pp. 4687–4694 (2006), http://www.opticsinfobase.org/abstract.cfm?URI=oe-14-11-4687.

5. R. Völkel, C. Ossmann, T. Scharf, H.P. Herzig, and R. Dändliker, "Optical Microsystems for Imaging," in *Electronic Image Capture and Publishing*, Proc. EOS/SPIE 3410, Vol. 13, 1998.
6. J. Duparré, P. Schreiber, A. Matthes, E. Pshenay-Severin, A. Bräuer, A. Tünnermann, R. Völkel, M. Eisner and T. Scharf, "Microoptical telescope compound eye," *Opt. Express*, Vol. 13, pp. 889–903 (2005), http://www.opticsinfobase.org/abstract.cfm?URI=oe-13-3-889.
7. A.W. Lohmann, "Scaling laws for lens systems," *Appl. Opt.*, Vol. 28, pp. 4996–4998 (1989).
8. R. Völkel, K.J. Weible, and M. Eisner, "Kommerzialisierung von Mikro-Optik," *DGaO-Proceedings*, ISSN: 1614-8436, 2006, http://www.dgao-proceedings.de/download/107/107_h6.pdf.
9. H. Ottevaere, H.P. Herzig, M. Taghizadeh, T. Miyashita, K. Naessens, R. Cox, and H. Thienpont, "Comparative study of glass and plastic refractive microlenses and their fabrication techniques," SPIE Conference 5453.

Chapter 14

Metrology

14.1 Tactile Profile Measurement

B. Dörband

14.1.1 Basic assessment of the technology

Metrology comprises a pointwise sampling method and apparatus to determine the surface shape of an object. When applied to a plano, spherical, or aspherical surface of an optical element, suitable sensors must be used to avoid damaging.

Unlike surface-covering interferometric metrology, tactile profile measuring does not need any null systems or compensators. Coordinates of sample points are referenced against an intrinsic coordinate system, which is supplied by the machine after a calibration process.

- *Method of measurement*: Measured variables are x, z-coordinates (line scanning) or x, y, z-coordinates (surface scanning) of arbitrary points on the surface under test.
- *In-line*: Some diamond-turning machines used for the production of optical elements have built-in profile metrology.
- *Off-line*: The usual case is the off-line operating apparatus in a metrology room providing the necessary environmental conditions (cleanness, stable temperature and humidity, no vibrations or air turbulences).

14.1.2 Intended purpose of the technology

High-precision surface metrology can be applied universally to optical surfaces. The aim is to measure the shape (figure) as well as the roughness of optical surfaces in different states of the production process. Thus, not only polished but also a ground or lapped optical element of different materials should be measured. No *a priori* information about the shape should be necessary. Unknown elements should be measured as long as they fit into the measuring range of the apparatus.

The accuracy and resolution requirements should be comparable to the visual inspection methods in interferometry ($\lambda/10$ with $\lambda = 633$ nm). For roughness metrology, subnanometer resolution normal to the surface as well as sampling rates between 1 μm and 1 mm are required.

14.1.3 The technology's typical features

- Sequential sampling of arbitrary coordinate points on an optical surface in different states of production;
- No *a priori* information about the surface needed;
- No additional surface-specific compensating elements needed;
- Shape measurement with accuracies of $\sim\lambda/10$ ($\lambda = 633$ nm);
- Roughness measurement with subnanometer resolution and sampling rates between 1 μm and 1 mm;
- Medium duration of measurement (minutes);
- Metrology room environment preferable;
- Different contacting sensor types can be used (ruby ball, diamond tip) with different forces;
- Noncontacting or quasi-noncontacting sensor types are available, such as optical, pneumatic, or atomic force sensors, thus diminishing the risk of damaging the surface under test.

14.1.4 Description of process

14.1.4.1 Operating resources

Cross-section metrology (2D)

- Two-coordinate measuring machine;
- Lens holder;
- Lens adjustment and manipulation (lens holder on rotary table with x, y, z-translation adjustment plus x, y-tilt adjustment);
- Computer to control the x-position of the driving unit and read out the z-position of the stylus or sensor unit;
- Software for machine control;
- Software for adjustment compensation, calibration, error correction of
- sampled data sets;
- Data processing including comparison of results with design data, documentation, and storage of data.

Surface metrology (3D)

- Three-coordinate measuring machine;
- Lens holder;
- Lens adjustment and manipulation (lens holder on stage with x, y, z-translation adjustment plus x, y-tilt adjustment);

- Computer to control the x, y-position of the driving unit and read out the z-position of the stylus or sensor unit;
- Software for machine control;
- Software for adjustment compensation, calibration, error correction of sampled data sets;
- Data processing including comparison of results with design data, documentation, and storage of data.

14.1.4.2 Layout/test setup

In Fig. 14.1, the principle of a cross-section metrology (2D) is shown.

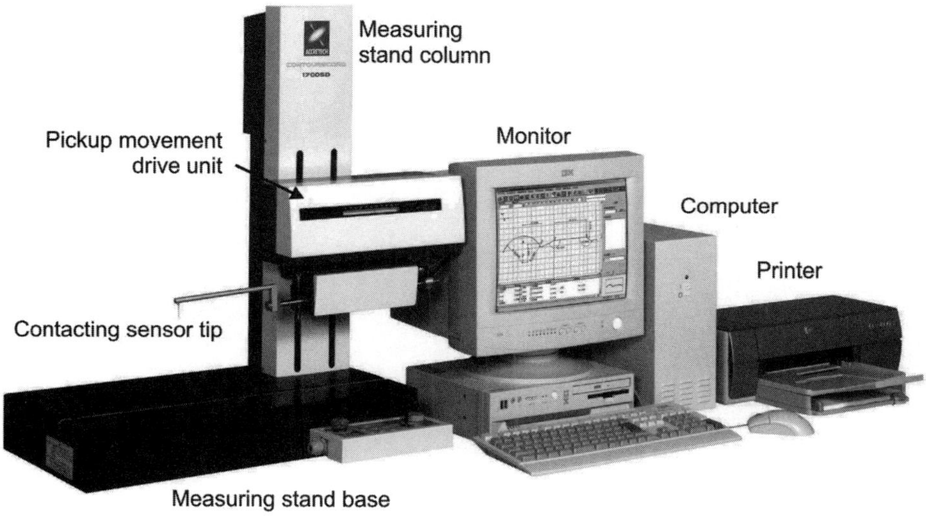

Figure 14.1 Principle of a cross-section metrology (example from Accretech/Surfcom).

14.1.4.3 Mode of operation

In Fig. 14.2, the mode of operation is shown.

14.1.5 Versions (state of the art)

Tactile asphere metrology can be classified as 2D and 3D scanning systems. Sensors can be classified as contacting and noncontacting. Contacting sensors differ in the type of tip being used (i.e., ruby ball, diamond tip) and the applied contacting forces. Noncontacting sensor types can be optical, pneumatic, or atomic force sensors.

Examples of cross-section metrology (2D) include the Mahrsurf 120 (Fig. 14.3) and the Surfcom 3000A (Fig. 14.4). An example of surface metrology (3D) is the UPMC Carat from Carl Zeiss (Fig. 14.5).

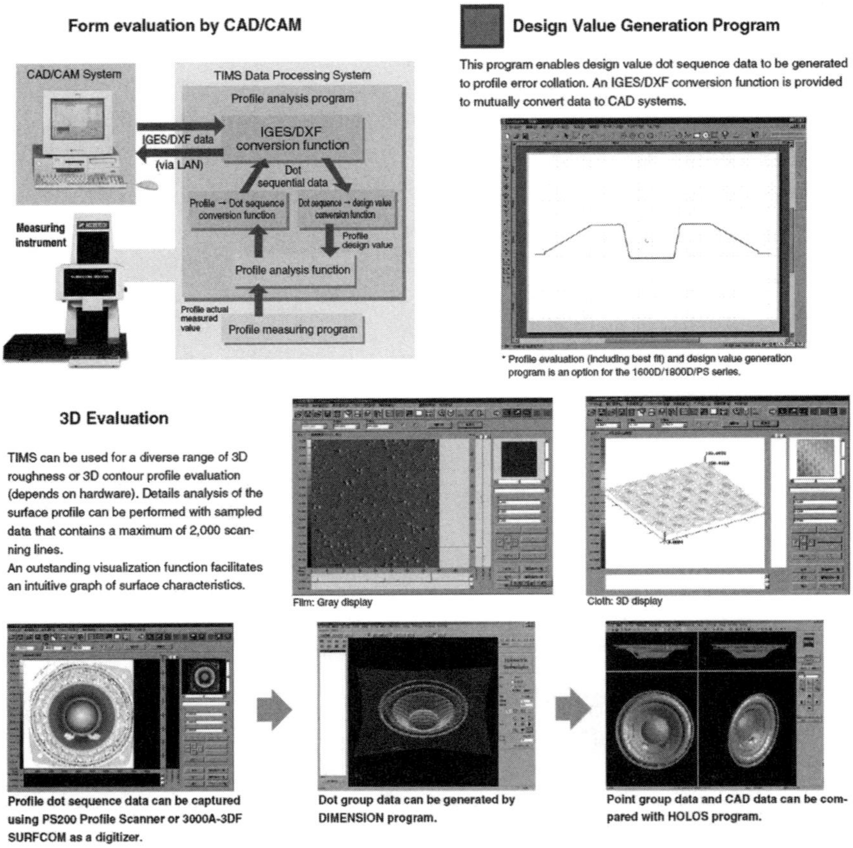

Figure 14.2 Mode of operation/sc diagram (example from Accretech/Surfcom).

Figure 14.3 Mahrsurf 120 from Mahr GmbH.

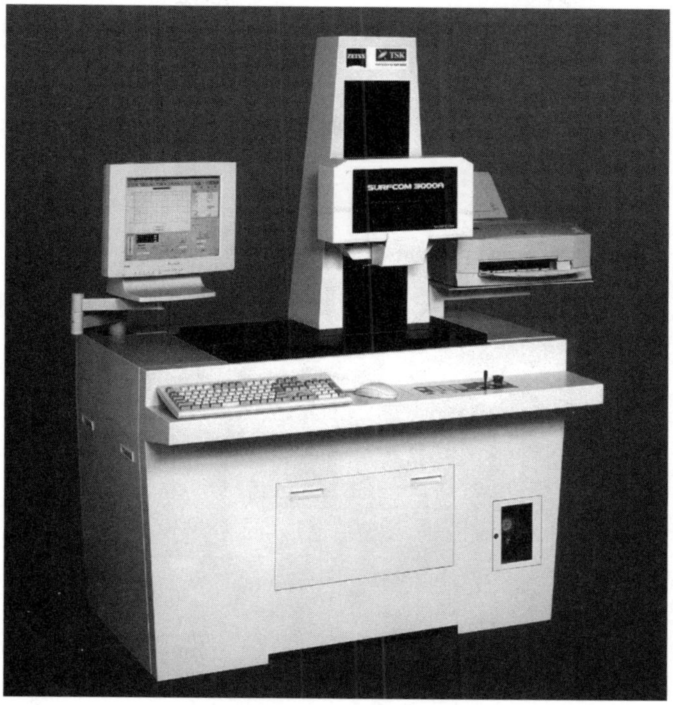

Figure 14.4 Surfcom 3000A from Accretech.

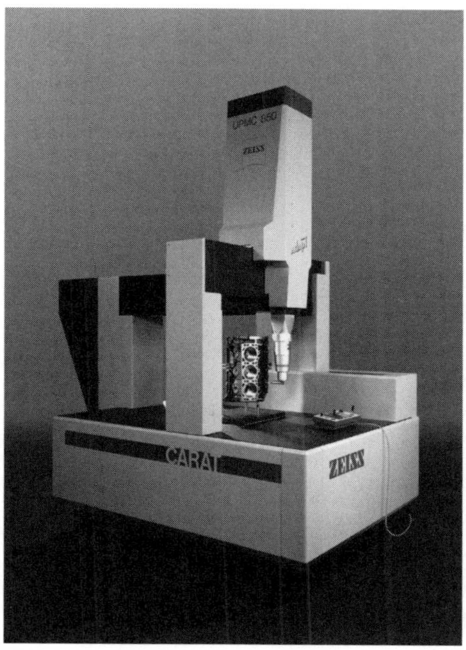

Figure 14.5 3D coordinate measuring machine UPMC Carat from Carl Zeiss.

14.1.6 Data for tactile profile measurement (Tables 14.1–14.7)

14.1.6.1 Cross-section metrology (2D)

Table 14.1 Typical operation parameters.

Parameter	Minimum	Maximum
Duration of measurement for Ø 100 mm	50 s	17 min
Measurement rate	0.1 mm/s	2 mm/s
Resolution in z	0.8 nm	3.2 nm
Repeatability in z	25 nm	50 nm
Range in z	6 mm	12.5 mm

Table 14.2 Machinable dimensions.

Geometrical dimension	Minimum	Maximum
Diameter	1 mm	200 mm
Sag in z	0 mm	38 mm

Table 14.3 Cost drivers.

Cost factor	Volume
Investment (total machine typically)	120,000$

Table 14.4 Limits of the technology.

Parameter	Limiting value	Reason for limitation
Slope of surface	Max. 60 deg, measurement of semisphere not possible	Mechanical construction principle
Accuracy	Calibration process	Accuracy only as good as calibration sphere

14.1.6.2 Surface metrology (3D)

Table 14.5 Typical operation parameters.

Parameter	Minimum	Maximum
Duration of measurement for Ø 100 mm	5 min	Several hours
Measurement rate	0.01 mm/s	20 mm/s
Resolution in z	3 nm	10 nm
Repeatability in z	30 nm	50 nm
Range in z	35 mm	90 mm

Table 14.6 Machinable dimensions.

Geometrical dimension	Minimum	Maximum
Diameter	1 mm	400 mm
Sag in z	0 mm	90 mm

Table 14.7 Cost drivers.

Cost factor	Volume
Total machine typically	800,000$

Table 14.8 Limits of the technology.

Parameter	Limiting value	Reason for limitation
Accuracy	No limitation for regular CMM like UPMC Carat Calibration process	Accuracy only as good as calibration sphere

14.1.7 Links

Producer/supplier of auxiliary materials

Taylor Hobson Ltd.: http://www.taylor-hobson.com/
Mahr GnbH Göttingen: http://www.mahr.de/
Ocean Optics (Long Trace Profilometer): http://www.oceanoptics.com/
Panasonic Industrial Europe GnbH: http://www.panasonic-industrial.com/
Carl Zeiss Industrielle Messtechnik GnbH: http://www.zeiss.de/imt
Accretech: http://www.accretechusa.com/

Research and development:

PTB Braunschweig: http://www.ptb.de/

14.2 Interferometry

H. J. Tiziani

14.2.1 Basic assessment of the technology

For the analysis of polished optical surfaces to a shape accuracy of some nanometers rms or less, interferometry is the most appropriate measuring tool. Interferometric testing of polished optical components and systems is well established. It delivers in one measurement typically 10^4–10^6 measurement points. Highly aspherical systems, however, produce fringe patterns with fringes too numerous and too dense to be recorded and interpreted. For detection, the Nyquist frequency limitation requires a minimum of two pixels for a fringe on the detector, typically a CCD, which sets an upper limit on the measurable wavefront slope. In addition, typical interferometric setups are corrected and calibrated only for a small deviation from the null setup; that is, additional errors are introduced if the interferogram shows more than a few fringes.

The use of a so-called Null optic converts the system into a Null test configuration. A Null corrector compensates the asphericity of the surface under test; hence, the interferogram shows only the deviations from the ideal aspherical shape. In such a Null test setup, the null optics is of vital importance and needs to be calibrated for precision measurements. It defines the ideal shape of the asphere and thus has to be considered as part of the reference. With careful design and calibration of the test setup, the accuracy for testing aspherical surfaces reaches that of spherical surfaces, that is, a few nanometers or better, even though the deviation from the best sphere can be some 100 μm or more.

The efforts and price of a Null optic are prohibitive for a single test object or small series. Alternative interferometric non-Null methods are therefore under development for aspheric testing. They include

- Multiple wavelength interferometry,
- Sub-Nyquist interferometry,
- Stitching methods, with and without mechanical movements, and
- Adaptable Null optics.

Some of the most promising are described in Sec. 14.2.10.

14.2.2 Intended purpose of the technology

Interferometry is the standard method for high-precision measurement of polished optical surfaces like plane, spherical, and—by using a Null corrector—also for aspherical surfaces. It is capable of measuring the surface geometry, microroughness, as well as material homogeneity and birefringence. Interferometry can be used to test components or the whole system, both in reflection as well as in transmission. In addition, interferometry can be used for lens thickness or separation measurements.

14.2.3 The technology's typical features

Interferometry uses the wavelength of light as scale (a typical wavelength used for shop testing is 632.8 nm, of the HeNe laser). It is the standard method for high-precision measurement of polished optical surfaces. An accuracy of shape deviation of up to 1 nm or even better can be achieved. It is capable of measuring a 2D map of optical path differences (now typically about 1000×1000 up to 2000×2000 data points) in one measurement. The optical path differences as well as the surface deviations from a reference or material properties are mapped using known setups and calibration data. For automated fringe analysis, different programs are available [9].

It should be noted that the surface under test is usually compared to a known reference surface. For testing aspherical surfaces, however, no reference surface usually exists. Instead, the wavefront of the test arm is sculptured by a Null optics such that a perfect asphere under test retroreflects the incident wavefront. It is obvious that the accuracy of the Null optics limits the accuracy of the test.

Aspherical surfaces with up to a few 100 μm deviation from the best-fitting sphere can be measured, and an accuracy of 1 nm and less can be obtained. The local slope limits the application of the technique. Furthermore, calibration is important for high-accuracy measurements (Section 14.2.4.6).

14.2.4 Description of process

14.2.4.1 Operating resources

For testing aspherical surfaces with interferometric methods, the interferometer needs to be equipped with a refractive or diffractive Null corrector. Both types of Null corrector are expensive to design and to fabricate and need to be calibrated. Computer generated holograms (CGH) have become the state-of-the-art procedure for testing aspherical surfaces. Design and production of a CGH is time consuming. Only recently has the developing and production time of a CGH been reduced to one week. The cost and time requirement might become an issue for prototypes and small size lots.

14.2.4.2 Test set-up

Twyman–Green, Fizeau, or Mach–Zehnder interference arrangements for two-wave interference are frequently used for optical testing. The most used interference arrangements for testing aspherical surfaces are based on the Twyman–Green or Fizeau principles. Fizeau setups are almost the common path and therefore quite robust with respect to disturbances in the measuring setup [2, 9].

In classical interferometric testing, the test object can be a plane or spherical surface, or an optical system. For testing an aspherical surface, the Null optic needs to be designed such that the rays according to geometrical optics hit the test surface nearly normal. Figure 14.6 shows three typical test beam setups (right) based on a Twyman–Green interferometer (left).

In the test arm, the Null compensator is placed as shown in Fig. 14.6(b), where a CGH is used as a Null corrector for testing an aspherical surface. In Fig. 14.6(c), a collimating lens frequently used for testing spherical surfaces is added in addition to the CGH in order to take care of some power. Using this kind of hybrid system is even more appropriate because it reduces the requirements with respect to fringe density in the CGH and thus decreases the sensitivity of fabrication, centering, and position errors. In Fig. 14.6(a), the setup for testing an aspherical system in transmission is shown; the CGH is used in reflection.

In a Fizeau arrangement, the reference beam is reflected back at a beamsplitter, which could be combined with the CGH in Fig. 14.6(b).

Care needs to be taken in the accurate positioning and centering of the CGH as well as the surface under test. It was therefore found very helpful to generate an

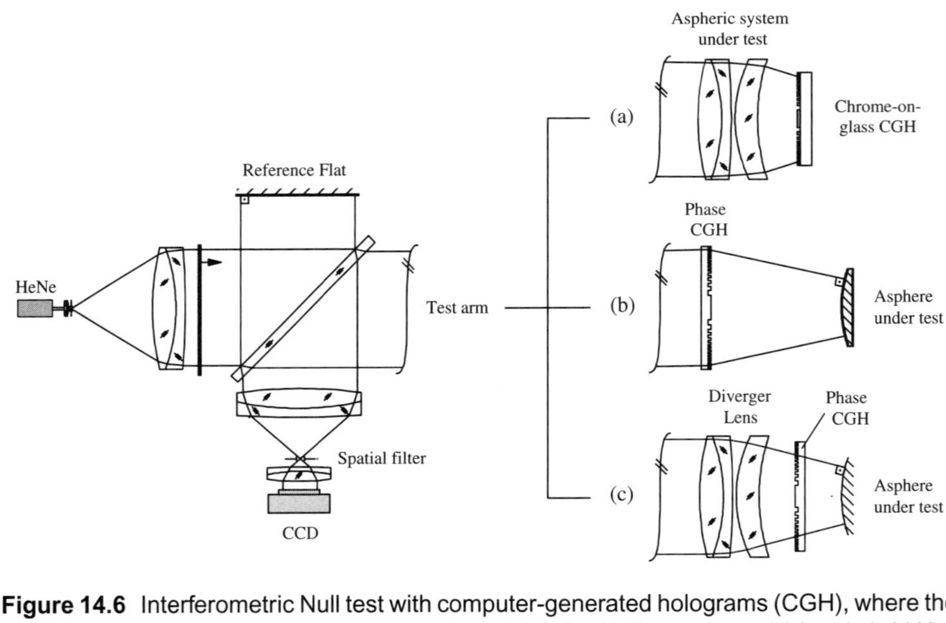

Figure 14.6 Interferometric Null test with computer-generated holograms (CGH), where the CGH is applied (a) as diffractive master, (b) as diffractive Null system, and (c) as hybrid Null system.

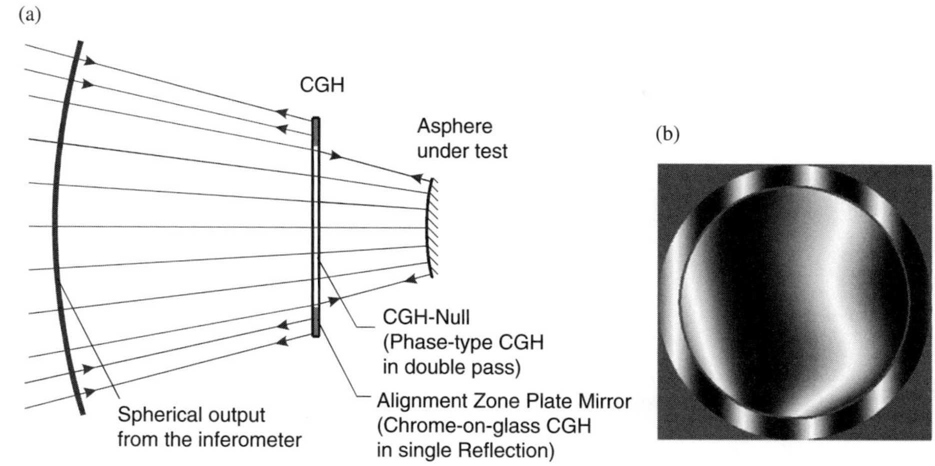

Figure 14.7 (a) Multiple function CGH with a ring-shaped reflection hologram (amplitude hologram) for positioning, alignment, and centering. The test hologram is a phase hologram. (b) A small lateral decentering is present in the interference pattern.

additional reference CGH on the same substrate on the periphery, for instance for centering and positioning the CGH (Fig. 14.7). Furthermore, additional holograms for calibration can be added.

After the interference patterns are obtained, the interference fringes are analyzed using one of the well-known techniques for fringe analysis [2, 9], and the deviation from the perfect required shape of the asphere is determined. Essential is an error analysis and calibration of the whole test system including the Null lens (Fig. 14.8). By taking the mentioned precautions, the accuracy of the measurement can be nearly the same as for spherical surfaces, that is, some nanometers or less for high-precision systems used for lithography, even though the deviation from the best sphere can be some 100 μm.

A test setup based on a Twyman–Green principle is depicted in Fig. 14.6. The test procedure is shown schematically in Fig. 14.8. At first, the test configuration with the Null lens needs to be designed (1), and the Null lens (such as the CGH) has to be manufactured with very high precision and tested (2). The Null lens, a CGH in Fig. 14.8, is introduced and accurately aligned (3). The alignment is facilitated by special alignment structures on the CGH. Calibration of the measuring system is also important [4–6].

14.2.4.3 Testing procedure and precautions

The test piece should be imaged onto the CCD chip in order to avoid disturbing fringes caused by diffraction at the edges or ripples. However, the test piece is often not flat; hence, it could be useful to image the edge onto the CCD. When designing the test system, care needs to be taken regarding aberrations such as distortion. For aspherical Null testing, a geometrically distorted image of the testpiece on the

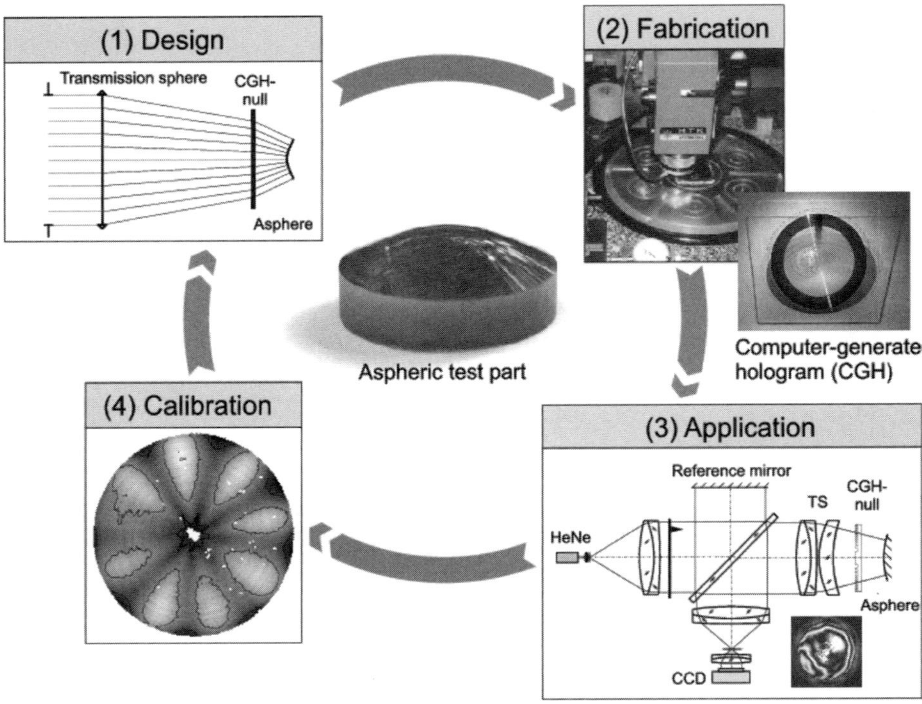

Figure 14.8 Operation for the application of CGHs.

CCD often occurs. Thus, care needs to be taken in design and calibration in order to compensate for errors due to distortion or field curvature.

After the interference patterns are obtained, the interference fringes are analyzed using one of the well-known techniques, and the deviation from the perfect required shape of the aspherical surface is determined [9].

14.2.4.4 Alignment and centering of the CGH

Alignment and centering are very important for high-precision measurements. It should be noted that an aspherical surface does not possess the high symmetry of a spherical surface. Therefore, more degrees of freedom need to be taken care of than for spheres, and centering tolerances, for instance, are more severe.

Recently, additional holograms for alignment and centering have been used. In our case, additional alignment holograms are generated on the periphery of the CGH. The manufacturing of the auxiliary holograms should be of the same substrate and manufactured practically at the same time as the test CGH in order to reduce systematic errors in the testing and calibration procedure [3–6]. In Fig. 14.7, the alignment hologram is a reflecting amplitude hologram placed at the periphery, which makes the positioning, alignment, and centering of the hologram much simpler.

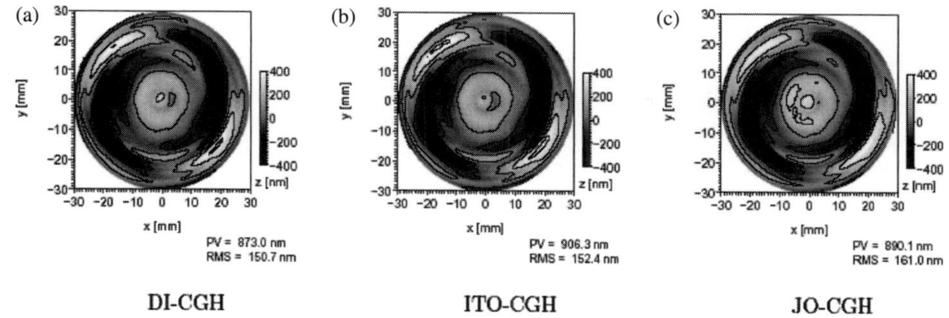

Figure 14.9 A comparison of results of a measured aspherical surface obtained with three different CGHs produced with different technologies at different locations.

14.2.4.5 Comparison of measurements on an aspherical surface using different CGHs

Results of measurements with three different CGHs from different sources using different writing techniques (E-beam and laser direct writer) for the same aspherical surface are shown in Fig. 14.9. The comparison was carried out by Jenoptic as part of a common project. The CGHs were provided by Diffraction International (DI) USA, Institut für Technische Optik (ITO), and Jenoptik (JO). Even though no special precautions were taken, a good agreement of the comparisons was obtained.

14.2.4.6 Calibration of the measuring system

In high-precision interferometry, systematic errors can be as high as or even higher than the accepted tolerance. Therefore, calibration is a necessary requirement. Calibration can be done by comparing the unknown wavefront with an absolutely known reference. However, to establish and measure a standard master can be difficult. Alternatively, interferometric self-calibration techniques can be used to determine the systematic errors in the interferometric setup in an absolute manner. The basic idea is to cancel the different error contributions originating from the measuring system. There are different techniques to be used; the "N-position test" and the "three-position method" will be discussed briefly.

The "N-position test" is used to compensate the asymmetric contributions by measuring N times under different azimuthal orientations by angular increments of $2\pi/N$. Typically, $N = 12$ is an appropriate compromise with respect to measuring time and accuracy. An averaging of the N measurements leads to a cancellation of the asymmetric contributions. The difference of a single measurement compared with the average of all the measurements leads to the nonsymmetric error components of the testpiece.

The "three-position method" was proposed by Jensen [4]. The method is mainly used for testing spherical wavefronts. Combining three measurements, two at confocal (azimuth zero and 180 deg) and one at cat's eye positions, the wavefront error of a spherical surface and that of the interferometer can be obtained in an absolute manner. Additional improvement can be reached by including the N-position test. By testing aspherical surfaces interferometrically, the three-position test needs to be modified.

Calibration by testing aspherical surfaces with a CGH as a Null lens

By using a CGH as a null lens, the phase of the wavefront shaping is encoded in the CGH, and the amplitude is considered to be constant. However, several composite errors may be present in CGHs. They can be divided into pattern errors of the hologram, a figure error of the hologram substrate, and a profile depth error of the diffractive structure.

It was found to be useful to use a property already known by the calibration of a linear grating in autocollimation configuration where the errors of the diffractive structures in the complementary diffractive orders change the sign only. Therefore, by subtracting the measured wavefronts pixel by pixel for the complementary orders, the distortion error is obtained.

The wavefront error of the diffracting structure in the first diffraction order is therefore given by

$$W_{\text{PD}-1} = 1/2(W_1 - W_2),$$

where W_1 and W_2 are the phase errors in the plus and minus first order, respectively, which are different in sign only. The hologram error is, however, part of the error budget. The error of the substrate, for instance, remains unknown.

A combination of the conjugate diffraction order measurement principle with the previously discussed three-position test leads to a solution. Therefore, a reference spherical wave generated by the CGH can be verified as follows: In addition to the cat's eye position, the wavefronts are measured intra- and extra-focal, W_1 and W_3 at azimuth angles 0, and W_2 and W_4 at azimuth angles of 180 deg with respect to W_1 and W_3 in the same positions, but with the azimuth angles of 180 deg with respect to W_1 and W_3 (Fig. 14.10a).

Calibration by multiplex CGHs

CGHs offer important additional degrees of freedom in the design as well as in an application status. Using multiplex encoding techniques by practically simultaneous generation of structures in the hologram leads to the possibility of producing several optical functions by a single diffracting element. This technique is now used for calibration. By coding two holograms in the CGH, for instance, two reconstructions, namely a spherical wave as reference for calibration purposes and the compensating wavefront (Null lens) are adapted to the aspherical surface under test. With the spherical wavefront the error analysis occurs. It should be noted that both holograms are generated practically at the same time; therefore, the aberrations are practically the same. Hence, the errors detected with the spherical reference

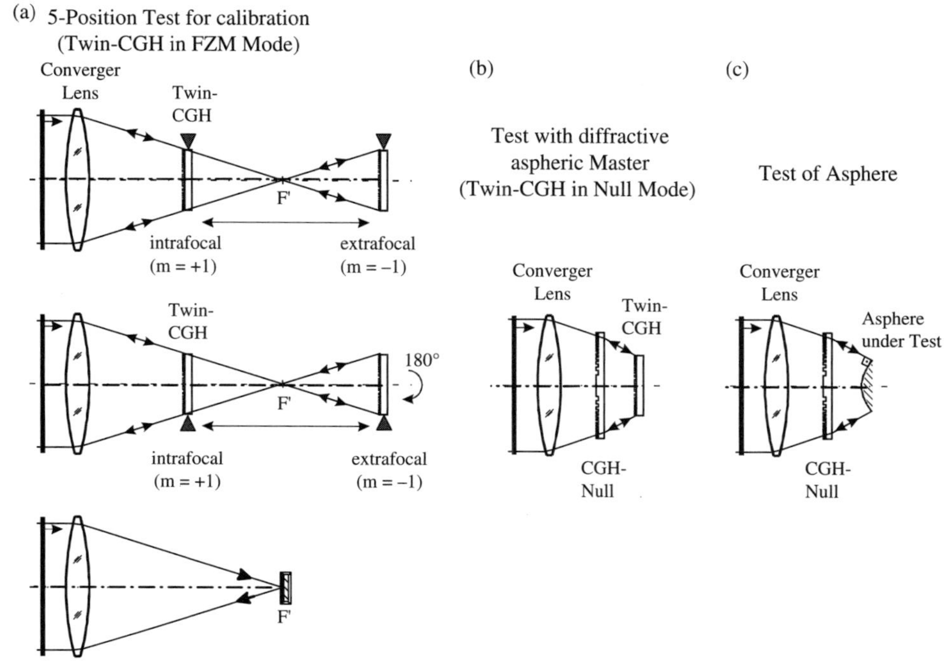

Figure 14.10 Absolute calibration of test configuration for testing an aspherical surface in reflection.

wave will be used to compensate for the wavefront adopted for the aspherical surface under test (Null lens). The new possibilities are very useful for calibration. Different encoding possibilities can be adapted depending on the applications. For instance, if the phase function of the test wave is symmetrical, the auxiliary phase function can have the same symmetry. It should be remembered that an additional hologram for adjustment and centering only can be added.

In Fig. 14.10(a), a calibration procedure is shown together with the measurement of an aspherical surface. At first, the interferometric setup is calibrated by using the twin hologram (five-position test as discussed above) [Fig. 14.10(a)]. In Fig. 14.10(b), the twin CGH in the Null mode is compared with the CGH-Null [Fig. 14.10(c)]. The remaining errors are compensated in the real measurements of the aspherical surface. Therefore, seven interference patterns need to be analyzed for an absolute calibration of the whole interferometric setup, including the CGH [4–7].

The concept discussed above opens the door for an overall calibration of any CGH-Null test setup. The absolute test for a rotationally symmetric aspherical surface has been shown. In the case of free-form aspheres, there is no preferred symmetry either in the test piece or in the CGH-Null lens. To apply the procedure outlined above, three phase functions (two mutually orthogonal oriented linear gratings and an arbitrary aspherical phase function) need to be encoded simultaneously in the multiplex CGH. For a complete calibration of an interferometric Null test setup for free-form aspheres, seven measurements are required: five calibration measurements to obtain the error of the diffractive master and that of the plane output wave of the interferometer and one calibration measurement for the test setup with the calibrated multiplex CGH operating in the aspherical mode. Finally, the measurement of the actual free-form asphere is performed.

14.2.5 Data for interferometry (Tables 14.9–14.11)

Table 14.9 Typical operation parameters.

Parameter	Minimum	Maximum
Duration of measurement	1 min (without calibration)	6–20 days (lead time for Null corrector)
Measurement rate	1 min/surface	
Precision of measurement	$\lambda/4$	$\lambda/20$ (1 nm for high precision)
Geometrical dimension of object measured	2 mm	Concave: several meters Convex: somewhat limited by CGH diameter Standard: <140 mm, can be up to 250 mm

Table 14.10 Cost drivers.

	Cost factor	Volume
Equipment	Interferometer	100,000€
Additional components	CGH, necessary for each type of asphere	9,000€

Operating supplies are not a relevant cost factor for interferometry.

Table 14.11 Limits of the technology.

Parameter	Limiting value	Reason for limitation
Aspherical shape	Caustic region	CGH must be positioned outside caustic region
Diffraction orders	Stray light due to parasitic diffraction orders	Aspheres requiring not realizable setups for control of diffraction orders

14.2.6 Conclusions

Testing aspherical surfaces with Null lenses, in particular with CGHs, is state of the art. Care needs to be taken with respect to positioning and centering of the CGH. Additional holograms for centering and calibration are very useful; they reduce errors and save time. Calibration is an important aspect of precision measurements. Null lenses are, however, expensive, and their manufacture is time consuming. Additional flexible methods are needed, especially for single surfaces and prototype testing.

14.2.7 Status

Testing aspherical surfaces and systems with CGHs is well established. However, the production and implementation of CGHs is time consuming and expensive, especially for a single surface or prototypes or small lots. Therefore, alternative measuring techniques that are more flexible are needed in the future. A stylus instrument on a special machine can be used. Some optical techniques are under investigation and will be discussed below.

There are different candidates such as stitching of interferograms taken by defocusing or displacement of the test surface or tilting it in order to investigate zones where the fringe density is such that it can be analyzed. The different interferograms need to be fitted together with very high accuracy. An overlap of typically 50% may be needed to minimize the stitching errors. Alternatives are techniques based on deflectometry or the use of longer wavelengths, such as using a CO_2 laser ($\lambda = 10.6\,\mu m$). Multiwavelength techniques, adaptive optics (like adaptive membranes), adaptive mirrors, and liquid crystals are other alternatives to be considered and will be discussed below. The Hartmann or the Shack–Hartmann wavefront sensor will be discussed in more detail separately.

14.2.7.1 Multiwavelength interferometry

To reduce the sensitivity of surface measurements, an IR wavelength with $\lambda = 10.6\,\mu m$ can be used. Although laser is available, high-resolution detectors with a comparable performance to those obtained with CCDs, in the visible, are not yet available or too expensive. Two or more neighboring wavelengths can be used, leading to beat wavelengths. The beat wavelength for two wavelengths λ_1 and λ_2 is

$$\Lambda = \frac{\lambda_1 \lambda_2}{\lambda_1 - \lambda_2}.$$

As an example, the interferograms of an aspherical surface are shown in Fig. 14.11(a) for $\lambda_1 = 822\,nm$ and in Fig. 14.11(b) for $\lambda_2 = 812\,nm$. The synthetic wavelength is $\Lambda = 66.7\,\mu m$. Fringes for the single wavelengths are too dense and cannot be analyzed. By using both wavelengths, producing the synthetic wavelength, three fringes are observed, as seen in Fig. 14.11(c). The wavefront can be analyzed pixelwise, as shown in Fig. 14.11(d) from which the shape of the surface is obtained. For accurate measurements, the interference pattern of the single wavelength can be used by taking into account (pixelwise) the results obtained with the synthetic wavelength. Care needs to be taken with respect to dispersion.

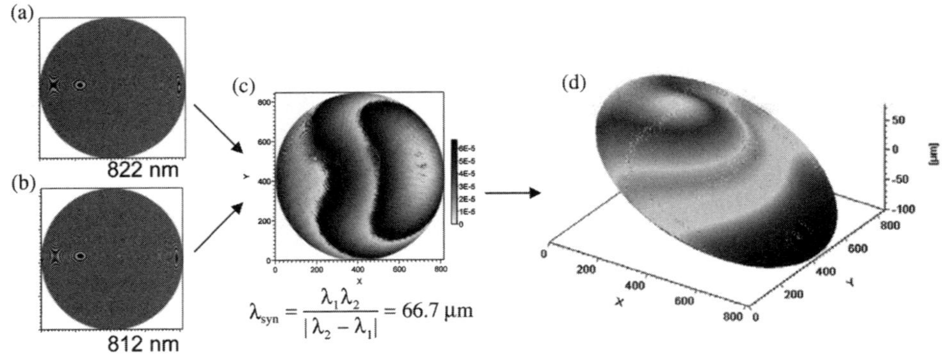

Figure 14.11 Principle of multiple wavelength interferometry.

14.2.7.2 Deformable membrane mirror

To correct wavefronts or adapt the wavefront to an aspherical surface, a deformable membrane mirror can be useful, as long as the local slope is not too high. The principle of a membrane mirror is shown in Fig. 14.12(a). A thin (about 1 μm) Si-nitride membrane mirror coated with Al is stretched over an array of electrodes. Due to electrostatic forces, the membrane deforms when applying an electrostatic field between membrane and electrodes. There are different configurations for the electrodes, circular or hexagonal, as shown schematically in Fig. 14.12(b), where a hexagonal structure is shown, whereas in Fig. 14.12(c), a ring shape structure is shown.

Because the computer-controlled membrane deformation is smooth, the wavefront deformation is smooth too. Instead of Al-coatings, dielectric high-reflectivity coatings are possible. Due to the small mass, the membrane mirror has a fast response time of about 1 ms.

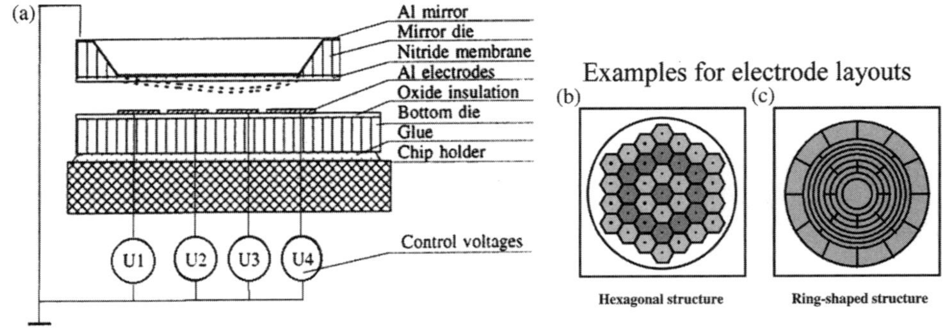

Figure 14.12 (a) Principle of membrane mirror, (b) hexagonal electrode configuration and (c) ring-shaped electrode structure.

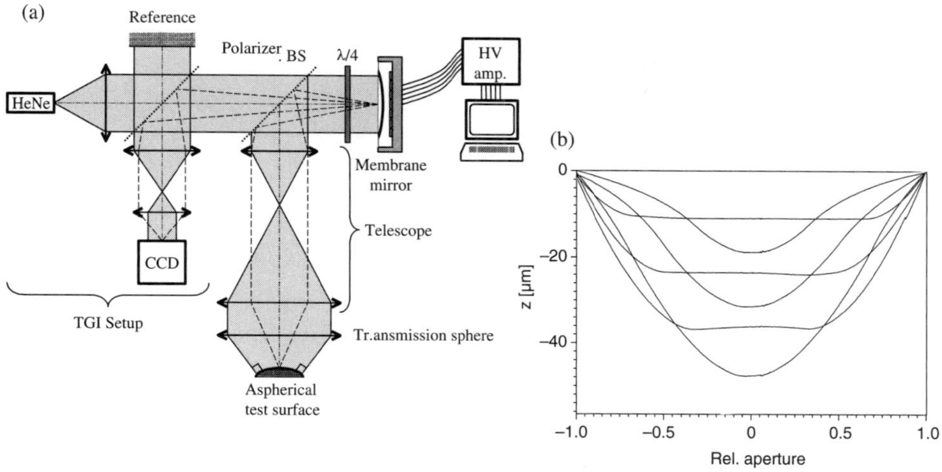

Figure 14.13 (a) Arrangement for testing an aspheric surface by the use of a membrane mirror to reduce the fringe density (b) shows deformations of the membrane for different voltages applied.

The wavefront deformation depends on

- The material constants and stress of the membrane,
- The gap between electrodes and membrane, and
- The voltage applied.

Material constants and stress depend on the fabrication process. The gap between electrodes and membrane defines the maximum deflection of the membrane. The deflection should not exceed 30% of the gap width. For small deflections, the deflection is proportional to the square of the applied voltage.

Membrane mirrors with a diameter up to 50 mm were used in our laboratory. Some experimental results obtained with a membrane mirror with a diameter of 25 mm and a gap width of 170 μm are shown in Fig. 14.13(b). For driving the 37 channels of the mirror with a voltage up to 700 V, HV amplifier electronics were developed. Deflections of more than 30 μm PV were obtained with a shape reproducibility of less than 50 nm.

14.2.7.3 Testing configuration with a membrane mirror

In Fig. 14.13(a), a dynamic membrane mirror for testing an aspherical surface is integrated into a Twyman–Green interferometer setup [7]. The polarized beam in the test arm passes the polarizing beamsplitter and the quarter wave, and it is reflected back from the membrane mirror surface and passes the retardation plate again. The quarter-wave plate rotates the polarization by 90 deg. The returning beam from the aspherical test surface passes the membrane mirror again, which leads to a deliberately introduced wavefront deformation that is four times the deformation of the membrane. For calibration of the membrane mirror, the quarter wave plate

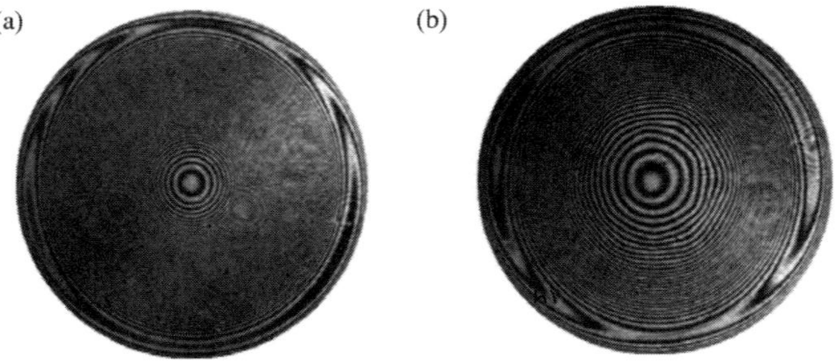

Figure 14.14 Reduced fringe density with a membrane mirror (a) with the membrane not activated and (b) with the activated membrane mirror, leading to a partial wavefront adaptation.

is rotated in order to directly observe the membrane mirror. Figure 14.13(b) shows the deformation of the membrane by applying different voltages to the electrodes. Figure 14.14(a) shows an interference pattern without and Figure 14.14(b) with wave compensation.

14.2.7.4 Stitching interferometry, with stitching by dynamically tilted reference wave

For resolving high-density interference patterns, a stitching technique can be used in order to select a small portion of the interference pattern at a time. By combining the different overlapping portions of the interferograms, the entire field wavefront is obtained. There are different methods for selecting and adding the different portions of the wavefront. A lateral shift is used by QED [10], and an axial shift is introduced by Knechal (sold by Zygo) [11, 12].

A very promising technique for flexible wavefront measuring of otherwise not resolvable interference patterns is a flexible tilted reference or target wave [8]. This approach does not include any mechanical movements but uses adaptive optics for optical stitching. The technique is in the process of being developed for testing aspherical surfaces. The principle shown in Fig. 14.15 is based on a Twyman–Green interferometer. The main difference is a dynamically and flexibly tilted reference wave. The key element is a switchable point source array consisting of an addressable LCD display and a microlens array together with an array of pinholes. Details of the key element are shown in Fig. 14.16. Together with a collimating lens, the switchable point source array defines plane waves with well-defined selectable tilt angles.

A plane wave is falling obliquely on the liquid crystal (LC) and is focused by microlenses in the pinhole plane, but it is missing the pinholes all together (Fig. 14.16). By generating a grating in the electrically addressed LC elements, the first diffraction order deviates the selected beam (or beams) to pass the appropriate pinhole. The point source is now active and leads to a tilted reference wave after the collimating lens. Furthermore, a phase shift for the automated fringe analysis is introduced by laterally shifting the displayed grating proportional to the shifted grating. The amplitude of the reference wave can be adapted by the diffraction

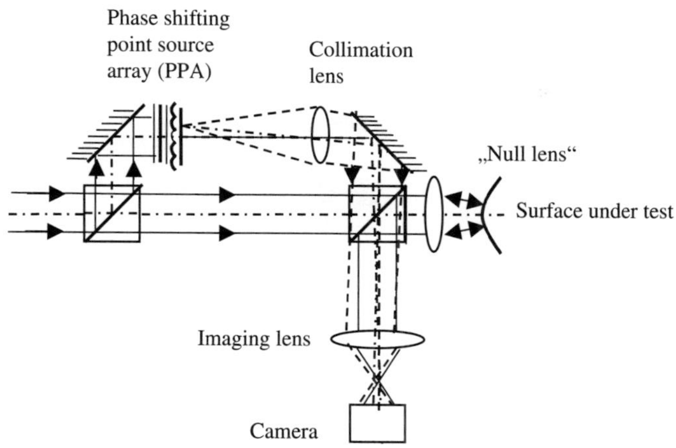

Figure 14.15 Principles of a dynamic reference beam.

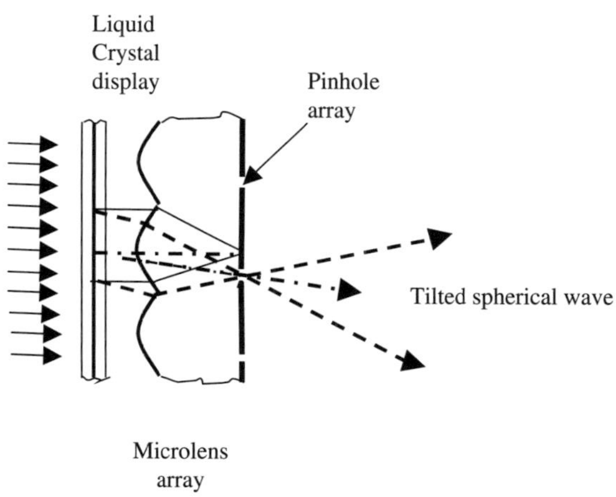

Figure 14.16 Detail for generating the tilted reference wave and phase shifting.

efficiencies of the generated phase grating independently for the different point reference sources.

The LCD used was from CRL OPTO and had 1024 × 768 pixels with a pixel pitch of 36 μm. From the microlens array, 70 × 70 LCD pixels were used in front of each microlens. Gray tone structures with eight levels were produced by direct writing into photoresist with a laser writer and transferred into fused silica substrate by dry etching (RIE). The array of pinholes was etched into a black chromium layer as the absorbing layer with a separate mask. Figure 14.17 shows the first experimental results of the new stitching procedure. On the left are single phase measurements of a test surface and on the right the unwrapped wavefront with a PV value of 270λ. In a comparison with simulated results, after some calibration of the experimental setup, the rms value agreed by λ/14. The accuracy can be further improved [8]. It turns out that it is even more appropriate to introduce tilts into the object wave instead of the reference wave. Work is underway at the Institute of Technical Optics, University of Stuttgart [13].

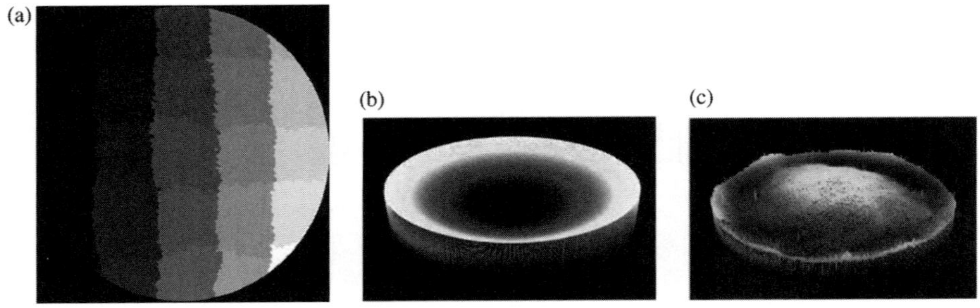

Figure 14.17 Phase measurement of a defocused reference surface with 5 × 5 tilts. (a) The stitching areas are defined using the phase-shifting contrast. (b) The result after unwrapping of 5 × 5 interferograms shows a measured wavefront of more than 270λ. (c) After subtraction of the known wavefront, a residual error of λ/14 rms remains.

Adaptive and active optical elements can be very useful for beam shaping and wavefront measurements. New active optical elements may come on the market to improve speed, field size, and resolution.

14.2.8 Further reading

1. J.C. Wyant and V.P. Bennett, "Using computer generated holograms to test aspheric wavefronts," *Appl. Optics*, Vol. 11, pp. 2833–2839 (1972).
2. J.E. Greivenkamp and J.H. Bruning, "Phase shifting interferometry," in *Optical Shop Testing*, 2nd edition, Malacara, D., pp. 577 (1992).
3. B. Dörband and H.J. Tiziani, "Testing aspheric surfaces with computer generated holograms: Analysis of adjustment and shape errors," *Appl. Optics*, Vol. 24, pp. 2604–2611 (1985).
4. A.E. Jensen, "Absolute calibration method for laser Twyman-Green wavefront testing interferometers," *Journal of the Optical Society of America*, Vol. 63, No. 10, pp. 1313 (1973).
5. S. Reichelt, H.J. Tiziani and H. Zappe, "Self calibration of wavefront testing interferometers by use of diffractive elements," *SPIE Proc.*, San Diego (2006).
6. S. Reichelt, C. Pruss and H.J. Tiziani, "Absolute interferometric test of aspheres by use of twin-generated holograms," *Applied Optics*, Vol. 42, pp. 4468–4479 (2003).
7. C. Pruss and H.J. Tiziani, "Dynamic null lens for aspheric testing using a membrane mirror," *Optics Communications*, Vol. 233, pp. 15–19 (2004).
8. J. Liesener and H.J. Tiziani, "Interferometer with dynamic reference," *Proc. SPIE*, Vol. 5252, pp. 264–271 (2004).
9. D.W. Robinson and G.T. Reid (eds) *Interferogram Analysis-Digital Fringe Pattern Measurement Techniques*, IOP Publishing Ltd, Bristol (1993).
10. P. Murphy, J. Flieg, G. Forbes, D. Miladinovic, G. De Vries and S. O'Donohue, "Subaperture stiching interferometry for testing mild aspheres", *Proc. SPIE*, Vol. 6293, 62930J, pp. 1–10 (2006).

11. M.F. Kuechel, "Absolute measurement of rotationally symmetrical aspheric surfaces", US Patent Nos. 6,781,700, 6,972,849 and 6,879,402.
12. M.F. Kuechel, "Absolute measurement of rotationally symmetrical aspheric surfaces", OSA Optical Fabrication and Testing, Rochester, OFTub5, pdf (2006).
13. E. Garbusi, Ch. Pruss, J. Liesener and W. Osten: "New technique for flexible and rapid measurement of precision aspheres", *Proc. SPIE*, Vol. 661629, (2007).

14.2.9 Links

Suppliers for CGHs

- E-beam technology (typically up to 140 mm in diameter)
  - JENOPTIK Laser, Optic, System GmbH: http://www.jenoptik-los.com/
  - Diffraction International: http://www.diffraction.com/
- Laser writing (up to 280 mm in diameter, customized substrates)
  - Institut fur Technische Optik (ITO): http://www.uni-stuttgart.de/ito/

14.3 Wavefront Sensor (Shack–Hartmann)

H. J. Tiziani

14.3.1 Basic assessment of the technology

The Shack–Hartmann wavefront sensor offers an interesting alternative to interferometry for testing optical systems and components. It is a geometric test with a lenslet array to be designed in order to be adapted to the system or surface under test with respect to sensitivity and dynamic range.

For the wavefront sensor first proposed by Hartmann, a set of holes selects a set of rays from which its intersection in selected planes (near the image plane) is detected. The analysis of the intersection of the rays with two separated planes perpendicular to the optical axis leads to the ray aberrations. The small holes proposed by Hartmann are replaced by microlenses (refractive or diffractive) in the Shack–Hartmann sensor.

The sensitivity depends on the geometrical configuration of the whole measuring system as well as on the resolution of the CCD or CMOS detector and the wavelength used. Care needs to be taken with respect to calibration. An absolute calibration is required when used in optical metrology for testing aspherical surfaces.

14.3.2 Intended purpose of the technology

The Hartmann and Shack–Hartmann sensors have been used for many years, especially in astronomy to measure wavefronts and in feedback control systems for adaptive optics. Image performance was measured even in the presence of air turbulence in astronomy. In addition, eye aberrations were measured with the wavefront sensor.

The Shack–Hartmann wavefront sensor is very appropriate for measuring the shape of a wavefront. Currently, the Shack–Hartmann wavefront sensor is frequently used in optical metrology and optical testing because of its robustness, easy application, and flexibility in being adapted to different testing configurations and sensitivity.

14.3.3 The technology's typical features

The basic principle of the Shack–Hartmann wavefront sensor was first suggested by Roland Shack in 1971 as an extension of the more traditional Hartmann test. In the Shack–Hartmann test, the wavefront to be analyzed is imaged on a regular array of microlenses. Each microlens focuses its portion of the wavefront area onto a position-sensitive detector in an array. The average wavefront tilt across each microlens aperture results in a lateral shift of the perspective focal spot. An unaberrated wavefront produces a regular array of equidistant spots [indicated by the crosses in Fig. 14.18(b)], whereas an aberrated wavefront produces a distorted spot pattern. Mapping the overall spot pattern distortions produces a map of the wavefront slopes, and an integration of these slopes will reconstruct the wavefront under test. The primary requirement for the analysis is that each spot in the detected pattern must be associated with the specific microlens that produced it.

For an aberrated wavefront, the phase gradient incident on a perfect microlens leads to a lateral shift of the spot in the x- and y-directions by Δx and Δy. For a focal length of the microlens f, the mean wavefront slopes across the microlens

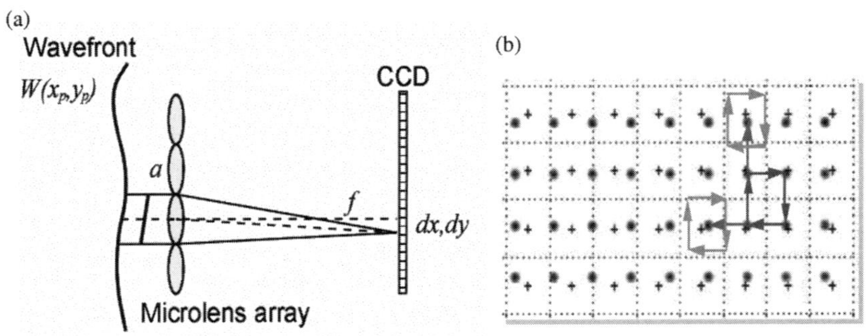

Figure 14.18 Principle of the Shack–Hartmann wavefront sensor. (a) A basic setup. (b) An array of spots on a CCD array partly displayed from its virtual center due to aberrations. The crosses indicate the virtual spot positions for a perfect wavefront.

aperture in x- and y-directions are given by

$$\langle L\rangle = \frac{\Delta x}{\sqrt{\Delta x^2 + \Delta y^2 + f^2}} \quad \text{and} \quad \langle M\rangle = \frac{\Delta y}{\sqrt{\Delta x^2 + \Delta y^2 + f^2}}.$$

The angle brackets indicate an average of the slope across the aperture of the microlens. By integration of the slope over the pupil, the wavefront is obtained, from which Seidel and Zernike polynomials, for instance, can be calculated. Furthermore, the spots of the microlenses should be diffraction limited, but aberrations in the field of the microlenses may lead to a shift of the centroid. This is particularly true when the Shack–Hartmann sensor is used for measuring wavefronts with a strong deviation from a plane or spherical reference wavefront.

For highly aspherical wavefronts, for instance, the spots will no longer stay inside their subaperture regions. Spots can walk off the corresponding detector element [CCD pixel in Fig. 14.18(a)]. In addition, the spots may be blurred due to the wavefront variation over the subaperture (the microlens). A possible solution to solve both problems is the use of adaptive, dynamic microlenses generated by electrically addressed liquid crystal arrays, which will be discussed in Sec. 14.3.9.

14.3.4 Description of process

The Shack–Hartmann wavefront sensor measures the local wavefront slopes within a subaperture in the pupil defined typically by the microlenses of a lens array. The positions of the Shack–Hartmann spots on the detector(s) in the focal plane are analyzed. Their lateral locations relative to the calibrated lateral spot positions of a perfect wavefront are proportional to the local wavefront slope. The spot positions on the detector need to be calibrated before the measurements are started. With respect to the microlenses, they can be refractive or diffractive. Refractive elements can be produced by a melting process, for instance. The diffractive elements are typically binary elements, allowing fabrication by using lithographic techniques at least for the master, and they can be duplicated afterwards. For high diffraction efficiency, three-dimensional phase structures need to be generated, and gray tone technology can be used. The microlens diameter is typically a few hundreds of micrometers but can be as small as 50 μm [1–3].

14.3.4.1 Adaption of the wavefront under test to the Shack–Hartmann sensor

The Shack–Hartmann wavefront sensor in its elementary state consists of a two-dimensional array of positive microlenses of selected focal length together with a detector such as a CCD (charge-coupled device) in the focal plane. However, a more general application includes an imaging system to relay the wavefront to be tested onto the pupils of the microlenses, while at the same time presenting a nominally collimated beam to the microlens array, as shown in Fig. 14.19.

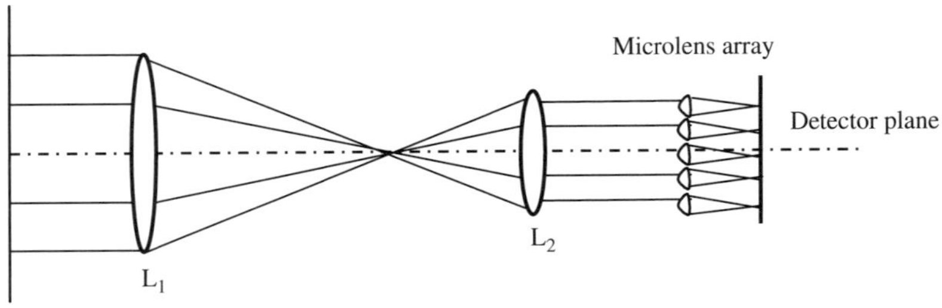

Figure 14.19 Principle for imaging the wavefront under test onto the microlenses.

For the imaging system, the exit pupil of the system under test should be imaged by the lens systems L_1 and L_2 onto the pupils of the microlenses. For relatively weakly aberrated wavefronts, caustic is not a problem. In Fig. 14.19, the imaging system is optimized for a plane wavefront as a reference or infinite conjugate. For an aspherical surface with a best-fitting sphere as a reference, a slightly modified setup is used. Care needs to be taken in order to avoid clipping of rays or distortion. For finite conjugates of the pupil planes, an appropriate lens system needs to be designed.

The Shack–Hartmann sensor is an attractive method for testing aspherical surfaces. The microlens array can be designed to take advantage of the trade-off between sensitivity, accuracy, and dynamic range. The dynamic range is, however, limited but can be extended by proper system design. It should be noted that calibration of the sensor system is very important [4].

A possible setup for testing an aspherical surface is shown in Fig. 14.20. The laser beam is guided by a monomode fiber and expanded and adapted to the aperture of the surface to be tested. At the same time, the surface under test should be imaged onto the pupil plane of the microlenses as shown schematically in Fig. 14.20. It can be taken care of by the appropriate lens design of L_3 and L_4. An additional lens system, or Null lens system comparable to the Null lens, described in Sec. 14.2 for testing aspherical surfaces, could be required as indicated in Fig. 14.20.

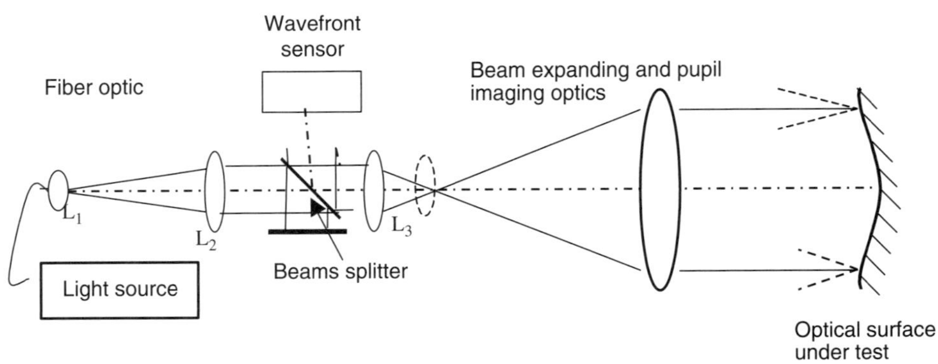

Figure 14.20 Test setup for the aspherical surface to be tested.

14.3.5 Data for wavefront sensor (Shack–Hartmann) (Tables 14.12–14.15)

Table 14.12 Typical operation parameters.

Parameter	Minimum	Maximum
Duration of measurement	1 min, without calibration	5 min with calibration
Measurement rate	1min/surface	
Precision of measurement	$\lambda/4$	$\lambda/20$
Spatial resolution	100 μm	1500 μm
Resolution of tilt	1 μrad	6 μrad
Wavelength	$400\,\text{nm} \leq \lambda \leq 1200\,\text{nm}$	

Table 14.13 Geometrical dimensions.

Geometrical dimensions	Minimum	Maximum
Diameter	2 mm	100 mm depending on lateral resolution and adaptive optics

Table 14.14 Cost drivers.

	Cost factor	Volume
Equipment	Wavefront sensor	15,000–35,000€
Additional components	Optical setup, pupil adaptation	500–2000€

Operating supplies are not a relevant cost factor for wavefront sensors.

Table 14.15 Limits of the technology.

Parameter	Limiting value	Reason for limitation
Local slope/microlens v	$2f\,v \leq$ pixel size	Ambiguity

The diameter of the microlenses of the lenslet array should be chosen such that each subaperture experiences local tilt only. The pixel size should be chosen to optimize the dynamic range. The actual spot shift should not drift into a neighboring subaperture.

14.3.6 Conclusions

A long-standing goal for testing aspherical surfaces is testing without a "Null lens," as discussed for interferometric testing. The Shack–Hartmann sensor can be used without a Null lens for weak aspherical surfaces. For testing more complex and stronger aspherical surfaces (strong local slope), further investigations are needed. One possibility to extend the range of applications is discussed in Sec. 14.3.7.

14.3.7 Status

By replacing the static microlens array by an electrically addressed light modulator, a more flexible adaptive Shack–Hartmann sensor is obtained. The liquid crystal display consists of individually addressable pixels, which are used to generate an array of Fresnel microlenses. Parameters such as focal length aperture size can be adapted in a wide range. Single synthesized microlenses (40 × 40 pixels for instance) can be activated and deactivated in order to localize the spots in the appropriate fields on the detector as well as to compensate curvature in an individual aperture.

The advantages of the adaptive Shack–Hartmann sensor using LCD elements include the following:

- Ambiguity problems can be avoided because individual microlenses can be turned on and off, leading to a correspondence of the appropriate microlens and spot.
- Design parameters, such as focal length, geometry, size, and number of microlenses, can be adapted quickly.
- If the incident wavefront across the Hartmann sensor is so strongly deformed that its gradient over one microlens diameter is no longer linear, then the corresponding spot would be severely aberrated and thus difficult to measure. However, knowing the wavefront deformation by some *a priori* means would allow tuning the aberrations of each microlens to suppress the nonlinear part of the wavefront.

It should be noted that a simulated microlens for the results in Fig. 14.21 and Fig. 14.22 consisted of 40 × 40 LC elements. Results obtained by correcting the local curvature when testing a progressive eye glass are shown in Fig. 14.21. For

Figure 14.21 Focal spots of a progressive eye glass without correction.

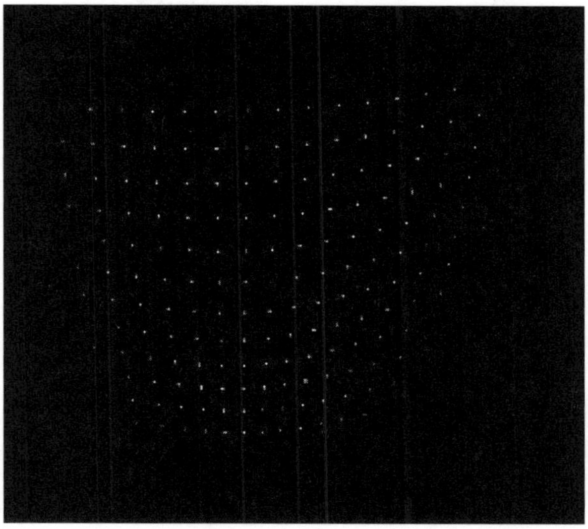

Figure 14.22 Result of corrected microlens array of the progressive eye glass.

comparison, Fig. 14.22 shows the spots without correction of the local curvature. The spots are partly blurred. With correction of the local curvature, the spots are relatively sharp and in a plane. Their positions are no longer difficult to measure accurately. To correct for distortion in the analysis is not a problem.

14.3.8 Further reading

1. B. Platt and R.V. Shack, "Lenticular Hartmann screen," *Optical Science Center Newsletter*, Vol. 5, No. 1, pp. 15 (1971).
2. J. Pfund, N. Lindlein and J. Schwider, "Non null testing of rotationally symmetric aspheres: A systematic error assessment," *Appl. Opt.*, Vol. 40, No. 4, pp. 439–446 (2001).
3. N. Lindlein, J. Pfund and J. Schwider, "Expansion of the dynamic range of a Shack Hartmann sensor by using astigmatic microlenses," *Opt. Eng.*, Vol. 39, pp. 2220–2225 (2000).
4. J.E. Greivencamp, D.G. Smith and E. Goodwin, "Calibration issues with Shack-Hartmann sensors for metrology applications," *SPIE Proceedings*, St. Etienne (2004).

14.3.9 Links

Suppliers:

- OPTOCRAFT GmbH, http://www.optocraft. de/
- Imagine Optic, http://www.imagine-optic.com/

- Wavefront Science, 14810 Central Ave SE, Albuquerque, New Mexico 87123, USA; Tel. +1 (505) 275 4747, Fax +1 (505) 275 4749; e-mail: info@wavefrontsciences.com; http://www.wavefrontsciences.com/

14.4 Surface/Microstructure Inspection

A. Duparré

14.4.1 Basic assessment of the technology

For the inspection of surface microstructures of ultraprecision optical surfaces, height sensitivities in the nanometer and subnanometer range are required. Scanning force microscopy and optical profilometry (white light/phase shift profilometry) are appropriate tactile and optical methods, respectively, for local measurements. For local as well as measurements across small areas that are relevant for optical function, light-scattering methods are the preferred techniques. The main established techniques for light-scattering measurements of surfaces are angle-resolved scattering (ARS) and total integrated scattering. For aspherical surfaces, ARS is preferred because of its higher flexibility regarding the adaptation of the measurement procedures to the surface curvature.

There are two methods of measurement:

- *In-line, pointwise*–optical profilometry, tactile profilometry, pointwise light scatter measurement, and
- *Off-line, areawise*–angle-resolved light-scattering techniques, optical profilometry with stitching methods.

14.4.2 Intended purpose of the technology

Light scattering measurement techniques have been developed, in particular for smooth optical and nonoptical surfaces as well as thin-film coatings to determine their optical losses, microstructures, roughness, defects, and contaminations. Volume scattering and related material inhomogeneities can also be investigated.

14.4.3 The technology's typical features

ARS of optical components is determined by means of an arrangement where a goniometer arm carrying a photodetector rotates about the sample, which is illuminated by a well-collimated light beam. This measurement arrangement is divided into the following parts: radiation source, beam preparation system, sample positioning system, and detection system. The majority of such instruments operate at a laser wavelength in the visible spectral region (typically 632.8 nm, He–Ne laser). The essentially possible wavelength range, however, extends into the infrared and ultraviolet spectral regions. Using radiation sources other than lasers is possible, but it significantly confines the sensitivity of the measurement. For areal measurement of aspherical optics, due to the influence of curvature, the detection angle of the scattering signal is no longer identical to the real scattering angle. Therefore, this deviation must be compensated for in order to avoid falsified determination of the microstructural data from the measurement. This compensation can be accomplished through mechanical adaptation by the sample positioning system such that, for each effective local curvature, the sample normal at the illumination point remains fixed with respect to the direction of illumination. Alternatively, the compensation can be performed on a software basis that calculates the locally varying detection angles into the correct scattering angles.

14.4.4 Description of process

14.4.4.1 Operating resources

Operating the system under cleanroom conditions (or normal laboratory environment with a flow box) is recommended to avoid disturbing influences of scattering from contaminations. Which cleanroom class is needed depends on the particular specifications of the surface finish of the aspherical test object. Class 10,000 can be sufficient for a variety of applications; for microstructure height sensitivities in the subnanometer range, however, Class 1000 is recommended.

14.4.4.2 Layout/test setup

A typical test setup is schematically shown in Fig. 14.23. The laser radiation is guided through the beam preparation system, consisting of various optical and

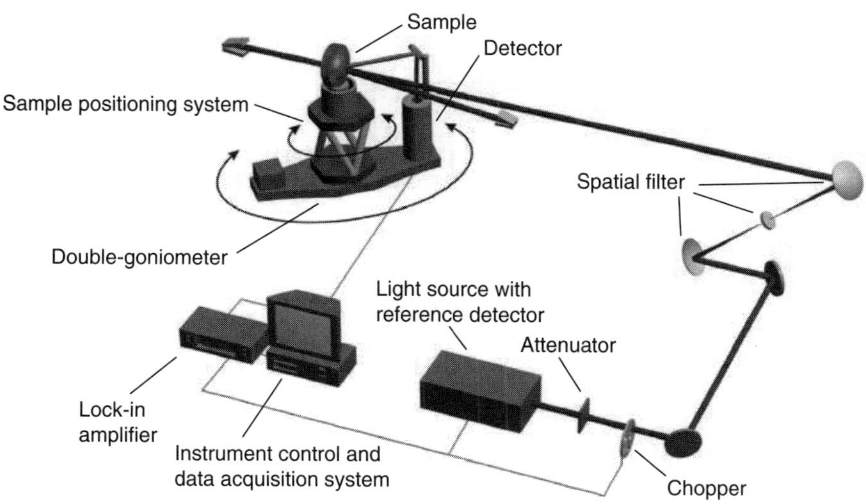

Figure 14.23 Schematic layout of a setup for ARS measurement.

mechanical elements in order to provide the needed beam quality. The sample holder and the scatter signal detector are each mounted on a precision goniometer. The sample positioning system is fixed on the inner goniometer arm; the outer goniometer carries the detector system. The detector can be revolved 360 deg around the sample with 0.01 deg resolution. The dynamic range of the arrangement comprises 12 orders of magnitude, which is necessary for the microstructure inspection of well-polished surfaces.

14.4.4.3 Mode of operation/sc diagram

The basic parts of the test procedure are demonstrated in Fig. 14.24. Prior to the measurement of the aspherical element, the instrument signature is recorded and the calibration is performed.

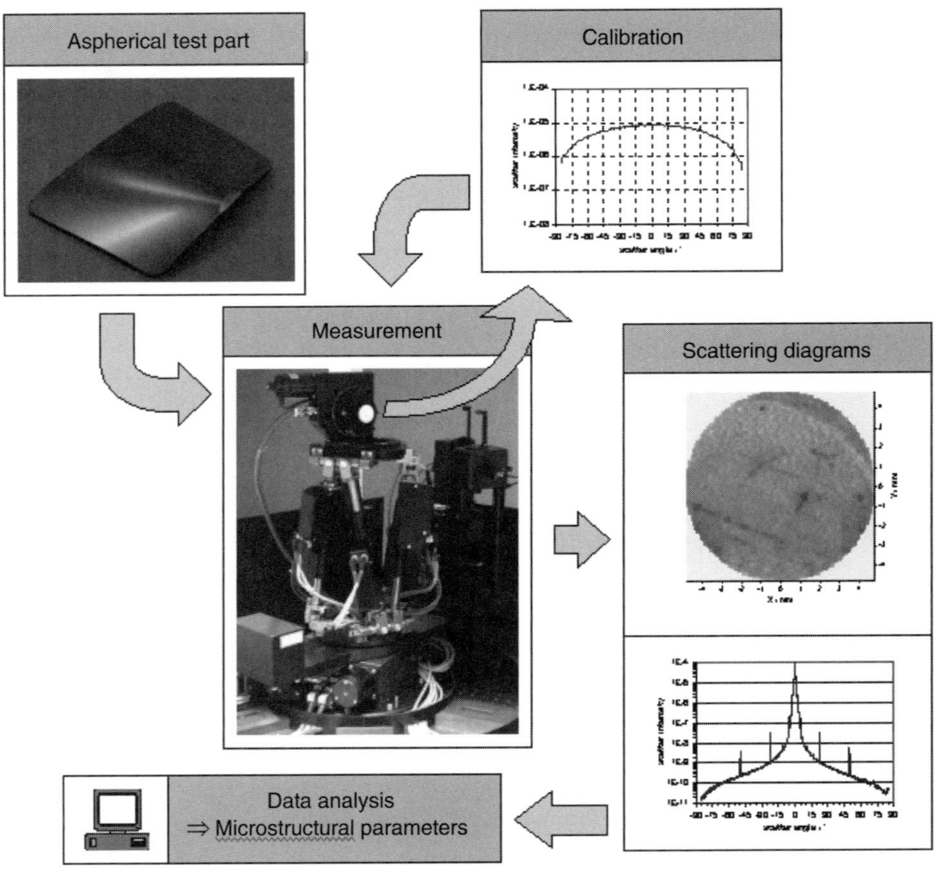

Figure 14.24 Schematic of operation for microstructure testing.

The signature is measured while keeping the sample holder empty. The incident beam is directed through the center of the sample holder, and the scatter signature is taken around 360 deg. The resulting ARS distribution is compared with the required signature, and fine alignment is carried out until the signature does not exceed the required low levels at any angle. This is important for ensuring a dynamic range of 12 orders of magnitude, which is necessary for the measurement of well-polished aspherical surfaces. Afterwards, calibration is performed by measuring a plane, all-diffuse reflectance standard sample with a Lambertian scatter behavior (Fig. 14.23) for the white sample in the sample holder of the instrument and corresponding measurement curve. This measurement is carried out for normal incidence in the backscattering hemisphere (-180 deg to $+180$ deg).

For ensuring appropriate measurement of the aspherical test part, the surfaces (front as well as back side) need to be carefully cleaned to prevent influences of the measured curves by contamination-induced scattering.

The aspherical test part is mounted onto the sample positioning system and aligned such that the illumination point on the sample surface is located on the goniometer axis. The sample is turned around the illumination point until the predefined illumination conditions (polar and azimuthal angles of incidence) are realized. The measurement, as demonstrated in Fig. 14.24, is divided into two parts. In the first step, the surface finish quality is tested with respect to homogeneity and polishing defects like scratches, holes, and so on. Therefore, surface mappings at selected detection angles are registered (Fig. 14.24, "scatter diagram," upper picture). The illumination conditions need to be chosen such that (for transparent samples) specular reflexes from the back side of the asphere cannot enter the detector system. The deviation between the detection angle and the real scattering angle must be counterbalanced for each effective local curvature, either mechanically or on a software basis.

In the second step, full scattering distributions in the backscattering hemisphere are recorded for the defect-free sample parts for the inspection of the homogeneous microstructure with regard to roughness, groove distances for diamond-turned surfaces, and so on (Fig. 14.24, lower picture). For the inspection of singular defect structures, scattering curves are measured accordingly at the disturbed sample parts.

14.4.5 Data for surface/microstructure inspection (Tables 14.16–14.19)

Table 14.16 Typical operation parameters.

Parameter	Minimum	Maximum
Duration of measurement	10 min (without calibration)	5 days
Precision of measurement	1 nm	0.1 nm

Table 14.17 Machinable dimensions.

Geometrical dimension	Minimum	Maximum
Diameter	4 mm	400 mm
Height	–	200 mm
Weight	–	5 kg

Table 14.18 Cost drivers.

	Cost factor	Volume
Equipment	Setup	150,000€
	Infrastructure (clean room or flow box)	10,000€
Operating supplies	Expert	2 days per month
	Technician	3 days per month
	Consumables (calibration standards, cleaning chemicals)	5000€

Table 14.19 Limits of the technology.

Parameter	Limiting value	Reason for limitation
Curvature	Radius of local curvature >10 spot size	Detection angle ambiguity

14.4.6 Status

Scattering measurements are very suitable techniques for the inspection of the surface microstructure of high-quality optical surfaces. For application to aspheres, compensation for local curvature must be carefully taken into account. In addition to ARS, future research and development will also be directed to adapting total scattering procedures to the specific requirements of aspheres, because of their attractive capability for time- and cost-efficient operation. Furthermore, microstructure inspection of aspheres can benefit from a well-tailored combination of scattering techniques with local scanning force microscopy as well as optical profilometry.

14.4.7 Further reading

1. J. Stover, *Optical Scattering: Measurement and Analysis*, 2nd ed. SPIE Optical Engineering Press, Bellingham, WA, 1995.
2. A. Duparré, "Scattering from surfaces and thin films," in *Encyclopedia of Modern Optics*, Guenther, R. D. et al. (eds) Elsevier, Oxford, 2004, pp. 314–321.

3. A. Duparré, J. Ferre-Borull, S. Gliech, G. Notni and J. Bennett, "Surface characterization techniques for determining the root-mean-square roughness and power spectral densities of optical components," *Applied Optics*, vol. 41, pp. 154–171, 2002.

14.4.8 Links

Producers/suppliers of auxiliary materials

- Diffuse reflectance calibration standard: Spectralon®, supplied by Labsphere, http://www.labsphere.com

Chapter 15

Coating Technologies

15.1 Coating Design

U. Schallenberg

15.1.1 Basic assessment of the technology

Coating design involves generating the design of an optical coating for a particular application and the conditions of manufacturing. It is part of the manufacturing process of optical coatings itself. It gives the layer sequence for the production run, assists the monitoring of the deposition process during the run, and enables a reverse engineering after the run. It is also particularly part of any R&D in optics in general and goes hand in hand with the optic design itself (Fig. 15.1). Necessarily, it has to use corresponding thin-film design software.

15.1.2 Intended purpose of the technology

Optical coatings consist of thin films of different materials and thicknesses. Thin films are deposited onto a substrate, and the optical application of such a coating is based on the interference effect of light within the thin films and between the ambient media. The propagation of light as an electromagnetic wave is described by Maxwell's equations. Under defined conditions, there are solutions that describe the interference effect exactly. Various recurrent calculating techniques exist to evaluate the effect of optical coatings, but the most versatile method is based on the matrix formulation of Maxwell's equations. Each individual thin film is described by a characteristic matrix including refractive index, extinction coefficient, and physical thickness of the film in dependence on the wavelength of light. The optical coating is described by the product of all the characteristic matrices and the vector whose elements depend on the refractive index of the substrate.

There are only less explicit equations to obtain the reflectance, transmittance, absorptance, and phase changes on reflection or transmission of an optical coating,

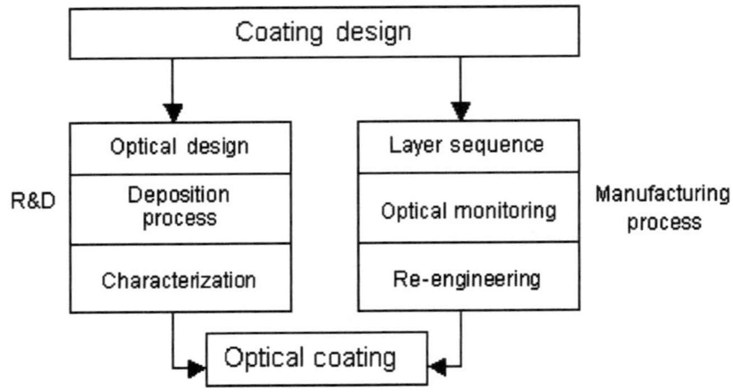

Figure 15.1 Coating design as basis of both the R&D process and the manufacturing process of an optical coating.

and computer programs are needed to perform these evaluations. In this sense, coating design is strongly connected with progress in computer hardware and software. It is not necessary to develop a corresponding program if someone is interested in performing coating design, because a lot of commercial thin-film design software exists. However, computer-assisted coating design does not reduce the need for comprehensive knowledge, skill, and experience in thin-film optics—both in theory and in practice.

15.1.3 The technology's typical features

Coating design creates a sequence of layers or thin films. It is possible to use the same coating design for different coatings (e.g., at different applications by changing the angle of incidence) in different spectral regions by changing the materials, and with different parameters by using higher orders of phase thickness. On the other hand, the same application can require different designs if different substrates or different manufacturing methods are used.

Based on the intended optical application, coating design distinguishes coating types with respect to the beams created by the coating and the energy balance regarding reflection, transmission, absorption, and scattered loss (Fig. 15.2). If only the transmitted beam is used, the coating is called a "filter," if only the reflected beam is used, the coating is called a "mirror," and if only the absorption within the coating is used, the coating is called an "absorber." If both the transmitted and the reflected beam are used, the coating is called a "beam splitter": Splitting into different spectral regions gives a "dichroic" (filter or mirror), splitting into different polarization states gives a "polarizer," and splitting into beams with different phases gives a "phase shifter." The scattered loss both in reflection and in transmission has a negligible value in most cases.

In many optical applications, there is an undesired reflected beam, which can be reduced using an "AR coating" (antireflecting). An "attenuator" changes the ratio of intensity from a reflected to a transmitted beam. A "nonpolarizing beam splitter" shows no polarization effect of the reflected or transmitted beam. In general, the

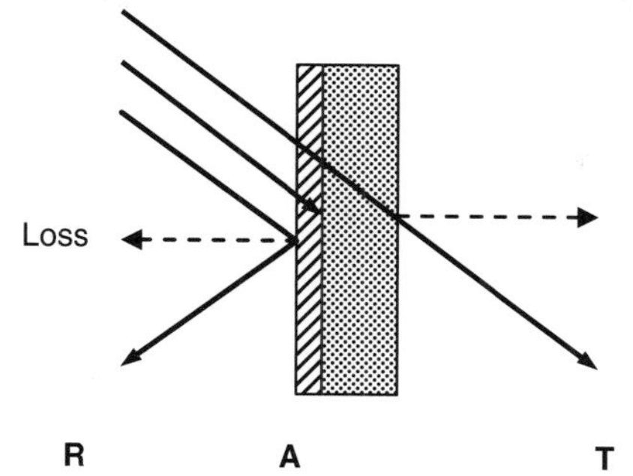

Figure 15.2 Schematic diagram of beam splitting in reflection R, transmission T, and absorption A, and stray light losses at an optical coating (hatched) on a transparent substrate (dotted).

term "filter" is also used for the optical coating itself, often completed to the term "interference filter."

15.1.4 Description of process

Optical coating design uses a special notation to define the coatings themselves. The different materials used for the coatings are named by capitals A, B, C, and so on. Because most optical coatings consist only of two different materials, one with a high refractive index and the other with a low refractive index, the capitals H and L are used for two-material coatings. The thickness of each individual film is given by its optical thickness or its absolute physical thickness.

Usually, the optical thickness is normalized with respect to a reference wavelength in a fraction of $1/4$. A layer with an optical thickness of exactly $1/4$ of the reference wavelength is called a quarter-wave layer (QWL) having a normalized optical thickness of 1. A two-material sequence of QWLs gives the basic design of most optical coatings.

Often, a group of two or more layers are repeated within the coating, which is schematically given by a sequence in brackets with an exponent announcing the number of repetitions. The notation itself starts with an abbreviation for the substrate and ends with a term indicating the incident medium, for example

$$\mathrm{BK7(0.5L\,1H\,2L\,1H\,0.5L)^{21}air.}$$

The individual films within the coating are counted, starting at the substrate or starting with the first film deposited onto the substrate, respectively, or they are counted starting at the incident media with the layer that sees the incoming beam at first.

15.1.4.1 Packages of thin-film design software

Coating design needs thin-film design software consisting of packages for input, analysis, refinement, synthesis, and manufacture assistance (Fig. 15.3). There are

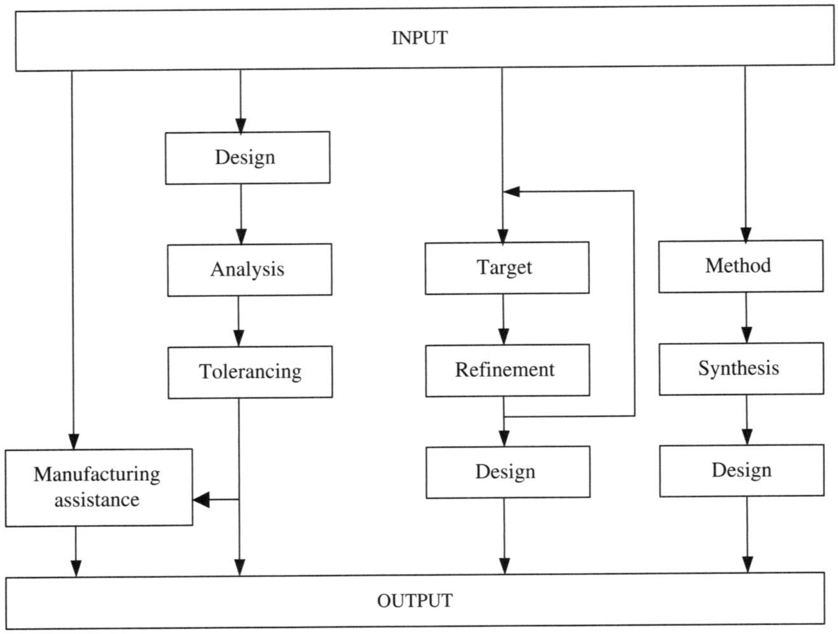

Figure 15.3 Flowchart of different ways of using thin-film software tools.

different ways of using the diverse software tools. Which tool is used for which task depends usually on the requirements for the coating design and the familiarity of the designer with the program.

15.1.4.2 Input

The input is given by the environment, the design itself, and the material definition including index extraction. The environment refers to the refractive index of the incident medium and substrate, a wavelength-dependent illumination or detector, and the angle of incidence of the incoming beam. The design itself is prompt, as an alphanumerical string considering the corresponding notation rules and the defined materials. The design program converts the string into the layer sequence, and a refractive index versus thickness plot illustrates the index profile.

Material definition is a database of materials suitable for use as thin films in an optical coating. Each material is defined by its refractive index as a complex value, with a real part as the refractive index itself and an imaginary part as the extinction coefficient. Usually, the given complex refractive index is listed in its wavelength dependence. The material data files are included in the material software package, or they can be called from the software or the material vendor.

15.1.4.3 Analysis

The analysis program performs evaluations based only on the given design, with the design parameters of layer number, layer type, and layer thickness. Analysis is used to check a design under different parameter values or different environments, to refer or to compare designs of a database or a reference, or to modify designs by material substitution.

Because different coatings are possible for the same application, a method to select an optimal design solution is to analyze the design performance if small errors in thickness or refractive index are created. The checked errors have their origin in tolerances of thickness and refractive index during the manufacturing process, which gives this analysis tool the name "tolerancing." Multiple numerical simulations of the design performance can be performed using a random error generator or using schematic deviations. Finally, a worst-case consideration indicates error values that yield a coating out of the target design.

15.1.4.4 Refinement

To obtain a desired coating performance, a target has to be defined. A broad range of performance values can be used as targets simultaneously, for example, reflectance versus wavelength and versus angle of incidence. The refinement program modifies a given design in its layer sequence and in the thickness of each individual layer to reach the targets as well as possible. Also, the refractive index of each layer material can be used as a refinement variable. Constraints can be defined concerning the layer thickness, layer number, and available refractive indices. A merit function is used to drive the changes in the refinement parameters and to indicate when a satisfactory solution has been reached.

The refinement program itself uses known optimization procedures similar to those that have been developed for the solution of general engineering problems. Because there is no one optimization method that works best on all problems, the coating designer can choose different procedures for different problems, but usually this has to be done with the aid of his own experience. The final value of the merit function, and the process time to get this value, may be parameters to select good start designs and optimal refinement algorithms for further design works.

15.1.4.5 Synthesis

Because the refinement program only varies given variables of the design, the result of refinement depends strongly on the start design. To overcome this problem, there is a synthesis program to find a design that matches the specified performance only when target and constraints are given. Synthesis creates an optimal design by adding layer by layer into the design, and thickness and refractive indices of the layers are so refined.

There are various methods of synthesis implemented in most of the thin-film software, such as comprehensive search, gradual evolution, needle technique, minus-filter approach, flip-flop design, and inverse Fourier transformation. In contrast to refinement procedures, thin-film syntheses do not require specific starting designs and were developed solely for solving thin-film problems. However, synthesis requires special practices in handling of the different methods, because different problems are best solved by different methods. Furthermore, different synthesis methods yield different types of solutions to the same problem, and particular attention has to be paid to the creation of the targets.

15.1.4.6 Manufacturing assistance

Special thin-film software tools have the ability to guide the designer through the manufacturing process depending on the method of thickness monitoring and the deposition process used. Monitoring methods are based directly on the change in transmittance or reflectance of a coated substrate. Different direct optical monitoring techniques exist in which light of only one wavelength, several optimum wavelengths, or a whole spectral range are used. For each method, calculations can be performed that assist in determining the exact point where the deposition of individual layers should be terminated.

A more satisfactory approach is to use real-time design analysis. If the individual layers are measured for thickness and refractive index during the deposition process, it seems possible to adjust the thickness of the remaining layers of the coating if any error at an individual layer is indicated by the monitoring system. Such a completely automatic coating process is not yet implemented in a commercially available coating plant, but it is in development and demonstrates the future in thin-film manufacturing.

If the coating is manufactured, a re-engineering program based on the refinement program can be used to identify real errors in thickness and refractive index made during the manufacturing process. This gives feedback for the evaluation, both of the chosen design and the quality of the manufacturing process used.

15.1.4.7 Output

Output of the program tools take the form of different plots as spectral characteristics (e.g., reflectance or transmittance versus wavelength with angle of incidence as a parameter), as angle characteristics (e.g., transmitted or reflected phases versus angle of incidence with wavelength as parameter), or as thickness characteristics (e.g., transmittance or absorptance versus physical thickness with wavelength and angle of incidence as parameters). Some applications require an analysis of color in reflection or transmission under a given illumination. The CIE color specification system and common illuminations are implemented in any analysis program, and different chromaticity coordinates can be used as output plot.

Other outputs illustrate special design methods as vector or admittance diagrams, or the electric field distribution within the coating. Special performance parameters as group delay or group delay dispersion are also available in most of the programs. Naturally, parameter units can be changed easily, for example, length units in nanometers or micrometers, spectral units in nanometers or electron volts, and reflectance or transmittance in percent or decibel.

15.1.4.8 Features of different software offers

Essential thin-film software products are listed in the Links section stated below. Each of the suppliers offers yearly updates and telephone support for its product. All the programs include packages for analysis, refinement, synthesis, and manufacture assistance, but there are special features:

- *FilmStar* has a BASIC module as an environment for customizing the program use and a MEASURE tool that can be used to drive different types of spectrometers directly by the thin-film software.
- *OptiLayer* is characterized by the needle optimization technique. This powerful synthesis method is also part of other software programs, but the *OptiLayer* supplier is the developer of the method. *OptiLayer* software also uses multi-threaded computations. This allows the user to interact with the software without interrupting computations.
- *TFCalc* is very easy to use and is a cost-effective solution if a lot of workstation has to be provided with thin-film software.
- *The Essential Macleod* includes a vStack as a special tool for the calculation and optimization of a set of nonparallel coatings, for example, an all-face coated prism. There are also tools for designing DWDM filters and induced-transmission filters. *Essential Macleod* designs can be directly exported to the ZEMAX optical design software.
- *Film-Wizard*™ combines optimization and synthesis design capabilities with powerful tools for analyzing spectroscopic and ellipsometric data, modeling thin film layers, and performing regressions to determine actual thicknesses and indices of multilayered thin-film structures.

15.1.5 Further reading

1. H.A. Macleod, *Thin Film Optical Filters*, Institute of Physics Publishing, Bristol and Philadelphia (1996).
2. A.J. Thelen, *Design of Optical Interference Coatings*, McGraw-Hill, New York (1989).
3. H.K. Pulker, *Coatings on Glass*, Elsevier, Amsterdam (1984).
4. J.D. Rancourt, *Optical Thin-Films: User Handbook*, SPIE Press, Bellingham (1996).
5. Philip W. Baumeister, *Optical Coating Technology*, SPIE Press, Bellingham (2004).

15.1.6 Links

Further information on thin-film software:

- *FilmStar*: http://www.ftgsoftware.com
- *OptiLayer*: http://www.optilayer.com

- *TFCalc*: http://www.sspectra.com
- *The Essential Macleod*: http://www.thinfilmcenter.com
- *FilmWizard*: http://www.sci-soft.com

15.2 Electron-Beam Evaporation

M. Falz

15.2.1 Basic assessment of the technology

The technology essentially comprises the generation of atoms (or molecules) in a source that is heated resistively or by an electron beam. These atoms are transported to the substrate through a process chamber evacuated to high vacuum. The atoms are then condensed on the substrate and film growth is enabled.

15.2.2 Intended purpose of the technology

The optical properties of components (transmission, reflection, absorption) are affected precisely by means of resistance or e-beam evaporated films.

15.2.3 The technology's typical features

The technology includes evaporation of metals or compounds (oxides, fluorides, and so on) under high-vacuum conditions. By means of resistance-heated evaporators (coils, boots) or e-beams, the evaporating materials vaporize and are deposited on the substrates.

15.2.4 Description of process

15.2.4.1 Operating resources

- Vacuum chamber (diameter from 250 mm up to more than 2000 mm);
- Vacuum pumps (prevacuum pump or system in combination with high vacuum pumps such as diffusion-, turbo-, or cryopump);
- Glowing unit (plasma cleaning of substrates);
- Heating (front- or back-side heating for raising the substrate temperature);
- Evaporator (films of high purity with e-beam evaporator, less sophisticated layers with resistance evaporators);
- Substrate holders (palettes, calottes, or planetary systems);
- Shutters (start/stop of the deposition process and film thickness correction);
- Plant control (type: SPS or PC control);
- Measuring devices (quartz or optical monitoring for film thickness control).

15.2.4.2 Mode of operation

The parts to be coated are arranged in the upper section of a vacuum chamber on flat or slightly curved substrate holders. The vessel is evacuated to a high vacuum of lower than 10^{-4} mbar by use of combinations of powerful mechanical and high-vacuum pumps.

The evaporating sources are arranged on the bottom of the vessel. Depending on the technology, resistance or e-beam sources are used. E-beam evaporators are preferred to be used for films of high purity and, furthermore, for nonevaporable materials and high-temperature materials.

The performance of the evaporation sources is computer-controlled.

Shutters start and complete the deposition process. Normally, the deposition rate and the film thickness are measured by the quartz crystal method. In order to improve adhesion and the mechanical properties of the deposited films, the parts are plasma-treated by a glowing process prior to the actual coating step as well as preheated.

The common optical evaporation materials like Ti_3O_5 and SiO_2 are normally evaporated under a gas inlet (oxygen). A parallel operation of two evaporation sources is possible without problems.

15.2.5 Versions (state of the art)

Improvement of film quality by raising energy of vapor particles:

- PVD-IAD (ion-assisted physical vapor deposition),
- PVD-IP (ion plating),
- PVD-APS (advanced plasma source),
- PVD-magnetron sputtering,
- PVD-ion beam sputtering.

15.2.6 Data for electron-beam evaporation (Tables 15.1–15.7)

Table 15.1 Coating materials.

Material	Application
Metals (Al, Ag, Cu, etc.)	Mirrors
Oxides (TiO_2, HfO_2, Ta_2O_5, SiO_2, etc.)	Interference filter, antireflection films
Fluorides (YbF_3, MgF_2, etc.)	Interference filter, antireflection films
Suboxides (CrO_x, SiO_x, etc.)	Adhesion/absorbing layers

Table 15.2 Antireflection system.

Reflection	Wavelength range	Remarks
$\leq$0.5%	420–680 nm	Per side

Table 15.3 Interference filter.

Maximum transmission	Blocking	Full-width-at-half-maximum
80%	0.01%	10–20 nm

Table 15.4 Typical operation parameters.

Parameter	Minimum	Maximum
Evaporation rate	0.05 nm/s	10 nm/s (metals)
Residual gas pressure	5×10^{-6} mbar	10^{-8} mbar
Reactive gas pressure	5×10^{-5} mbar	3×10^{-4} mbar

Table 15.5 Machinable dimension (in dependence on system diameter).

Geometrical dimension	Minimum	Maximum
Calotte	Ø ~ 100 mm	Ø ~ 2000 mm
Palette (plan)	Ø ~ 100 mm	Ø ~ 2000 mm
Planetary system	Ø ~ 100 mm (calotte)	Ø ~ 700 mm (calotte)

Table 15.6 Process measurement technique.

Measured variable	Method of measurement
In-line	
Deposition rate, film thickness	Quartz crystal frequency measurement
Transmission, reflection	Spectral photometry
Temperature	Thermocouple, pyrometer
Off-line	
Transmission, reflection	Spectral photometry
Polarization	Polarizer
Color	Spectral photometry
Mechanical/chemical properties	Various
Conductivity	Four-point probe

Table 15.7 Limits of the technology.

Parameter	Limiting value	Reason for limitation
Thickness uniformity	±0.5%	Distribution of vapor type of substrate holder
Minimum thickness	1 nm	Metrology
Maximum thickness	100 μm	Film stress

15.3 Ion-Assisted Deposition (IAD)

U. Brauneck

15.3.1 Basic assessment of the technology

The technology essentially comprises the generation of atoms (or molecules) in a source that is heated resistively or by an electron beam. These atoms are transported to the substrate through a high vacuum. The atoms then condense on the substrate and film growth occurs during concurrent bombardment with energetic ions accelerated from an ion gun.

15.3.2 Intended purpose of the technology

The technology has been developed for the production of temperature-stable and dense coatings with excellent climatic resistance.

15.3.3 The technology's typical features

IAD includes a densification of a coating during the growth process by ion bombardment.

15.3.4 Description of process

15.3.4.1 Operating resources

- The material is evaporated by an electron beam or evaporation boat (usually metal-oxide or metal).

- An ion source or plasma source is used to produce reaction gas ions and/or noble gas ions. In some versions of the process, the evaporated material is partly ionized (plasma-assisted deposition and ion-plating) in the vapor phase and can therefore direct kinetic energy towards the substrates via a bias voltage. In all versions, the film is bombarded by reactive gas ions and noble gas ions directly. These ions transfer momentum to the surface atoms and improve the mobility of these atoms. Free positions in the growing atom layers can so be filled, and loosely bound particles may be sputtered away. Voids and subsequent shadowing effects (typical of conventional evaporation) are avoided.

15.3.4.2 Mode of operation

The mode of operation is described in Fig. 15.4.

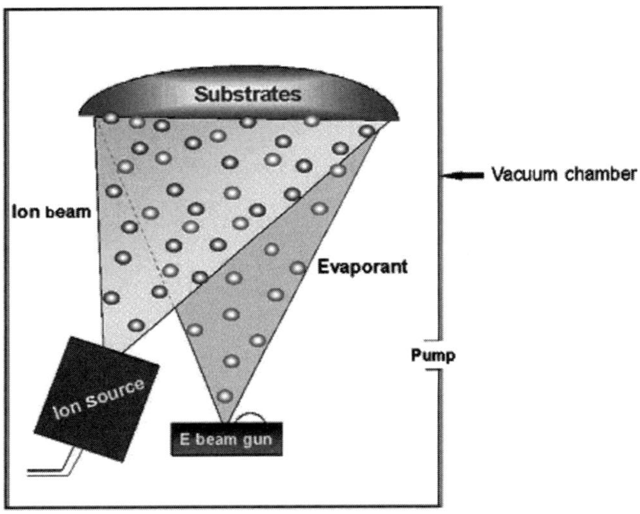

Figure 15.4 Sketch of principle of ion-assisted evaporation.

15.3.5 Versions (state of the art)

- IAD (ion-assisted deposition),
- Plasma-IAP,
- Ion plating.

15.3.6 Data for ion-assisted deposition (Tables 15.8–15.14)

Table 15.8 Coating materials.

Material	Application
SiO_2	Vis/NIR coatings
TiO_2	Vis/NIR coatings
Ta_2O_5	Vis (+partly UV) coatings
Nb_2O_5	Vis (+partly UV) coatings
HfO_2, ZrO_2	UV + Vis coatings

Table 15.9 Antireflection system.

Reflection	Wavelength range
Per side 0.5%	420–680 nm

Table 15.10 Interference filter.

Maximum transmission	Blocking	Full-width-at-half-maximum
80%	0.01%	10–20 nm

Table 15.11 Typical operation parameters.

Parameter	Minimum	Maximum
Evaporation rate	0.1 nm/s	>2 nm/s
Residual gas pressure	5×10^{-6} mbar	10^{-8} mbar
Reactive gas pressure	8×10^{-5} mbar	2×10^{-4} mbar

Table 15.12 Machinable dimensions (in dependence on system diameter).

Kind of substrate holder	Maximum geometrical dimension of filter	Maximum geometrical dimension of antireflection system
Calotte	~120 mm	~400 mm
Plate	~400 mm	>1000 mm

Table 15.13 Process measurement technique.

Measured variable	Method of measurement
In-line	
Transmission, reflection	Monochromator, spectral photometer
Ellipsometric angles	Ellipsometer
Off-line	
Transmission, reflection, absorption, and derived values	Photometry
Color (+derived values, temperature, R_a, and so on)	Photometry
State of polarization	Polarizer, analyzer
Laser damage threshold	Laser damage setup

Table 15.14 Limits of the technology.

Parameter	Limiting value	Reason for limitation
Thickness homogeneity	±0.5%	Form of evaporation "cloud"
IAD of fluorides	e.g., R_{vis}<0.2%	Decomposition of fluoride-component

15.3.7 Links

Further reading:

- http://www.uni-stuttgart.de/izfm/lehre/DunS_Auf.pdf

Plant engineering and construction:

- http://www.leyboldoptics.com
- http://www.unaxis.com
- http://www.dentonvacuum.com
- http://www.satis-vacuum.com

Toolmakers and suppliers of auxiliary materials:

- http://www.merck.de
- http://www.gfe-online.de
- http://www.umicore.com

Research and development:

- http://www.iof.fhg.de

15.4 Ion Plating (IP) Deposition

H.K. Pulker, A. Wälti

15.4.1 Basic assessment of the technology

The technology involves generation and ionization of atoms (or molecules) in a source that is heated by an electron beam and bombarded by a plasma beam accelerated from a low-voltage, high-current electron source. The ionized atoms are transported to the substrate through a low-pressure plasma environment. Condensation of the atoms on the substrate and film growth occurs in a reactive gaseous environment, with concurrent bombardment with energetic ions accelerated from the ionized vapor source.

15.4.2 Intended purpose of the technology

The technology has been developed for the technology production of dense, amorphous, abrasion-resistant, and environmentally stable films with excellent optical and functional properties.

15.4.3 The technology's typical features

A low-voltage, high-current electron source operating in a low-pressure argon atmosphere produces a dense plasma by electron impact into the coating material vapor cloud above an anodic crucible and into an introduced reactive gas. Depositing accelerated positive ions of the coating material, the working and reactive gas, generated by the repulsive force of the anode and a self bias potential of the insulated substrates, form well adherent, dense, stoichiometric, and generally constrained amorphous and smooth films on large-area substrates. Film properties are controlled by gas composition, pressure, and electrical data in a batch-coating system.

15.4.4 Description of process

15.4.4.1 Operation resources

- Vacuum box coater system of various chamber and substrate holder sizes (BAP-800 substrate holder, 800 mm);
- Different rotating substrate holder systems (palette, calotte, planetary);
- E-beam evaporation sources;

- Substrate heating device;
- *In situ* quartz crystal (rate) and optical (thickness) monitoring;
- Integrated fully automated control system; and
- Low-voltage, high-current electron source operated in argon with an e-beam crucible as the anode.

15.4.4.2 Mode of operation

The reactive IP process is performed in a specially designed automatic box-type plating system, the BALZERS BAP-800, shown schematically in Fig. 15.5. It can be used for deposition of single-layer and multilayer oxide coatings onto unheated glass and other unheated substrates. The evaporations are made by two special 270 deg deflection-type electron-beam evaporators. The starting materials, metals or suboxides, form electrically conducting melts. Very effective ionization and activation of the evaporating coating material atoms and the admitted reactive gas molecules results mainly from a low-voltage (typically 60 V), high-current (typically 40–50 A) electron beam of a non-self-sustaining thermionic arc discharge plasma directed to the anodic crucible of the e-beam evaporators. The cathode of the non-self-sustaining arc discharge is a resistance heated Ta filament in a separate chamber kept under a relatively high argon pressure. The substrate holder is electrically insulated. In contact with the formed plasma cloud, the substrates receive a relatively high negative self-biasing potential of -15 to -25 V with respect to the plasma, which together with the repulsive forces of the positive ions from the anodic crucible (potential typically between 70 and 40 eV) mainly determines the impact energy of the film-forming ions.

The total pressure in the plant is in the low 10^{-3} mbar range. Film deposition is started and stopped by opening or closing moveable shutters in front of the electron

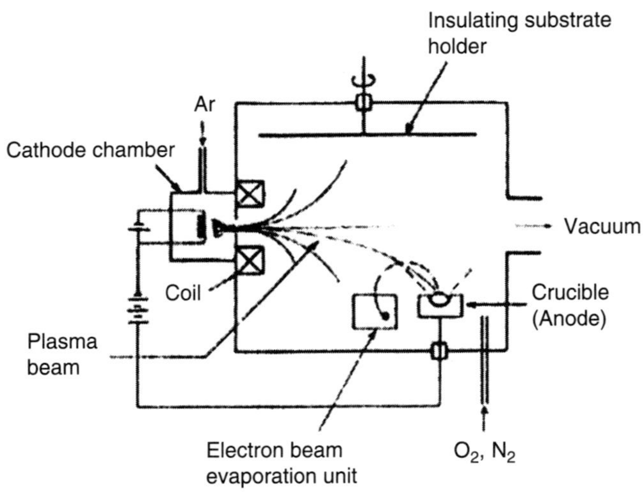

Figure 15.5 Operating principle of IP.

beam evaporators. Film thickness and deposition rate are controlled by optical thickness monitoring and oscillating quartz crystal monitoring.

Quarter-wave films in the visible are deposited within 2–5 min. A uniformity in thickness distribution of <1% over the whole substrate holder of 800 mm can be achieved with a simple static correction shield.

15.4.5 Data for ion plating deposition (Tables 15.15–15.22)

Table 15.15 Coating materials.

Material	Application
SiO_2, Ta_2O_5	AR coatings
SiO_2, Ta_2O_5, HfO_2, Al_2O_3	Interference filters 0.193–5.0 μm
SiO_xN_y, Si_3N_4, TiN	Decorative coatings
Ag, Al, Au, Ta, Nb, Ni, Si	Mirrors, absorbing layers
TiN, Ni	Antistress layers

Table 15.16 Versions (state of the art).

Coating machine	Substrate holder diameter
Balzers BAP	800 mm
Balzers BAP	1050 mm
Balzers BAV-1250	1250 mm
Evatec BAP-800	800 mm

Table 15.17 Antireflection system.

Reflection	Wavelength range
<0.2%	0.2–3 μm

Table 15.18 Interference filter.

Maximum transmission	Blocking	Full-width-at-half-maximum
>90%	<0.1%	1 nm

Table 15.19 Typical operation parameters.

Parameter	Minimum	Maximum
Process temperature	40 °C	280 °C
Residual gas pressure	2×10^{-4} mbar	1×10^{-3} mbar
Reactive gas pressure	8×10^{-4} mbar	1×10^{-3} mbar

Table 15.20 Machinable dimensions (in dependence on system diameter).

	Minimum	Maximum
Geometrical dimension	800 mm	1200 mm

Table 15.21 Process measurement technique.

Measured variable	Method of measurement
In-line	
Deposited mass	Quartz crystal frequency measurement
Transmittance, reflectance	Spectrophotometry
Temperature	Thermocouple
Off-line	
Conductivity	Four-point probe
Mechanical and chemical properties	Various
Transmittance, reflectance	Spectrophotometry

Table 15.22 Limits of the technology.

Parameter	Limiting value	Reason for limitation
Minimum thickness	1 nm	Metrology
Maximum thickness	100 μm	Coating stress
Homogeneity (m^2)	0.5%	Evaporation plume
Homogeneity (cm^2)	0.01%	Evaporation plume

15.4.6 Links

For further information see

- http://www.evatecnet.com

For further reading, plant engineering and construction, toolmakers and suppliers of auxiliary materials, and research and development.

15.5 Advanced Plasma Source (APS)

H. Hagedorn

15.5.1 Basic assessment of the technology

The technology involves generation of atoms (or molecules) in a source that is heated resistively or by an electron beam. The atoms are transported to the substrate through a low-pressure plasma environment. The ionized atoms condense on the substrate, and film growth occurs in a reactive gaseous environment, with concurrent bombardment with energetic ions accelerated from an advanced plasma source (APS).

15.5.2 Intended purpose of the technology

The technology is relevant for creating dense, shift-free coatings with superb layer qualities for optical and functional coatings.

15.5.3 The technology's typical features

A high-output plasma source delivers an additional plasma bombardment in an evaporation process. On large substrate areas, the film properties can be enhanced in terms of adhesion, density, morphology, and stoichiometry. The source concept allows an easy and long-term stable control of film properties in a batch coater system.

15.5.4 Description of process

15.5.4.1 Operating resources

- Vacuum coating system of various chamber sizes (diam. 700, 900, 1100, 1500 mm);
- Different substrate holder/rotation systems (palette, calotte, planetary, high-speed);
- E-beam or thermal boat evaporation;
- Additional substrate heating;
- Indirect or direct *in situ* monitoring system of thickness and optical properties;
- Integrated, fully automated control system (SYRUSpro); and
- Advanced plasma source.

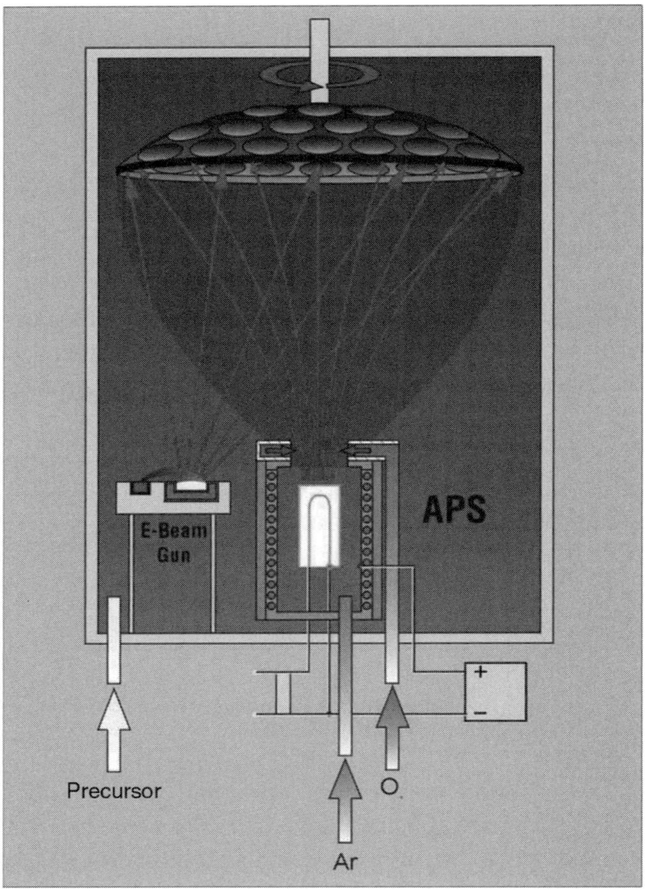

Figure 15.6 Operating principle of APS.

15.5.4.2 Mode of operation

A heated LaB_6 cathode, centered in the APS, emits an extremely high number of electrons (Fig. 15.6). An axial magnetic field hinders the electrons from reaching the water-cooled anode. A high-density plasma is created by using a rare gas and applying an appropriate DC voltage between anode and cathode. The plasma is extracted into the chamber and impinges the substrates. The source is on a floating potential. The voltage between source and ground is measured and controlled. The so-called bias voltage determines the ion energy and can be varied between 30 and 200 eV. Ion current densities between 300 and 600 $\mu A/cm^2$ can be achieved on the hole substrate area. Additional reactive gases and monomers can be used to change the film properties. For thin-film deposition, thermal or electron evaporation are used.

15.5.5 Data for advanced plasma source (Tables 15.23–15.30)

Table 15.23 Coating materials.

Material	Application
Oxides (SiO_2, TiO_2, Al_2O_3, Ta_2O_5...)	Interference filter: 0.193–8 μm
Suboxides (CrO_x, SiO_x)	Adhesion, absorbing layers
TCO (ITO)	Cold process, plastic compatible
Nitrides, oxynitrides (TiN, Si_3N_4, SiO_xN_y)	Decorative, mixed index
Fluorides (YbF_{3-})	Interference filter: 0.4–12 μm
Metals (Al, Ag, Mo)	Mirrors
Monomers	PECVD (hydrophobic, scratch protection)

Table 15.24 Versions available as state of the art.

APS I	1992–2000; scratch-resistant AR coatings for ophthalmic applications
APS II	1994–2003; temperature-stable band-pass filter
APS DWDM	1996–2002; narrow bandpass filter for telecom applications
APSpro	2003–; high-output version

Table 15.25 Antireflection system.

Reflection	Wavelength range (nm)
<0.15%	193–10,600

Table 15.26 Interference filter.

Maximum transmission	Blocking	Full-width-at-half-maximum
>96%	<0.1%	<1 nm

Table 15.27 Typical operation parameters.

Parameter	Minimum	Maximum
Process temperature	40 °C	380 °C
Residual gas pressure	8×10^{-5} mbar	8×10^{-4} mbar
Reactive gas pressure	1×10^{-4} mbar	4×10^{-4} mbar

Table 15.28 Machinable dimensions (in dependence on system diameter).

Geometrical dimension	Minimum	Maximum
Diameter	700	1500

Table 15.29 Process measurement technique.

Measured variable	Method of measurement
In-line	
Deposited mass	Quartz crystal frequency measurement
Transmittance, reflectance	Spectrophotometry
Temperature	Pyrometer, thermocouple
Off-line	
Conductivity	Four-point probe
Mechanical and chemical properties	Various
Transmittance, reflectance	Spectrophotometry

Table 15.30 Limits of the technology.

Parameter	Limiting value	Reason for limitation
Minimum thickness	1 nm	Metrology
Maximum thickness	100 μm	Coating stress
Homogeneity (m^2)	0.5%	Evaporation plume
Homogeneity (cm^2)	0.01%	Evaporation plume

15.5.6 Link

Further information:

- http://www.leyboldoptics.com

15.6 Magnetron Sputtering

B. Szyszka

15.6.1 Basic assessment of the technology

Atoms are generated due to sputtering of a metallic or compound target plate, which is the cathode of a glow discharge process. These atoms are transported to the substrate through a low-pressure plasma environment. Mostly neutral atoms condensate on the substrate. Film growth is in a reactive gaseous environment under concurrent bombardment with energetic species from the plasma.

15.6.2 Intended purpose of the technology

Magnetron sputtering was developed in the early 1970s. The first applications were the metallization of polymer parts within the automotive industry, the metallization of wafers for microelectronics, and the large area deposition of energy-efficient coatings on architectural glass. There are numerous further applications, major examples of which are coatings on polymeric web and display glass as well as coatings for data storage, such as CDs and hard disk drives.

15.6.3 The technology's typical features

Because the coating material is passed into the vapor phase by a mechanical (momentum exchange) rather than a chemical or thermal process, virtually any material is a candidate for coating. Films containing almost every solid element in the periodic table have been prepared by sputtering. Alloys and compounds can be sputter deposited while preserving their stoichiometry [1].

Magnetron sputtering is a production-proven, high-precision, and high-rate coating technology based on simple and rugged components. It is a unique technology for large-area coatings with a substrate dimension up to 18 m^2. For the optical industry, magnetron sputtering allows cost-effective, small-batch processes where only a few pieces have to be generated.

15.6.4 Description of process

15.6.4.1 Operating resources

A vacuum coating system, either a batch coater (footprint $\approx 1\ m^2$) or an in-line coater (footprint $\approx 5 \times 1\ m^2$, including PECVD hard coat) equipped with:

- Substrate holder (palette of planetary),
- Magnetron sputter sources,
- *In situ* process control techniques,
- Fully automated control system, and
- Facility for substrate cleaning and pretreatment.

15.6.4.2 Mode of operation

Sputtering is a process whereby material is dislodged and ejected from the surface of a solid target material as a result of momentum exchange associated with energetic particle impact. The impact of particles with energy of several 100 eV gives rise to a collision cascade in the target material, which results in the emission of a certain amount of target material at energies of a few eV, mostly as neutral atoms.

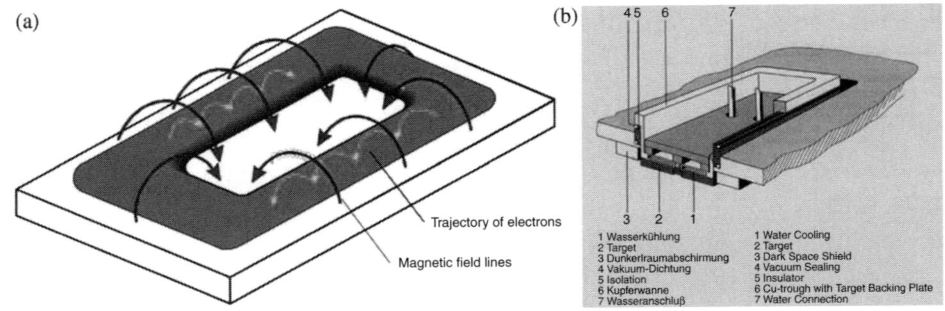

Figure 15.7 (a) The secondary electrons emitted from the target surface are trapped by the magnetic field due to the Lorentz force. The result is a plasma torus in front of the target. (b) Cross section of a planar magnetron cathode.

Glow discharge sputter deposition is a vacuum-based coating technology where the target operates as the cathode of the glow discharge. The sputtered atoms reach the substrates after several collisions in the gas phase. At the substrate, these atoms are deposited, and reactions with other gas atoms may occur. The result is the formation of either a metallic or a compound film.

The magnetron is a specially designed cathode of such a discharge device. It uses the effect of magnetic trapping of the electrons in order to confine the plasma close to the cathode. Due to the raised plasma density, a much higher ion current and deposition rate are possible. Furthermore, the pressure can be decreased, which gives rise to improved film properties because less scattering occurs in the gas phase. Also, the thermal load to the substrate is decreased. Magnetron sputtering therefore allows for coating on temperature-sensitive polymer substrates. These effects are achieved due to the use of a magnetic array behind the target plate, as shown in Fig. 15.7.

Compound film such as SiO_2 and Nb_2O_5 are usually deposited using reactive sputtering, where a pure target material (e.g., Si) is sputtered in a reactive atmosphere of Ar and O_2. At an appropriate oxygen partial pressure, stoichiometric SiO_2 films can be obtained at the substrate. Pulsed plasma excitation, either for single or dual magnetron sputtering, is performed in order to achieve sufficient process stability. Dielectric layers are deposited not only on the substrate but also on the target. On the target surface, these layers form a capacitor, which is charged by the impinging ion flux. For DC plasma excitation, a critical breakdown field strength is reached after a period of time. An arc discharge occurs that damages the substrate as well as the target. Using pulse power plasma excitation, the potential is either switched off or even reversed at frequencies of several tens of kilohertz. This allows for a neutralization of the surface charges prior to arc ignition.

15.6.4.3 Coating materials

Sputter processes are able to deposit virtually all materials with excellent layer properties, including adhesion, hardness, density, stability, absorption, stoichiometry, and phase composition. There is one exception, related to fluoride films, which cannot be deposited, to date, by sputter deposition. The reason for this is their strong tendency for dissociation, which gives rise to metal-rich fluorides when sputtered from a fluoride target. Even the addition of fluorine containing hydrocarbons does not seem to help. MgF_2 films with n below 1.38 have been deposited, but these films exhibit very poor adhesion.

A compilation of material data taken from [2] for sputtered films is shown in Table 15.31.

Table 15.31 Material data.

Material	Useful transmission range [nm]	Mid-range refractive index	References
Sputtered oxide optical coating materials			
SiO_2	200–8,000	1.47 @ 550 nm	Coleman 1974 [3] Pawlewicz 1982 [4] Pawlewicz 1986 [5]
SiO		1.9 @ 670 nm	Coleman 1974 [3]
GeO_2	280–8,000	1.60	Pawlewicz 1982 [4]
TiO_2	380–8,000	2.1–2.7	Coleman 1974 [3] Pawlewicz [4, 5] Kienel 1981 [6]
ZrO_2	240–8,000	2.05 @ 550 nm	Coleman 1974 [3] Pawlewicz 1982, 1986 [4, 5]
ZrO_2:CaO	250–8,000	2.10	Pawlewicz 1982 [4]
HfO_2	220–8,000	2.00	Pawlewicz 1982 [4]
HfO_2:Y_2O_3	220–8,000	2.00	Pawlewicz 1982 [4]
Ta_2O_5	350–8,000	2.06 @ 550 nm	Pawlewicz 1982, 1986 [4, 5] Kienel 1981 [6]
Nb_2O_5	320–8,000	2.20	Pawlewicz 1982, 1986 [4, 5]
Y_2O_3	220–8,000	1.90	Pawlewicz 1982 [4]
La_2O_3	250–8,000	1.76	Pawlewicz 1982 [4]
MgO	225–8,000	1.69	Pawlewicz 1982 [4]
Al_2O_3	<200–8,000	1.63 @ 550 nm	Pawlewicz 1982, 1986 [4, 5]
$Y_4Al_2O_9$	220–8,000	1.76	Pawlewicz 1980 [7]
$MgAl_2O_4$	225–8,000	1.61	Pawlewicz 1980 [7]
Sputtered nitride optical coating material			
Si_3N_4	250–9,000	1.95	Pawlewicz 1982, 1986 [4, 5]
Ge_3N_4	800–9,000	2.11	Pawlewicz 1982 [4]
BN	250–9,000	1.67	Pawlewicz 1982 [4]
AlN	220–	1.94	Pawlewicz 1982, 1986 [4, 5]
Sputtered semiconductor optical coating materials			
Si	1,000–9,000	3.60	Pawlewicz 1982 [4]
Si:H	700–9,000	2.0–3.6	Pawlewicz 1982 [4]
Ge	1500–9,000+	4.08	Pawlewicz 1982 [4]
Ge:H	1200–11,000+	2.88–4.08	Pawlewicz 1982 [4]
CdS	600–11,000+	2.52	Pawlewicz 1982 [4]
CdTe	900–11,000+	2.70	Pawlewicz 1982 [4]
ZnSe	500–20,000+	2.42	Pawlewicz 1982 [4]

15.6.5 Versions (state of the art)

Coaters can be designed either as in-line coaters or as batch coaters. For in-line coaters, a cycle time of about 60–100 s can be achieved for the deposition of the layer stack

CR39/Hard Coat (approx. 3.5 μm) by PECVD/4 Layer AR by MS/

Top Coat (by PECVD)

on four lenses on one side. Nb_2O_5 is used as high-index material and SiO_2 as low-index material. Both materials are deposited by reactive AC magnetron sputtering using double magnetrons. The deposition of hard coat and top coat by PECVD is a part of the process line [8, 9].

For batch coaters, the time cycle is of the order of 720 s for a four-layer AR stack on four lenses for coating on both sides. In that case, the hard coat has to be applied prior to the AR coating. The high-index and low-index films are deposited using the same Si target due to switching of the process gas from Ar/N_2 to Ar/O_2 mixture [10].

Other layouts utilize a coating drum or a rotary platter on which the lenses are mounted. They can be equipped similarly to an in-line coater with magnetrons for reactive sputtering, but they allow also for another mode of operation where thin metal films are deposited that are oxidized in the highly reactive plasma of an oxygen ion beam [11] or oxygen radical source [12].

Another emerging technology is a result of innovations in the field of sputter target manufacturing. Cost-effective and conductive ceramic targets suited for DC sputtering such as Nb_2O_{5-x} or TiO_{2-x} can be made by sintering. This approach allows for more robust deposition processes, because the complex dependencies of deposition rate and stoichiometry on target state (either metallic mode or compound mode, where reaction products have formed on the target surface) put a strong demand on process control techniques. For conventional reactive sputtering, the closed-loop, *in situ* control of reactive gas partial pressure is important for achieving sufficient reproducibility of rate and film properties. For ceramic target sputtering, there is no need for fast, closed-loop control, because only very little additional reactive gas is necessary for the formation of stoichiometric films. For many applications, these processes can be operated at a given power and time without need for further *in situ* control.

15.6.6 Data for magnetron sputtering (Tables 15.32–15.36)

Table 15.32 Antireflection system.

Reflection	Wavelength range
<0.2%	0.2–3 μm

Table 15.33 Interference filter.

Maximum transmission	Blocking	Full-width-at-half-maximum
>95%	<0.2%	<4 nm

Table 15.34 Typical operation parameters.

Parameter	Minimum	Maximum
Process temperature	40 °C	380 °C
Residual gas pressure	Typical 10^{-6} mbar	
Reactive gas pressure	Typical 10–100 mPa	
Deposition rate	0.1–10 nm/s	

Table 15.35 Machinable dimension.

Geometrical dimension	Minimum	Maximum
Diameter of optics	1 mm	5 m

Table 15.36 Process measurement techniques.

Measured variable	Method of measurement
In-line	
Oxygen partial pressure	λ-probe
Process gas partial pressure	Mass spectroscopy
Target state	Optical emission spectroscopy
Target state	Voltage and current; FFT analysis of voltage and current
Transmittance and reflectance	Spectroscopic photometry, flat substrates
Ellipsometry	Spectroscopic ellipsometry, flat substrates
Off-line	
Conductivity	Four-point probe
Structural, mechanical, and chemical properties	Various
Transmittance, reflectance, ellipsometry	Spectroscopic photometry and ellipsometry

15.6.7 Conclusions

Magnetron sputtering can be considered as a cost-effective production technology, in particular for mini batches where activated evaporation is not cost-effective. It is a precise and flexible technology, but ion beam sputtering performs even better if the small rate can be accepted. It is a unique technology if large substrate areas $\gg 1\ m^2$ have to be considered.

A strong impact on production processes for optical films will be realized due to emerging milestone technologies such as the high-power pulse magnetron sputtering as an ionized PVD technique. The modeling of vacuum and plasma processes is emerging as a detailed quantitative process model of magnetron sputter processes. This approach provides a powerful tool for developing robust processes and realizing customer-specific coatings.

15.6.8 Further reading

1. J.A. Thornton, "Recent advances in sputter deposition," *Surface Engineering*, Vol. 2, No. 4, pp. 283–292 (1986).
2. R. Herrmann and G. Bräuer, "DC and RF-Magnetron Sputtering," in *Handbook of Optical Properties I*, R.E. Hummel, K.H. Guenther (eds), CRC Press, Florida, pp. 135–187 (1994).
3. W.J. Coleman, "Evolution of optical thin films by sputtering," *Applied Optics*, Vol. 13, No. 4, pp. 946–951 (1974).
4. W.T. Pawlewicz, P.M. Martin, D.D. Hays, I.B. Mann, "Recent developments in reactively sputtered optical thin films," *SPIE* **325**, *Optical Thin Films*, pp. 105–116 (1982).
5. W.T. Pawlewicz, P.M. Martin, R.W. Knoll, I.B. Mann, "Multilayer optical coating fabrication by DC magnetron reactive sputtering," *SPIE* **678**, *Optical Thin Films*, Vol. II (1986).
6. G. Kienel, "Optical layers produced by sputtering," *Thin Solid Films*, Vol. 77, pp. 213–224 (1981).
7. W.T. Pawlewicz, D.D. Hays, P.M. Martin, "High band gap oxide optical coatings for 0.25 and 1.06 μm fusion lasers," *Thin Solid Films*, Vol. 73, pp. 169–175 (1980).
8. R. Beckmann, G. Deppisch, H. Hagedorn, H.-U. Hermann, T. Naumann, J. Pistner, "Beschichten transparenter Kunststoffe mit Kratzschutz und AR in einer kombinierten InLine-Anlage," *Vakuum in Forschung und Praxis*, Vol. 1, pp. 9–15 (2002).
9. http://www.leybold-optics.com/
10. http://www.satis-vacuum.ch/site/productsset.html
11. http://www.jdsu.com
12. http://www.shincron.co.jp/test/en/technical/index.html

15.7 Ion Beam Sputtering

D. Ristau

15.7.1 Basic assessment of the technology

Atoms (or molecules) are generated by sputtering from a target with energetic ions from a separate ion gun. These atoms are transported to the substrate through a high vacuum. The atoms are then condensed on the substrate to lead to film growth.

15.7.2 Intended purpose of the technology

Initially, ion beam sputtering (IBS) was developed for the coating of high-quality mirrors for laser gyroscopes, which are dependent on extremely low backscatter values below 10 ppm. In the course of further development of IBS, coatings with an outstanding optical quality and high laser-induced damage thresholds could be achieved.

15.7.3 The technology's typical features

IBS is a high-energy coating process where the material is sputtered by a separate ion gun from a target. The typical energy of the molecules deposited on the growing layer is in the range 5–20 eV. As a consequence, the layer structure of IBS is nearly ideal and can be considered as the best coating structure known today in optical thin-film technology.

15.7.4 Description of process

15.7.4.1 Operating resources

Three major components are necessary to implement an IBS process:

- An ion gun for the generation of energetic noble gas ions with defined energies in the range between 0.5 and 2 keV. Modern ion sources for IBS comprise a discharge chamber with an *Rf*-excited plasma. The ions are extracted from the plasma by a multigrid system and subsequently neutralized with an additional electron source in order to achieve a confined ion beam.
- A water-cooled target consisting of the deposition compound or the corresponding metal. The high-energy ions sputter the material from the target with a defined energy and spatial distribution. If compound targets are employed, an additional source of gas is necessary to compensate for the stoichiometric deficiency in the growing layer induced by the selective interaction of the ion beam with the different atomic classes.
- A rotating holder for the substrates, which are subjected to the flow of sputtered particles. In some systems, an additional flow of reactive or inert ions is directed onto the substrates to achieve an ideal stoichiometry and an improved microstructure.

The functional units are operated in a high-vacuum environment independently on the basis of only few parameters. Therefore, IBS is a very stable and reproducible process, which can be well controlled and guarantees low contamination rates of the growing layers. The substrates are subjected to low thermal loads and radiation doses.

15.7.4.2 Mode of operation

The operation principle of IBS is shown in Fig. 15.8. The ion source generates ions with a high kinetic energy in the range of 0.5–2 keV and a total current of around a few 100 mA. As a consequence of the ballistic interaction of the target surface with

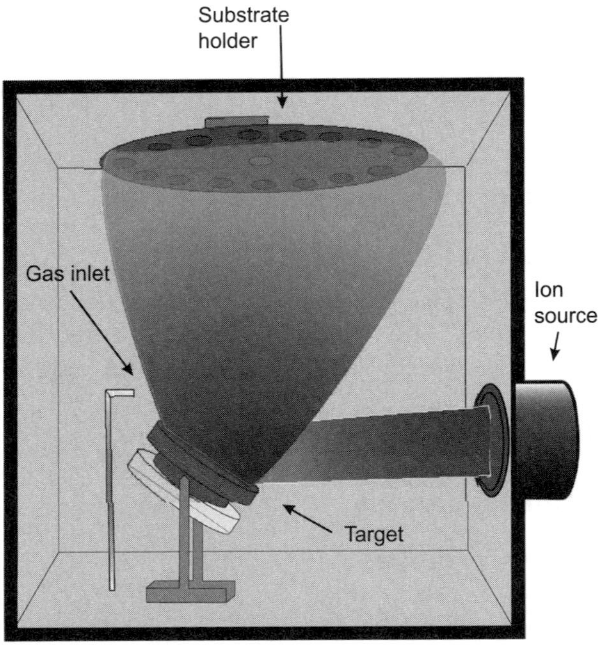

Figure 15.8 Operation principle of IBS.

incoming ions, deposition material is released into the vacuum environment in a specific spatial and energetic distribution. If the geometry of the process is adjusted correctly, most of the sputtered particles impinge onto the substrates forming the coating. The growing mechanism of the IBS coating is governed by the high energies of the particles (ranging between 5 and 20 eV), which allow for a dense and ideally amorphous microstructure of the deposited layers.

15.7.4.3 Coating materials used

The coating materials along with their applications are shown in Table 15.37.

Table 15.37 Coating materials.

Material		Applications
Metals	Al, Ni, Nb, Mo, Ge, Fe, Cu, Ag, Pd, Pt, Si, V, W, Zr	Optics, sensor technology, semiconductor technology, tribology, and so on
Oxides	Al_2O_3, HfO_2, Nb_2O_5, SiO_2, Ta_2O_5, TiO_2, Y_2O_3, ZrO_2	Optics, sensor technology
Fluorides	CaF, MgF_2, LiF, YF_3, LaF_3	Optics, sensor technology
Others	$YBa_2Cu_3O_{7-\delta}$, ITO, C, BN, TiC, DLC, SiO_xN_y, TiO_xN_y	Superconductivity, optics, tribology, display technology, and so on

15.7.5 Versions (state of the art)

Versions forming the state of the art are shown in Table 15.38.

Table 15.38 Versions forming state of the art.

For optical applications
• VEECO SPECTOR® ion-assisted ion beam sputtering systems; http://www.veeco.com/html/product_bymarket_proddetail.asp?ProductID=73 • Nordiko ion beam sputtering plants and ion sources; http://www.nordiko.com/html/optics.html • Shincron ion beam deposition and other deposition systems; http://www.shincron.co.jp/en/products/index.html • Oxford Instruments ion beam deposition systems; http://www.oxfordplasma.de/technols/ibs.htm • CCR Technology Copra Cube® ion beam sputtering system; http://www.ccrtechnology.de/pdf/CUBE.pdf • Roth und Rau UniLab® deposition systems; http://www.roth-rau.de/en/ionbeamsys_en_01.html

15.7.6 Data for ion beam sputtering (Tables 15.39–15.45)

All entries apply for a design wavelength of 1.064 nm for the coatings. Deposition material consumption of IBS is very low compared to conventional evaporation processes. The deposition time is directly dependent on the rate and the area to be coated.

Table 15.39 Antireflection system.

Reflection	Wavelength range
$<0.1\%$	0.2 to ~2 μm

Table 15.40 Interference filter.

Maximum transmission	Full-width-at-half-maximum
Loss <-0.5 dB	50 GHz HDWDM filter for telecommunication

http://www.veeco.com/html/product_bymarket_proddetail.asp?ProductID=73& MarketID=3

Table 15.41 High-reflecting system.

Reflection	Wavelength range
$>99.999\%$	0.5–1.8 μm

Table 15.42 Typical operation parameters.

Parameter	Typical value
Sputtering rate	0.1 nm/s
Residual gas pressure	$<10^{-7}$ mbar
Reactive gas pressure	10^{-4} to 10^{-3} mbar
Deposition temperature	Ambient to maximum of 350 °C

Table 15.43 Machinable dimensions (in dependence on system diameter).

Geometrical dimension	Minimum	Maximum
Diameter of optics	1 mm	500 mm

Table 15.44 Process measurement technique.

Measured variable	Method of measurement
In-line	
Deposited mass	Quartz crystal frequency measurement
Transmittance, reflectance	On-line spectrophotometry
Reactive gas composition	Mass spectrometer
Target surface emission spectra	On-line spectrophotometer
Substrate temperature	Thermocouple, pyrometer
Off-line	
Measurement according to ISO standards	

Table 15.45 Limits of the technology.

Parameter	Limiting value	Reason for limitation
Lateral homogeneity	0.1% on Ø 500 cm^2	Aperture of ion source low ion current
Deposition rate	Typically 0.1 nm/s	Low ion current
Minimum thickness	0.5 nm	Metrology
Maximum thickness	50 μm	Compressive stress

15.7.7 Conclusions

Ion beam sputtering can be considered as a reproducible high-end process for the production of optical coatings with outstanding quality and extreme complexity. The cost-efficiency of the process concepts is presently optimized to a level competitive with conventional coating techniques.

15.7.8 Further reading

D. Ristau, "Ion beam sputter coatings for laser technology," *Proceedings of Advances in Optical Thin Films*, SPIE 5963, pp. 13-1–13-12 (2005).

15.7.9 Links

Toolmakers and suppliers of auxiliary materials (only a selection):

- Targets
  A wide variety of target suppliers are active in the market. The following selection gives examples without any recommendation:

  http://www.cerac.com/
  http://www.fhr.biz/dienst_targets.php
  http://www.thinfilmproducts.umicore.com/
  http://metals.about.com/od/sputteringtargets/Sputtering_Target_Suppliers.htm

- Toolmakers
  See coating plant suppliers in Sec. 15.7.6.
- Research and development
  Major conferences:

  http://www.osa.org/meetings/topicalmeetings/OIC/
  http://www.svc.org/
  http://spie.org/boulder-damage.xml

  Advances in Optical Interference Coatings:
  http://www.leti.fr/
  http://www.lzh.de/

15.8 Plasma Impulse Chemical Vapor Deposition

S. Bauer

15.8.1 Basic assessment of the technology

Atoms (or molecules) are generated by pulsed plasma-assisted decomposition and the reaction of a chemical vapor precursor. These atoms are transported to the substrate through a low-pressure plasma environment. Atoms are condensed on the substrate, leading to film growth.

15.8.2 Intended purpose of the technology

Originally the technology was developed for coatings inside tubes. Currently, it is a major advantage of the technology that it enables inside coatings. The technology makes it possible to produce multilayer systems with a very quick change between layer materials. It is one of the most reliable CVD-based techniques for industrialized optical coatings and is currently used to produce millions of cold-light reflectors for general lighting, digital projection, and other applications. Moreover, the technology is used to produce scratch-resistant, antireflection, and decorative coatings on many different kinds of plastic substrates.

15.8.3 The technology's typical features

Coatings are deposited on preheated substrates in an evacuated chamber by decomposing chemical molecules (so called precursors) with the help of a microwave-induced plasma. This plasma is operated in pulsed mode (from several Hz to more than a kHz). This is why it is called a plasma impulse chemical vapor deposition (PICVD). The decomposed molecules, so-called radicals and ions, together with reactive gases like oxygen are deposited onto the substrates and form dense oxide layers.

15.8.4 Description of process

15.8.4.1 Operating resources

Utilizing PICVD technology requires at least the following resources (Fig. 15.9):

- A vacuum chamber (base vacuum $\leq 10^{-4}$ mbar) with a pumping unit for high gas throughput;
- A gas supply unit;

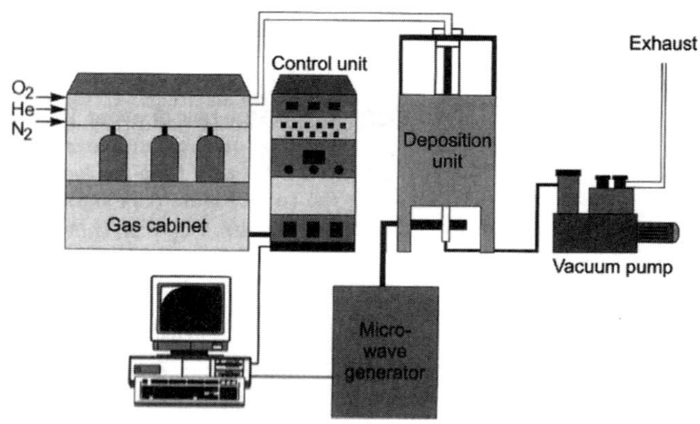

Figure 15.9 Components of a PICVD setup.

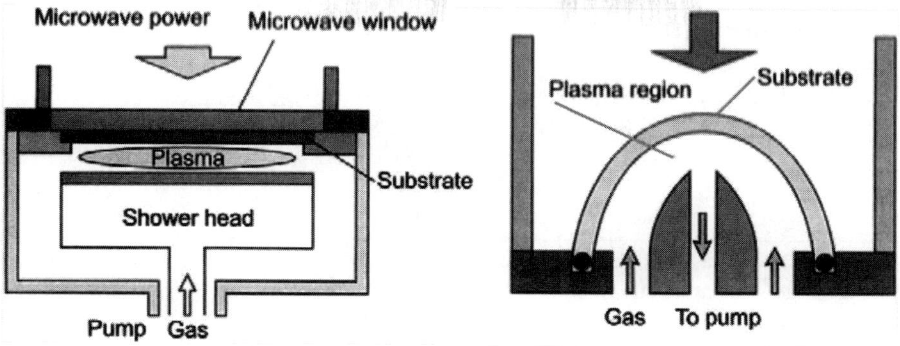

Figure 15.10 PICVD setups for planar and dome-shaped substrates.

- A microwave generator and a magnetron (usually operated at 2.45 GHz);
- A computer control unit for controlling valves, pumps, microwaves and process monitoring.

15.8.4.2 Mode of operation

A vacuum chamber is evacuated to a base vacuum of $\sim 10^{-4}$ mbar. Reactive chemicals (precursors) are transported into the deposition chamber by a carrier gas (usually oxygen or nitrogen). The precursors might be gases, liquids, or solids, which have to be evaporated in a specially designed gas cabinet, either by direct heating, evaporation, or bubbling. Mass-flow-controller units control the gas flows precisely. When the reaction chamber is filled with the gas mixture at a pressure in the mbar range, a pulsed plasma is ignited by applying microwave energy at a frequency of 2.45 GHz. According to the specific mode of operation, after the plasma pulse is switched off, the chamber is refilled with fresh gas and the next series of pulses is started. Every pulse generates a defined "film" thickness well below an atomic monolayer. Therefore, counting the number of pulses is in most cases sufficient for determining the desired layer thickness. According to Fig. 15.10, there are different setups for coating of planar substrates and three-dimensional embodiments.

15.8.4.3 Coating materials used

The coating materials along with their applications are shown in Table 15.46.

Table 15.46 Coating materials.

Material	Application
SiO_2	Optical/barrier
TiO_2	Optical/barrier
$SiO_xN_yC_z$	Scratch resistance, barrier
Nb_2O_5	Optical
Ta_2O_5	Optical

15.8.5 Versions (state of the art)

Versions forming the state of the art are shown in Table 15.47.

Table 15.47 Versions forming state of the art.

- For three-dimensional embodiments, usually a single-station concept, which is multiplied for large-scale production
- Batch reactors available for small ($<10 \times 10\,cm^2$) flat substrates
- Machine concept for inside coatings of container-like substrates
- Machine concept for double-sided coatings on lenses (especially for ophthalmic lenses) and flat optical substrates

15.8.6 Data for plasma impulse chemical vapor deposition (Tables 15.48–15.53)

Table 15.48 Antireflection system.

Reflection	Wavelength range
<1%	400–700 nm

Table 15.49 Interference filter.

Maximum transmission	Blocking	Full-width-at-half-maximum
>90%	>0.5%	Up to 4 nm

Table 15.50 Mirrors.

Maximum transmission	Wavelength range	Full-width-at-half-maximum
>95%	400–2000	Up to 2 nm

Table 15.51 Typical operation parameters.

Parameter	Minimum	Maximum
Deposition rate	Not limited (nm/min)	>1000 (nm/min)
Residual gas pressure	$>10^{-5}$ mbar	$<10^{-4}$ mbar
Reactive gas pressure	0.1 mbar	5 mbar

Table 15.52 Machinable dimensions (in dependence on system diameter).

Geometrical dimension	Minimum	Maximum
Flat substrates	Limited by sample holder	$<10 \times 30\,\text{cm}^2$
Container or container-like	>1 ml	$<1500\,\text{cm}^3$

Table 15.53 Process measurement technique.

Measured variable	Method of measurement
In-line	
In-line monitoring of layer stack	Optical monitoring
Monitoring of plasma emission	Optical monitoring
Gas pressure	Pressure transducer
Mass flows	Mass flow controller
Substrate temperature	Pyrometer
Off-line	
Transmission/reflection/absorption	Spectrometer
Thickness distribution	Scanning reflectometer
Refractive index	Ellipsometry
Composition of layers	SIMS, XPS, other surface analytics

Table 15.54 Limits of the technology.

Parameter	Limiting value	Reason for limitation
Thickness uniformity on large areas $>300\,\text{cm}^2$	Microwave applicator	Wavelength of microwave
No metal or conductive coatings	Microwave absorption	Availability of precursors and intrinsic shielding of microwave
Only oxide coatings	Availability of precursors	Precursors, feasibility

15.8.7 Status

Usually, layer thicknesses are determined by counting pulses with a known "rate" per pulse. For special filter applications, commercial optical on-line monitoring tools as well as self-assembled tools are used to control layer thickness. Plasma emission at specific wavelengths is controlled to ensure that a deposition of the desired materials has taken place (this is especially important for critical coatings, such as for pharmaceutical applications).

Commercial reflection and transmission spectrometers, as well as an ellipsometer, are used to measure layer thickness distributions and to determine n and k values of the films. Thin-film analytics contributes to a deeper understanding of the layer composition.

15.8.8 Conclusions

PICVD is a versatile chemical vapor deposition process for optical and functional coatings, yielding high-quality oxide coatings with very low absorption on glass and glass ceramics, as well as on many kinds of plastic substrates. The requirement for a vacuum in the range of only a few millibars reduces equipment costs and processing times. The resulting films have a high density and therefore excellent environmental stability. The PICVD process results in very high deposition rates up to more than 1000 nm/min. The pulsed mode reduces thermal load of the substrates to a minimum, which makes it suitable for coatings on temperature-sensitive substrates, such as plastics.

15.8.9 Further reading

1. H. Bach and D. Krause, *Thin Films on Glass*, Springer, Berlin, Heidelberg (1997).

15.8.10 Link

Schott AG: http://www.schott.com

Chapter 16

Assembly

16.1 Assembly of Spherical Lenses (Consumer Optics)

A. Bell

16.1.1 Basic assessment of the technology

All kinds of lenses assembled to doublets or triplets for optics consumers use this technology. The single components, lenses, and mechanical housings are manufactured using standard processes. The optical components can be directly assembled into housings without additional adjusting.

16.1.1.1 Assembly process

To correct color aberrations, most optical systems also need cemented optical elements, for example, for objectives, eyepieces, and erector systems. Cemented components are doublets, made of two singlets, or triplets, made of doublets and singlets.

16.1.2 Intended purpose of the technology

Assembly begins after manufacturing of singlets:

- Two singlets are cemented together to doublets.
- The single lenses and the doublets are mounted into mechanical housings (subassembling).
- The subcomponents are assembled into objective housings.

The optical cement has to meet a number of constraints: its refractive index should fairly match the refractive indices of both glasses used. It must not be absorptive, it must be thin (~10 μm), resistant to climatic conditions, and should not introduce any mechanical stress, which would cause gradients of the refractive index.

16.1.3 The technology's typical features

Typical steps include

- Alignment of the optical axes of two lenses to build the cemented doublets;
- Alignment of the optical axes of three lenses to build the cemented triplets.

In general, high quantity = lower cost.

The standard objective specifications allow all optical components, manufactured by standard procedures, to be directly assembled into the housings without additional adjusting steps.

For lenses, the following features apply:

- The cemented surfaces have to be polished thoroughly.
- The radius of the convex surface and of the concave surface of both single lenses to be cemented must be theoretically the same. Sometimes one modifies both radii slightly to keep the cement thickness constant across the diameter.
- The layer of the cement has to be even and thin.

For assembly, the following steps apply:

- Manufacturing of optical components and the mechanical housings in accordance with the tolerance specification.
- Assembly of single components into bodies in accordance with the tolerance specification.
- Assembly of bodies into housings without additional adjusting steps.

16.1.4 Description of process

16.1.4.1 Operating resources

- Considering the high requirements for cleanness of the cemented joint, the cementing is done in a flow-box (Fig. 16.1).
- The required equipment, tooling, and overhead materials are as follows:
  - *Equipment*: lamps, dosing device (ensures same size of cement drop), cementing centroscope with objective and a tie, table UV source of radiation to fix the lenses, inspection centroscope (Fig. 16.2), microscope, UV or electric hardening oven;
  - *Tooling*: cementing mount, centering mount;
  - *Overhead material*: cements (UV or two-component epoxy), spirit, ether, acetone, cotton on the stick, optical cloth (towel), and cork.

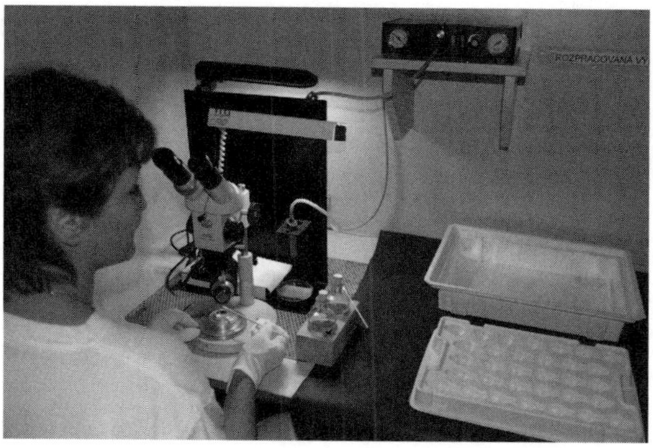

Figure 16.1 Flowbox with doser and cementing microscope.

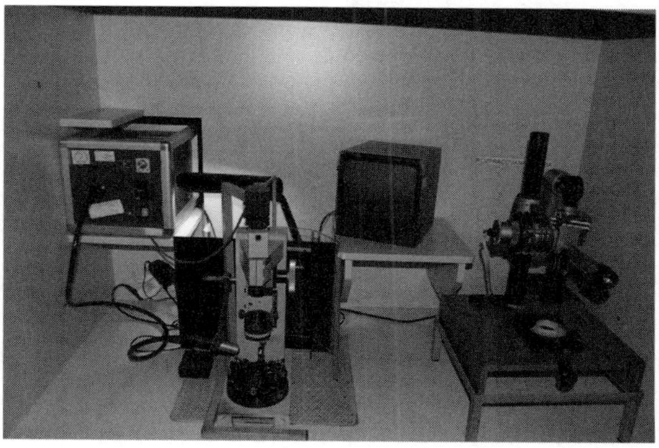

Figure 16.2 Workplace of cementing and measuring centroscope + UV source.

16.1.4.2 Mode of operation

Process steps

- Lens cleaning before coating;
- Pairing of lenses before cementing;
- Cementing
  - drop of defined amount of cement,
  - press to spread the cement and force out the air bubbles (inspection under the microscope);
- Align to optical axis with the help of cementing centroscope;
- Fixation of the achieved centered position;
- Cleaning of the edges (Fig. 16.3);
- Hardening of the cement;
- Inspection.

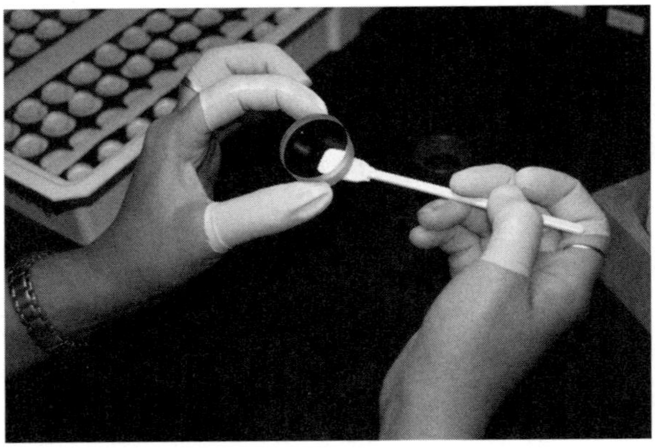

Figure 16.3 Manual cleaning of a lens.

16.1.5 Versions (state of the art)

- Cementing using UV cement;
- Cementing using mechanical fixers;
- Cementing using two-component epoxy cement;
- Cementing using microscope with monitor.

16.1.6 Data for assembly of spherical lenses (consumer optics)

16.1.6.1 Typical specifications reachable with this process

- MIL-PRF-13830B,
- MIL-STD-810 F,
- ISO 9022.

These are the most used standards to prove the quality and verify if the cemented parts are weatherproof.

16.1.6.2 Process measurement technique (Table 16.1)

During cementing, the cementing centroscope is used to unify the inclination of the optical and mechanical axis.

Table 16.1 Process measurement techniques.

Measured variable	Method of measurement
Axis	Centroscope
Centricity	Autocollimator
Cleanliness/air bubbles	Microscope/naked eye

For measuring, the measuring autocollimator is used, which measures the decentricity in a reflective way, or the transmittance centroscope, that is, a collimator in connection with a measuring microscope, which measures the decentricity after passing through the lens.

16.1.6.3 Cost drivers (Table 16.2)

The largest portion of the overall costs is the labor cost. The assembly time per doublets or triplets depends on the size and the shape.

Table 16.2 Cost drivers.

Cost factor	Volume
Material/glass type	10%
Labor cost	50%
Specification	10%
Inspection	5%
Machines/tooling	25%

16.1.6.4 Limits of the technology (Tables 16.3 and 16.4)

Table 16.3 Single lens tolerances.

Parameter	Limiting value	Reason for limitation
Radius	$\leq\lambda$	Machining
Form of surface	$\leq\lambda/10$	Machining
Center thickness of lens	± 0.02 mm	Process
Centering of lens	Decentric error 20″	Machining/radius

Table 16.4 Assembly tolerances.

Parameter	Limiting value	Reason for limitation
Single lens		
Tilting deviation $\Delta\varphi$	$\leq\pm 5'$	Method accuracy
Lateral Δr	$<\pm 50\,\mu\text{m}$	Method accuracy
Axial displacement Δz	$<\pm 50\,\mu\text{m}$	Method accuracy
Assembly group		Method accuracy
Tilting deviation $\Delta\varphi$	$\leq\pm 30''$	Method accuracy
Lateral Δr	$<\pm 50\,\mu\text{m}$	Method accuracy
Axial displacement Δz	$<\pm 50\,\mu\text{m}$	Method accuracy

16.1.7 Conclusions

Consumer optics, with good design, is only good enough with coatings to improve the light transmission. Each glass surface needs to be antireflective coated, which ensures increased light transmission of >99.4% per glass surface. Uncoated lenses lose, on each glass surface, 4% transmission.

To create a cost-advantageous bonding process for doublets, especially for big series, can create great savings on labor costs if the handwork is replaced with robotic supported cementing.

16.1.8 Link

- http://www.meopta.com

16.2 Assembly of Spherical Lenses (HQ Optics)

C. Gunkel, T. Sure

16.2.1 Basic assessment of the technology

The technology comprises final assembly and alignment of optical systems using subgroups of mounted lenses, such as cemented doublets or triplets.

16.2.2 Intended purpose of the technology

The technology has been developed for the production of high-quality diffraction-limited objectives starting from optical components.

16.2.3 The technology's typical features

The assembly contains the following working steps:

1. Alignment of the optical axis of two (or three) lenses to build the cemented doublets (triplets).
2. Fixing of the optical components with the mechanical mount to build the subgroups.
3. Assembly and adjustment of the subgroups to build the objective.
4. A fine-tuning process to minimize
   (a) the monochromatic geometrical residual aberrations, and if necessary,
   (b) the chromatic aberrations by adjusting of some specially assigned optical elements within in the objective.

Even the best technology available on the market does not have the potential to manufacture lenses and subgroups with tolerances tight enough to realize high-quality, high-numerical-aperture objectives (high-NA objectives); therefore, a fine-tuning and adjustment process is mandatory. In general, this must be done for all microscope objectives. The effort depends on the special requirements for the individual objective.

Based on the optic design, special algorithms for adjustment and fine tuning have been developed

- To minimize error propagation,
- To mark the critical subgroups that mostly influence the image quality,
- To define the linear region for optimization during the alignment process.

For the lenses, a typical deviation of the radius to its nominal value is about λ, surface form errors are about $\lambda/10$, and the surface roughness <1 nm rms. The deviation of the center of the radius vector (which defines the curved surfaces of the lenses) to the optical axis of the lens is $\leq 5\,\mu m$. Because some of the surfaces define the bearing surface for the next lens, the decentering errors will increase during the assembly of the various subgroups and be transferred to the objective. For objectives with high numerical aperture, the created overall decentering error will be too high to assure the required image quality. To reduce the decentering error, the centerline of the lens mount will be aligned to the optical axis of the lens. The realized accuracy is in the order of $2\,\mu m$.

16.2.4 Description of process

16.2.4.1 Operating resources

- Special machining tools for lens centering, cementing achromats, and gluing lenses into the mounts. Different kinds of interferometer setups working with dedicated software algorithms (built in house).

16.2.4.2 Layout

- Lens cleaning;
- Centering of lens surfaces, decenter error $\leq 5\,\mu m$ (deviation of surface radius vector to optical axis), optically controlled;
- Lens centering, optically controlled;
- Adjusting, centering of two lenses or more to build cemented achromats (doublets, triplets);
- Fixing lens into the mount
  - mechanical fixing, (O-ring, screwing ring),
  - gluing lens into the mount;
- Centering of subgroups, decenter error $\leq 2\,\mu m$ (deviation of optical axis to the centerline of the mechanical mount);

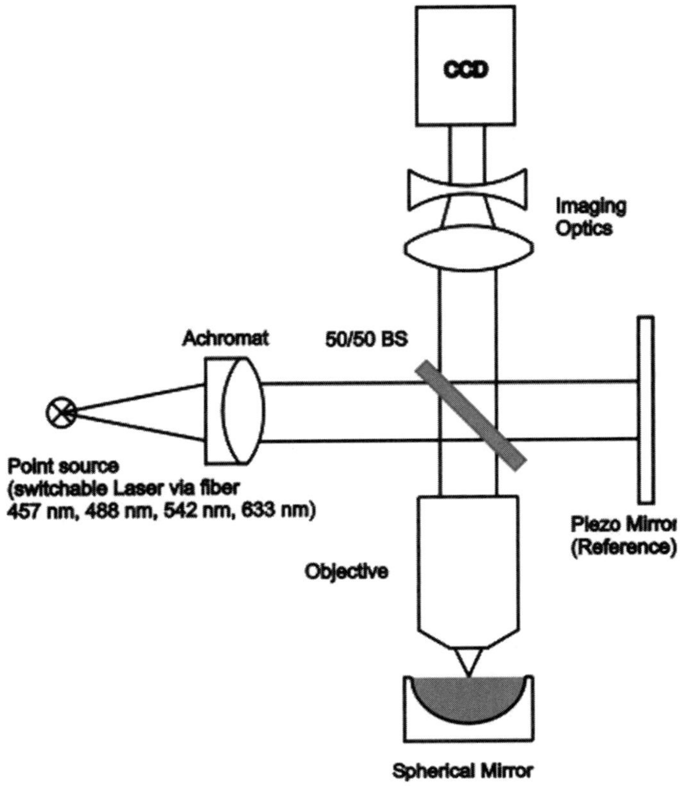

Figure 16.4 Typical arrangement of star-test equipment and interferometer.

- Assembly and adjusting of subgroups to build the objective;
- Fine-tuning process of complete objective (optical shop testing) regarding spherical aberration, coma, astigmatism, chromatic focal shift, with the help of a star-test setup and a quantitative interferometer (Fig. 16.4).

In general, different kinds of interferometer setups can be used to determine the wavefront aberration in the exit pupil of an optical system. A Twyman-Green interferometer is a common tool for evaluating microscope objectives. To determine the chromatic aberrations, a multiwavelength interferometer is available.

16.2.4.3 Mode of operation

Assembly of optical systems uses the subgroups without any additional adjustment procedure. It is typical for low-end optics like illumination systems, tube lenses, and so on, and is mostly done following steps 1 to 4 described in Sec. 16.2.3.

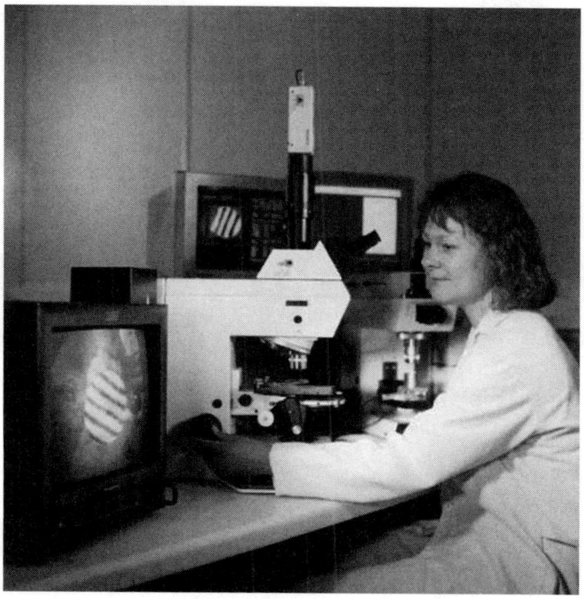

Figure 16.5 The principle setup of the Twyman–Green interferometer.

Optimization of monochromatic geometrical aberrations

A computer-controlled assembly of optical systems is used with a special fine-tuning procedure to minimize the monochromatically geometrical aberrations for on-axis, off-axis image points, typical for high-end optics, like microscope objectives with high numerical apertures. This is sufficient for optical systems where the tolerance band of the selected glasses, in particular the Abbe number, are tight enough to fulfill the requirements regarding chromatic aberrations like chromatic focal shift.

Optimization of chromatic and geometric aberrations

This is necessary for optical systems where even the selection of glasses with the tightest tolerances (1/1 glass) could not guarantee the required chromatic focal shift. For example, a $100\times$ oil-immersion objective with a numerical aperture of 1.4 used in a confocal microscope must have a chromatic focal shift less than 150 nm. Even if 1/1 glass is used, the possible maximum tolerances in one triplet will lead to a chromatic focal shift of 180 nm. Therefore, an adjustment and fine-tuning procedure to minimize the geometric as well as the chromatic aberrations, with the help of a multiwavelength interferometer, has to be carried out (Fig. 16.6).

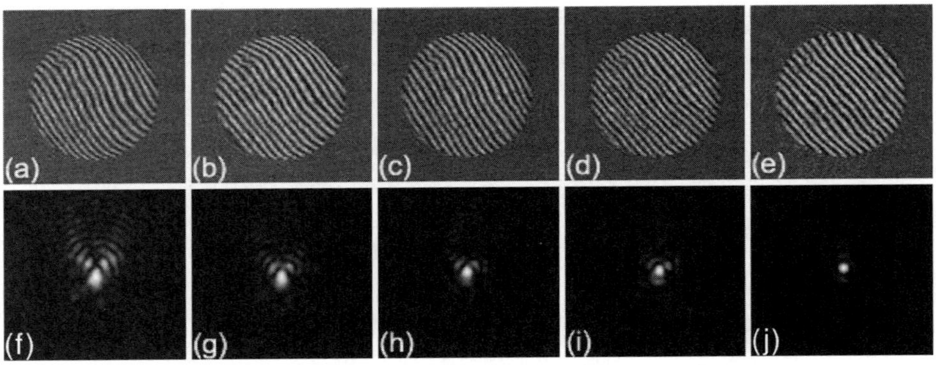

Figure 16.6 Interferograms (a) to (e) and the corresponding star-test pattern (f) to (j) during the fine-tuning process to minimize coma.

16.2.5 Data for assembly of spherical lenses (HQ Optics)

16.2.5.1 Typical specifications

The specification of the image quality of microscope objectives can be described by the Strehl-ratio, SR,

$$SR = \exp(-2\pi \text{rms})^2.$$

The rms value describes the wavefront deviation in the exit pupil of an objective from the ideal case, which will be a plane wave for infinity corrected optics or a perfect spherical wave with a defined radius of curvature for finite corrected optics. The wavefront deformation can be measured with an interferometer. To describe the measured wavefront, a set of 36 Zernike polynomials are fitted to the data. The coefficients, resulting from the fitting procedure, are used to calculate the rms value. For some special applications, it is not sufficient to specify only the overall Strehl-ratio; instead, the acceptable wavefront deformation has to be split regarding the symmetry to define the values for coma, astigmatism, and spherical aberration. The overall Strehl-ratio is then given by the product $SR = SR^*_{coma} SR^*_{ast} SR_{sph}$.

In 1949, Maréchal defined that objectives with SR > 80% will be called diffraction-limited. Today, the microscope objectives produced at Leica Microsystems have

- SR > 90% for serial production;
- SR > 95% for dedicated applications;
- SR > 98% for special customer requirements.

To explain the relevance of a change in the Strehl value of about 3%, we will take a look at an application where it is desired to measure the position of an edge by detecting the intensity of the line-spread function. The 50% value of the intensity of the line-spread function will define the edge position. Due to coma, the line-spread function will deviate from the ideal case. For a coma Strehl of 95%, this deviation will lead to a shift of the position of the 50% intensity value by about 65 nm. For applications in the semiconductor industry (where the position of structures of line width 135 nm, as for the Pentium 4 processor for example, has to be measured), this deviation could not be accepted. To increase the coma Strehl from 95% to more than 98% requires increasing the position accuracy of the shifting element, which is used to minimize coma aberration, from 1.5 μm to 300 nm. This means that increasing the Strehl values by about 3% requires enhancing the accuracy of the optical elements by about a factor of 5.

16.2.5.2 Process measurement technique (Table 16.5)

Table 16.5 Process measurement technique.

Measured variable	Method of measurement
In-line	
Lens decentering	Reflex image measurement
Image quality	Quantitative interferometry

16.2.5.3 Cost drivers (Table 16.6)

Table 16.6 Cost drivers.

Cost factor	Volume
Necessity of a repeatable fine tuning process to achieve the required image quality due to the residual errors of the single elements:	
1. Human resources	70%
2. Limits of the available technology	30%

16.2.5.4 Limits of the technology (Table 16.7)

Table 16.7 Limits of the technology.

Parameter	Limiting value	Reason for limitation
Radius deviation	$\leq\lambda$	Machining/measurement accuracy
Fine form error of lens surfaces	$\leq\lambda/10$	Machining/measurement accuracy
Center thickness of lenses/distance of air spaces	$\pm 1\ \mu m$	Machining/measurement accuracy
Surface roughness	≤ 1 nm rms	Machining/measurement accuracy
Centering of lens surfaces	Decenter error $\leq 5\ \mu m$	Machining/measurement accuracy
Centering of subgroups	Decenter error $\leq 2\ \mu m$	Machining/measurement accuracy
Position accuracy of shifting element	± 200 nm	Machining/measurement accuracy

16.2.6 Further reading

1. M. Born and E. Wolf, *Principle of Optics*, Pergamon Press (1993).
2. W.T. Wellford, *Aberrations of Optical Systems*, Adam Hiller Ltd, Bristol (1986).
3. D.W. Robinson and G.T. Reid, *Interferogram Analysis*, Institute of Physics Publishing (1993).

16.2.7 Links

- See Leica Microsystems homepage to download some publications
- http://www.oem-optics.com/website/sc_optics.nsf.

16.3 Assembly of Aspherical Lenses

C. Horneber, P. Karbe

16.3.1 Basic assessment of the technology

The described technology is utilized for the high-precision alignment of aspherical lenses. The procedure of positioning the lenses depends on the manufacturing process of the aspherical lenses.

16.3.2 Intended purpose of the technology

The precise alignment of an aspherical lens within the camera lens is particularly important, because its influence on the optical imaging quality is much higher than that of spherical lenses. Therefore, misalignment results in a much higher loss of imaging quality. This correlation gets even more important the larger the aperture of the camera lens is. Then the zones at the edge of the aspherical surface play an important role. These zones deviate greatly from the spherical shape and have a large influence on the optical imaging quality.

16.3.3 The technology's typical features

In contrast with spherical surfaces, an aspherical surface defines its own optical axis (Fig. 16.7). For this reason, one cannot always use techniques from spherical optics to adjust aspherical lenses. It is necessary to align the optical axis of the aspherical lens to the optical axis of the whole system. In principle, it is impossible to align these two axes only by rolling the aspherical lens on its banking surface. Errors occurring because of a wedge angle of the aspherical lens cannot be equalized at all. It is only possible to minimize their influence.

The procedure includes the following steps:

- Alignment of the aspherical lens;
- Fixture of the lens in its barrel;
- Assembly of the whole camera lens;
- Fine-adjustment by using shift elements.

16.3.4 Description of the process

The exact adjustment procedure depends on the manufacturing process and the geometry of the aspherical lens. A distinction is drawn between molded and polished aspherical lenses.

For an optimum alignment, it is required to avoid errors through wedge angles of the lens. The mechanical components have to be manufactured with high-precision barrels (e.g., h6/H6 according to DIN ISO 286 T1) and with minimum wedge angle (<1 arcmin).

16.3.4.1 Wedge error

The wedge error of an aspherical lens can be defined as in Fig. 16.7. The internal optical axis of the aspherical surface intersects the spherical surface in one well-defined point (Fig. 16.7, green point). There is an angle between a line defined by this point and the center of curvature of the spherical surface and the optical axis of the aspherical surface. This angle is called the wedge angle w. In contrast

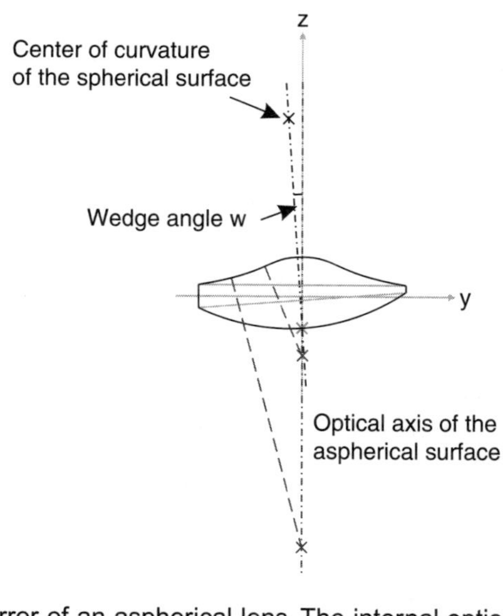

Figure 16.7 Wedge error of an aspherical lens. The internal optical axis of the aspherical surface intersects the spherical surface at the green point. The line from this point through the center curvature of the spherical surface defines the alignment of the sphere.

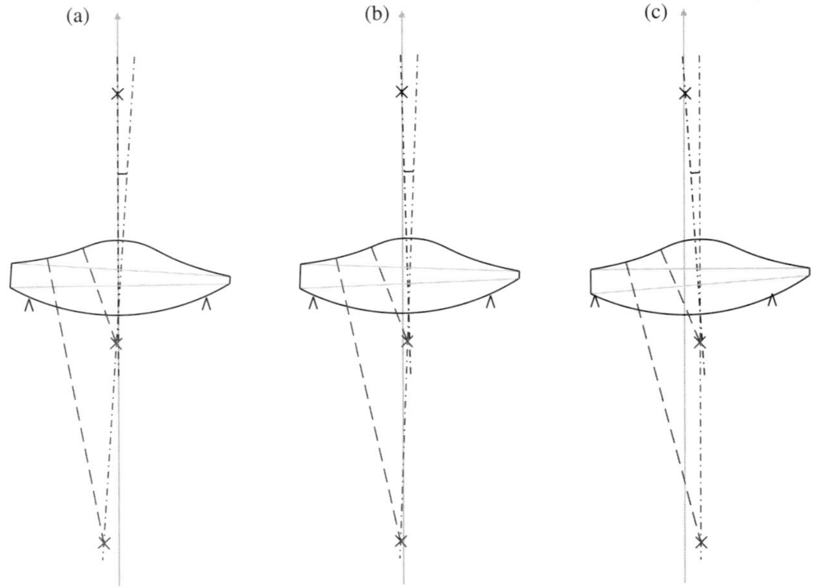

Figure 16.8 By rolling on the spherical surface, it is impossible to correct the misalignment of the wedge angle. (a) Aspherical lens is tilted; (b) the internal optical axis of the aspherical side is adjusted to obtain a mean tilting error. (c) Intern

to spherical surfaces, this error cannot be compensated by rolling the lens on its surface (Fig. 16.8). It is only possible to adjust the internal optical axis of the asphere parallel to the optical axis of the whole system. Then an offset of the aspherical lens perpendicular to its optical axis remains. For this reason, one has to pay attention especially to the given tolerances.

16.3.4.2 Molded aspheres

If the spherical side of an aspherical lens is quite flat, the lens can be simply supported on this side and then glued (Fig. 16.9). Thus, the aspherical lens is adjusted well within its tolerances. Unfortunately, the spherical side often shows a strong curvature (Fig. 16.10). Then the lens must be adjusted relative to its barrel before it can be glued. First, the exact position of the aspherical side has to be determined within the barrel. Then the aspherical side has to be adjusted by rolling the lens such that the internal optical axis of the aspherical side is parallel to the axis of the complete system. After gluing the aspherical lens, it is shifted in the lateral direction to receive the best adjustment for the lens. The shift between the optical axes is only a few micrometers. Typical tolerance values for the remaining wedge error are about 2 arcmin.

16.3.4.3 Polished aspherical lenses

For manufacturing aspherical lenses with the polishing method, a semifinished blank is used, whose spherical side is already tooled and polished. On this side, a

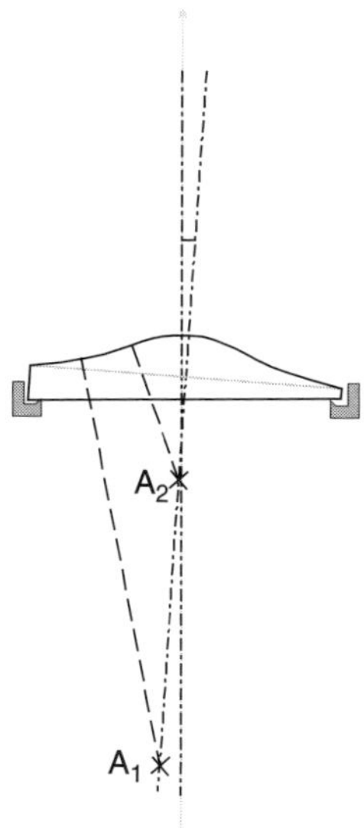

Figure 16.9 The wedge of the optical axis is not influenced by rolling if the aspherical lens has a only weakly curved spherical side.

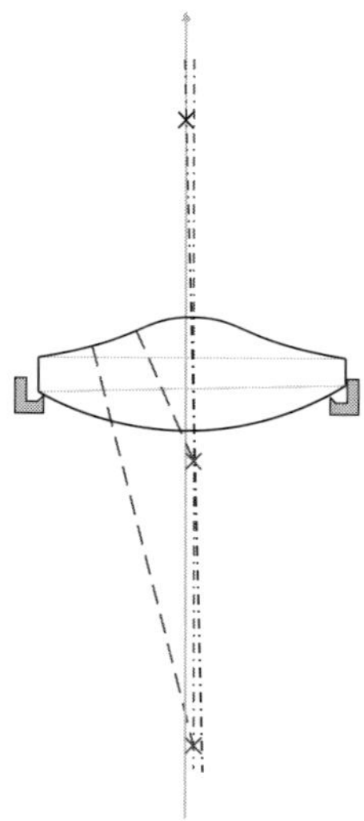

Figure 16.10 If the spherical side is curved more strongly, the internal optical axis of the asphere can be adjusted to the system. It remains a shift of the lens, which can be adjusted separately.

mandrel is glued, which is used for mounting in the molding and polishing process. This mounting ensures that the center of curvature of the spherical side is adjusted perfectly to the turning axis of the mandrel. Furthermore, the internal optical axis of the asphere is aligned with high precision to the turning axis of the mandrel and therefore to the center of curvature of the spherical side. We can achieve a very small wedge error with this method. This geometry is advantageous because the lens can be supported on its edge and is then automatically centered very well.

16.3.5 Versions (state of the art)

For the customer, the overall imaging quality of the complete system is paramount. The imaging quality can be described by the modulation transfer function (MTF) of the camera lens. As a rule of thumb, for acceptable tolerances we can say that the real MTF-values must not drop more then 10% below the theoretical values. To obtain the MTF, special measurement devices are used [1, 2].

For the evaluation of aspherical adjustment, different measuring methods are used. Deflectometers [3, 4] are suitable for measuring the tilt of the aspherical side down to a precision of 0.5 arcmin. They can be used for evaluating the given tolerances.

A further possibility for measuring imaging quality is the analysis of the point spread function. A point light source is imaged through the camera lens. From the image of this light source, one can draw conclusions with respect to different imaging errors (spherical aberration, coma, astigmatism). This method is called the star test [5].

For all these methods, it is important to measure the camera lens at different field angles. Only then is it possible to detect both tangential and sagittal imaging errors over the whole imaging field. If the measurement results differ at different angles, a centering error within the camera lens is suspected. However, it is impossible to distinguish with simple methods from which lens element or lens group the error results.

16.3.6 Data for assembly of aspherical lenses

16.3.6.1 Cost drivers

The largest contribution to the overall costs is manual work. The working time per lens strongly depends on the size and the shape of the asphere. It is roughly a few minutes. Automatization of the manufacturing process is possible but only cost-effective for mass production.

16.3.6.2 Limits of the technology (Table 16)

Parameter	Limiting factor	Reason
Radius deviation	$\leq 2\lambda$	Machine accuracy
Local irregularities	$\leq \lambda/5$	Machine accuracy
Center thickness	± 0.02 mm	Machine accuracy
Surface roughness	≤ 2 nm	Machine accuracy
Centering error of lenses	$\leq 5\ \mu$m	Machine accuracy
Centering error of lens groups	≤ 0.01 mm	Machine accuracy
Positioning shift element	$\pm 1\ \mu$m	Measuring accuracy

16.3.7 Conclusions

There is still a strong need for fast and high-precision measuring technologies for positioning aspherical lenses. To date, it has indeed been possible to adjust the lenses very precisely. However, for cost and time consumption reasons, it has been impossible to perform 100% process control.

16.3.8 Further reading

1. L. Baker (Ed.), "Selected papers on optical transfer function: measurement," *SPIE Milestone Series*, Vol. MS 60, SPIE Optical Engineering Press (1992).
2. B. Harand (Leica Camera AG), S. Rothe (OEG GmbH), "Universalobjektivprüfplatz für Foto- und Videoobjektive," *Photonik*, Vol. 3, S.54–57, AT-Fachverlag GmbH, Stuttgart (1999).
3. M. Knauer, J. Kaminski, and G. Häusler, "Phase Measuring Deflectometry: A New Approach to Measure Specular Free-Form Surfaces," in *Optical Metrology in Production Engineering*, W. Osten and M. Takeda (ed), Proc. SPIE, Strasbourg, France, Vol. 5457, pp. 366–376 (2004).
4. http://www.optik.uni-erlangen.de/osmin/research/papers.php
5. D. Malacara (Ed.), *Optical Shop Testing*, John Wiley & Sons, Inc. (1991).

16.4 Micro-Assembly TRIMO

B. Braunecker, L. Stauffer, A. Würsch

16.4.1 Basic assessment of the technology

New production technologies had to be developed to assure the required operating precision and stability within the preset cost limits. We describe in more detail a robot-driven mounting technology called TRIMO (three-dimensional miniaturized optical surface mounted devices). This is a new, flexible, and automated assembly technique for small optical components (maximum diameter for a lens is 2 mm) based on a laser reflow soldering technique. This technology can be compared to the electronic SMD technique, but applied to micro-optical devices.

16.4.2 Intended purpose of the technology

A significant part of the work in developing micro-optics devices is directed at solving the packaging aspect. The optical design very often requires an exact positioning of the optical components along the six degrees of freedom. Numerous techniques are now proposed for aligning (passively or actively) and fixing the optical components in the aligned position for long-term stability [1].

16.4.3 The technology's typical features

16.4.3.1 Standard holders

The interface between the optical element and the mounting plate is a standard metal holder (Fig. 16.11), produced by embossing a 0.1-mm invar sheet and consisting of a 2.6-mm-diameter round cup and two vertical arms, 2.5-mm long.

The holder cup is covered with a tin preform in preparation for the soldering process. Independently of the optical element, the holder shape and dimensions remain, which ensures a high repeatability of the assembly parameters.

Adaptation to the optics to be mounted is achieved with a steel submount, which is dimensioned and machined to fit into the holder and to accommodate the optical element. The submount is laser welded with the standard holder.

Some components such as lenses have to be inserted into the submount, limiting their size to a maximum of 2 mm in diameter. Mirrors and beamsplitters are flat cemented or soldered on the submount; therefore, their dimensions can be larger, for instance 4.5 mm (length) × 2 mm (height) × 1 mm (thickness). Because the optic parts are actively aligned, the optical elements can be premounted in the submount within the low tolerance of 100 μm. The mechanical parts thus have standard machining tolerances, which significantly reduce the cost of the final product.

16.4.3.2 Mounting plate

The mounting plate is typically a 1.5–2.5-mm-thick glass plate covered with a metallic grid-like film. During the reflow soldering process, a high-power diode laser beam is shot from below the plate and heats up the plate and the tin preform.

An adequate combination of plate material, grid metal composition, grid geometry surface, and preform solder alloy must be found for the laser reflow

Figure 16.11 Standard metal holder with a ball lens (left), a beam splitter (center) and a replicated lens (right).

process. The goal is to heat up the plate and the preform to a temperature slightly above the solder alloy fusion temperature [AuSn (80 : 20) melts at 280°C, SnAg (96 : 4) at 221°C, SnPbAg (62 : 36 : 2) at 179 °C].

The grid metal layer is thinner than 2 μm, and its dimensions are (75 × 75) μm^2 for the square openings, with a period of 130 μm. About 28% of the laser light is transmitted through a metallized 2-mm-thick sapphire plate.

During the initial TRIMO development work, the chosen plate material was sapphire. However, the high cost of the material and its machining difficulty limits its use. Nevertheless, due to its high tensile strength (300–400 MPa), sapphire is well adapted to peculiar applications such as soldering with the hard alloy AuSn (tensile strength 275 MPa). Substrates made of low-cost borosilicates or pyrex (tensile strength 69 MPa) can be used with other solder alloys such as SnPbAg (62 : 36 : 2) (tensile strength 45 MPa) or SnAg (96 : 4) (61 MPa). Borosilicate is easily available in large dimensions, and the sawing is easy.

16.4.3.3 Active alignment

The optical components mounted on the standard holder can be freely positioned on the whole surface of the mounting plate. They are aligned actively along six degrees of freedom and attached to the mounting plate in a "one laser reflow step" procedure. An assembly robot called "SIXTIFF," including a special mechanical gripper has been developed for the TRIMO application [3]. The robot linear resolution in x, y (horizontal), and z (vertical) is 0.25 μm, and the angular resolution is 1 μrad in θ_x and θ_y and 17 μrad in θ_z. The working volume is 100 × 100 × 50 mm^3 (Fig. 16.12). The robot is equipped with three trays for storing the parts. One special tray is dedicated to the module assembly. The parts are brought to a coarse position in a pick-and-place strategy.

Accurate alignment can be achieved because no mechanical contact exists between the holder and the plate before the attachment process (no stick & slip

Figure 16.12 View of the working volume of the assembly robot SIXTIFF.

effect). The optical elements are designed in such a way that a typical gap of 200 μm remains between the bottom of the tin preform and the plate. Depending on the height of the optical axis, as well as on the tolerances of the subcomponents, the gap can be reduced to almost zero or increased to 400 μm.

The fabrication tolerances on the elements are widely relaxed, because there is no restriction in the horizontal positioning of the parts and moderate restriction in the vertical placement. The solder allows the filling in a highly stable joint of gap up to 400 μm.

The gap is closed by the melted solder during the reflow process. A constant post-soldering shrinkage of about 7 μm has been measured for a 200-μm gap on a Pyrex glass plate. The shrinkage can be precompensated after the alignment.

The alignment concept of the optical components is already used in the early phase of any micro-optics module design. The miniaturization achieved with TRIMO requires a tight placement of the alignment sensors around the module. The typical sensors are cameras, position-sensitive detectors, wavefront sensors, radiometers, and so on. On the SIXTIFF robot, the installation of the alignment sensors close to the module is possible on the $x - y$ table, which has been widely extended outside of the working surface (Fig. 16.13). The sensors move simultaneously in x and y with the module, freeing the way to assembly process automation.

The technology TRIMO-SMD is particularly well suited for the fabrication of actively aligned, high-precision micro-optics modules. The global positioning errors can be kept within a small range independently of the numbers of components. Passive alignment with a large number of elements would require very tight subcomponent manufacturing tolerances in order to meet equivalent positioning performances (Fig. 16.14).

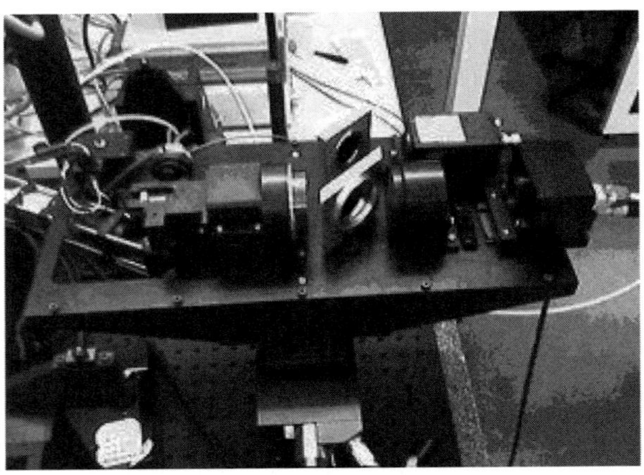

Figure 16.13 Alignment setup with a CCD camera and a wavefront sensor on the SIXTIFF for the assembly of a lidar module (on the left).

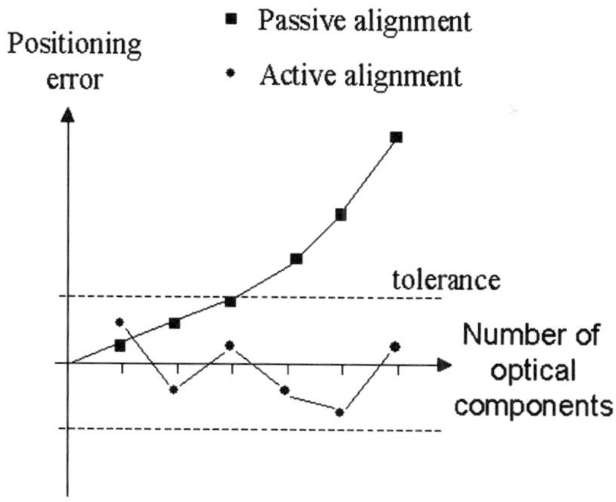

Figure 16.14 Schematic comparison of the expected positioning errors for active and passive alignments for a comparable number of components.

16.4.4 Description of process

The attachment procedure is based on a highly stable laser reflow soldering process between standard metal holders housing the optical components and a metallized transparent mounting plate. The active alignment in all six degrees of freedom of the optical holder is performed with a high-stiffness and high-resolution robot system. A schematic of the attachment principle is shown in Fig. 16.15.

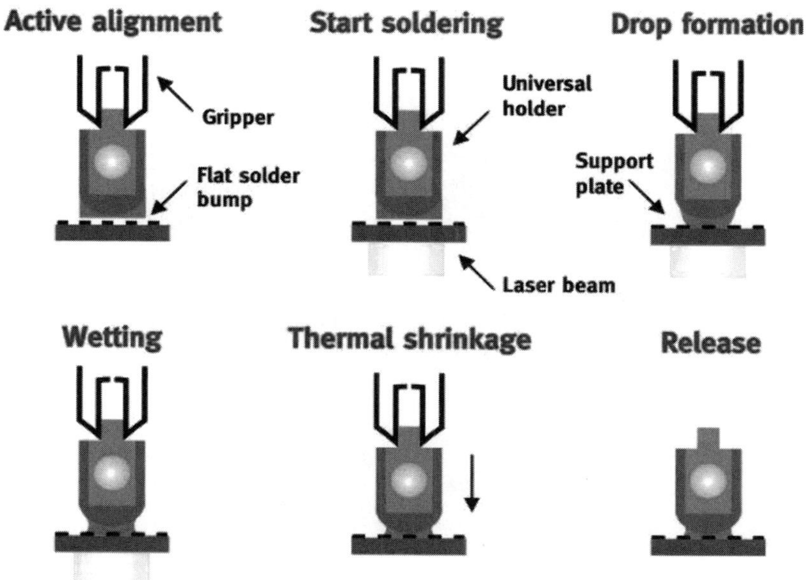

Figure 16.15 TRIMO-SMD attachment principle.

Because of the simple handling and high-precision properties, the new concepts allow both fast prototyping and mass production of micro-optical systems at low cost. High packaging density is achievable. The initial development of this technology has already been described [2].

16.4.5 Versions (state of the art)

16.4.5.1 Soldering accuracy

The precision of the active alignment depends first on the robot resolution and the sensitivity of the alignment sensors. However, it is the final position of the component reached after soldering that eventually matters.

The TRIMO attachment accuracy is specified as the difference between the measured position of the optical component still in the robot gripper and its final position after soldering on the mounting plate.

The lateral displacements in x and in z have been measured by capturing images of the holder head using a CCD camera. The rotations $\Delta\theta_x$ about x and $\Delta\theta_z$ about z were measured by capturing the images with a second CCD camera of a HeNe laser spot reflected on a mirror cemented on the optical submount. The test TRIMO holder was placed on a special tray, made of steel, and equipped with a large reference mirror and a reference metal arm (Fig. 16.16).

The images were processed off-line with a 2D least-square matching software program. We found with Pyrex substrates using a reinforced gripper, a repeatable soldering accuracy of 200 μrad (3σ) and 1 μm (3σ). With the Quartz plates, the results were 130 μrad (3σ) and 2 μm (3σ).

Figure 16.16 Soldering tray with reference arm and a reference mirror. On a metallized plate rests a soldered TRIMO test holder.

16.4.5.2 Joint stability

The TRIMO technology has been routinely applied for an industrial product by Leica-Geosystems in Heerbrugg for over two years. The TRIMO joints are controlled visually with a microscope on the side and from underneath the mounting plate. Moreover, the micro-optics modules are submitted to a temperature cycling up to 85 °C for 2 h for proofing the final alignment stability. Of more than 2000 solder joints, not one failed.

A quantitative measurement of the thermal stability has been performed on two demonstrators consisting of a 3-mm test mirror mounted on a TRIMO holder. The samples have been tested in autocollimation with a theodolite. A large reference mirror cemented on the mounting plate served as reference for deducting any drift of the test setup.

The parameters $\Delta\theta_x$ and $\Delta\theta_z$ have been measured at temperatures ranging between −42 °C and +63 °C. The tests showed that the measured angular deviations for both samples were within a tolerance circle of 100 μrad in radius (Fig. 16.17).

Due to the very light weight of the TRIMO components, a high resistance to vibration and shocks is expected. This could be demonstrated, for instance, with a very sensitive laser diode collimator equipped with a 60 μm cylindrical lens on a TRIMO mount. The stability of the collimation has been successfully tested after vibrations (10–500 Hz, 2 g, 2 h) and shocks (50 g, 3 ms, 3 ×, + and − directions).

16.4.6 Data for micro-assembly TRIMO

16.4.6.1 Cost drivers

The largest portion of the costs is related to the time required for active alignment. Position optimization with up to six degrees of freedom requires numerous iterations. Depending on the complexity of the alignment and whether the number of pieces to be assembled justifies it, the displacements of the robot can be fully automated.

The present design of the standard holder is quite complex. A reduction of the number of components would have a positive effect on the overall cost.

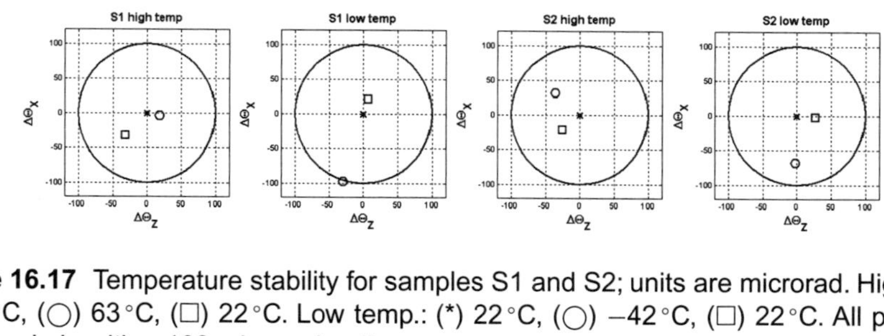

Figure 16.17 Temperature stability for samples S1 and S2; units are microrad. High temp.: (*) 22 °C, (○) 63 °C, (□) 22 °C. Low temp.: (*) 22 °C, (○) −42 °C, (□) 22 °C. All points are within a circle with a 100 microrad radius.

16.4.6.2 Limits of the technology (Table 16)

Parameter	Limits	Reason
Soldering accuracy (translation)	$\leq 2\,\mu m$ (3σ)	Soldering temperature and material
Soldering accuracy (rotation)	$\leq 200\,\mu rad$ (3σ)	Soldering temperature and material
Temperature stability (rotation)	$\leq 100\,\mu rad$ (3σ)	Material
Soldering time	Typically 2 s	Material thermal behavior
Size of micro-optics	≤ 2 mm	Standard holder design
Adjustment in x and y	Free	Size of the support glass plate
Adjustment in z	$\leq 400\,\mu m$	Soldering accuracy and solder volume

16.4.7 Conclusions

The optoelectronic packaging technology TRIMO-SMD combines different existing technologies (precision robotic assembly, soldering technologies, and glass wafer coating) to achieve miniaturization and flexibility in assembling micro optics. This flexibility allows us to go beyond what already exists, because this technology is not dedicated to one product but can be easily used the many optical assemblies without major changes by always using the same standard platform.

The TRIMO technology has made the jump from the feasibility demonstration to a real industrial product, and it is now ready to move to higher production volumes. Micro-optical components can now be assembled in a similar way to electronic SMD components.

The focus of future development will be to implement TRIMO in a larger palette of products and in other markets through licensing to other companies [4].

16.4.8 Further reading

1. A.R. Mickelson, N.R. Basavanhally, and Y.-C. Lee (Eds), *Optoelectronic Packaging*, Wiley (1997).
2. M. Scussat and A. Würsch, "An Innovative Flexible and Accurate Packaging Technique Suited to Fabricate Low Cost Micro Optoelectronic Modules," 50th ECTC Conference, Las Vegas, USA, May (2000).
3. A. Würsch, M. Scussat, "An Innovative Micro Optical Elements Assembly Robot Characterized by High Accuracy and Flexibility," 50th ECTC Conference, Las Vegas, USA, May (2000).
4. TRIMO web site: http://www.leica-geosystems.com/trimo-smd

16.5 CNC-Machined Monolithic Optics

B. Reiss

16.5.1 Basic assessment of the technology

Traditionally, an optical system is composed of several components, where each simple component serves a specific purpose. These components or distinct cemented subassemblies are mounted on/in a mechanical holder that afterwards is integrated into a higher system level. However, modern optical systems and instruments usually serve a variety of functions simultaneously. This results in ever more complex subsystems with ever greater numbers of single components contributing to increasingly critical requirements related to performance, precision, robustness, miniaturization, and also cost efficiency. The solution may lie in combining several optical and/or mechanical functions within one monolithic component. Thus, a improvement of quality, robustness, miniaturization, and cost efficiency at the same time will be achieved by the reduction of (Fig. 16.18; [1])

- The number of single parts within the system,
- The interfaces between mechanics and optics,
- The number of optically active surfaces,
- The number of cementing layers, and
- The adjustment work.

Besides molding technologies, which are primarily used for higher quantities and limited quality requirements, the combination of different computer numerical controlled (CNC) processing technologies opens up new ways of producing complex monolithic glass components. This CNC-machined monolithic optics does combine the intrinsic strengths of a monolithic design with the quality achieved in modern precision optics manufacturing of single refractive surfaces (Fig. 16.19).

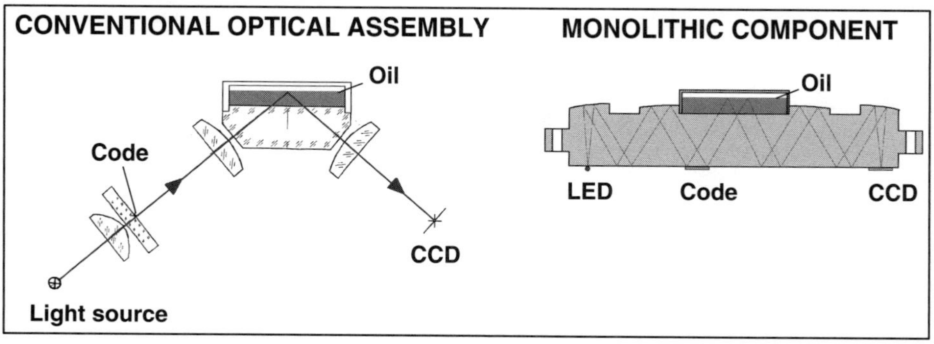

Figure 16.18 Tiltsensor: comparison of conventional optical assembly and monolithic component [2].

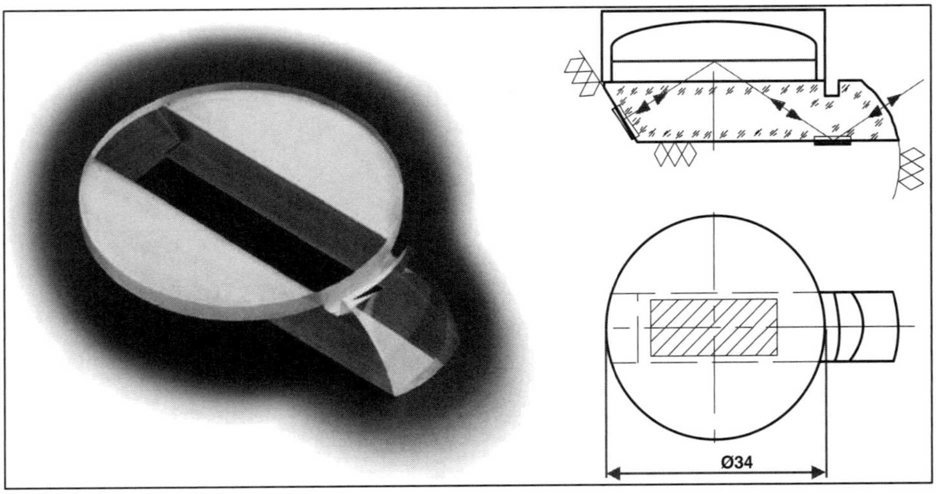

Figure 16.19 Tiltsensor: monolithic component: Integration of planar parallel plate, rhombic prism, and lens.

Furthermore, besides the different planar, spherical, aspherical, and/or free-form shaped optical surfaces, mechanical mounting and/or reference elements can be integrated within the same workpiece (Fig. 16.20).

The focus in this contribution is on the integration of several refractive surfaces and mechanical elements within one component, as this is the core contribution of CNC-machined monolithic optics. At SwissOptic AG, the possibilities of integration have been developed further by combination with etched as well as replicated diffractive surfaces and the application of defined partial filter coatings down to 8 μm.

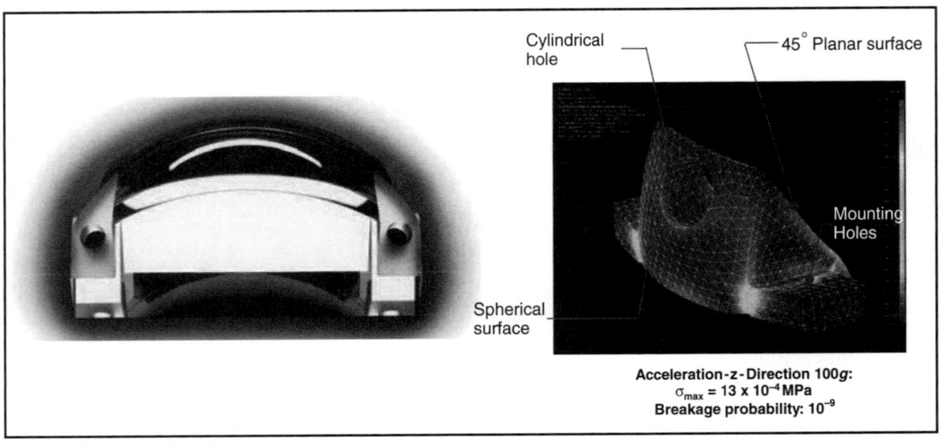

Figure 16.20 Monolithic component: integration of mechanical mounting and reference elements.

16.5.2 Description of process

Manufacturing CNC-machined monolithic optics is a combination of traditional and modern polishing processes with a central high-precision shaping step within the overall process chain. This central shaping step is basically equipped to measure the position and orientation of the semifinished product on the machine itself, according to already existing mechanical and/or optical reference surfaces. After measurement in the same setup, the shape of the remaining single surfaces is, in relation to the current workpiece coordinate system, ground into the semifinished product. Also, the fine-grinding of the optical surfaces is done in the same setup in such a way that, afterwards, only polishing is necessary (Fig. 16.21).

In this way, angular accuracies of up to 5″ and translational accuracies of up to 3μm can be achieved between the distinct surfaces. These are also demanding values for many traditionally mounted approaches. Regarding the accuracy and the quality of the single refractive surfaces, the specific polishing processes applied (CMP, CCP, FJP, IBF, and so on) are decisive and, consequently, there is no backlog in single surface quality compared with traditional systems. However, numerically controlled polishing technologies, like subaperture robotic polishing, are advantageous. They offer the possibility of defined local polishing and, when integrated into appropriate machine configurations like a six-axis robot, they provide the flexibility to polish the remaining surfaces in only one additional setup.

Because the combination of the different processing capabilities and the outlay of the process chain is defined by the specific product itself, it is not possible to give a standard process description for CNC-machined monolithic optics.

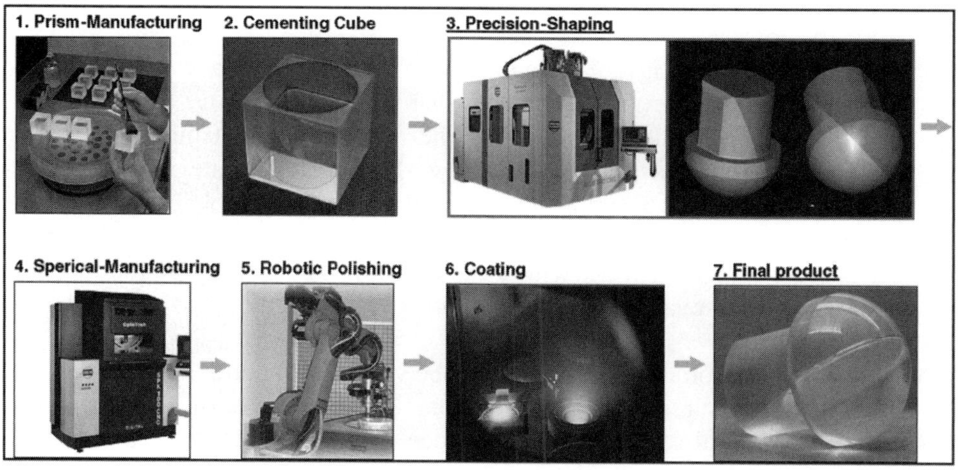

Figure 16.21 Example of process chain for CNC-machined monolithic optics.

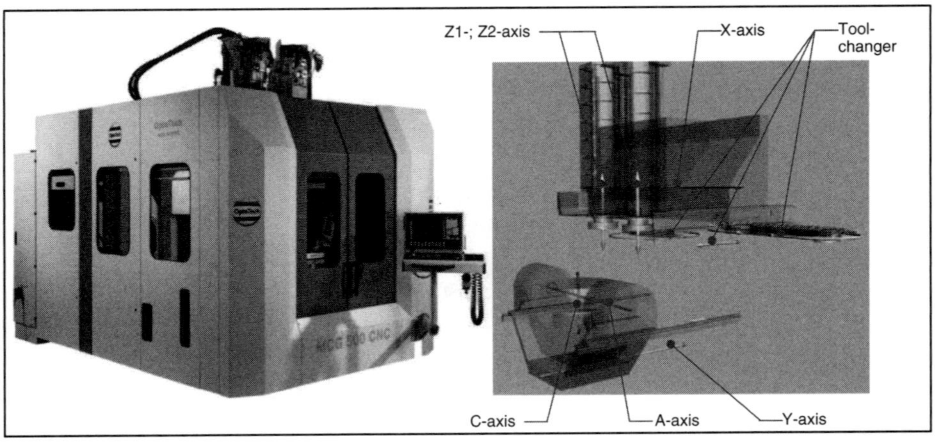

Figure 16.22 High-precision multicenter grinding system (courtesy of OPTOTECH GmbH, Germany).

However, the central precision shaping step is based on the following core elements (Fig. 16.22):

- The precision shaping is done on a five-axis CNC-machine (x-; y-; z-translational axis, A-; C-rotational axis). This opens up the possibility to calculate and conduct any coordinate system transformation according to the actual coordinate system of the component itself within the machining center. At the same time, modern torque-drives do offer combined drilling and simultaneous three-dimensional milling capabilities with an angle accuracy of approximately 2 arcseconds. By this means, it is possible to create flat, cubic, rotational (sphere; asphere), and/or free-form elements within one machine setup. In any case, the tool can be moved tangentially to the intended surface, thus eliminating existing tool-form errors and deviations through changing force vectors.
- By using an automatic tool-changer, several processing steps can be integrated within one machine setup, such as shaping, fine-grinding, and prepolishing. To eliminate the inaccuracy of the automatic toolchanger ($\leq 1\ \mu m$ in all three dimensions), there are two working spindles integrated ($z1$-, $z2$-axis), which allows leaving the last applied grinding tool in its respective working spindle while doing the preshaping with the other working spindle. In this way, one can work within the position repeatability of the machine itself and the measurement repeatability of the integrated measurement systems.
- By the integration of a mechanical probe system and optical position also as alignment measurement capabilities, one can determine the coordinate system of the actual semifinished workpiece within the machine. Furthermore, in-process automatic iterative process control also as statistical process control (SPC) can be conducted, thereby eliminating the transfer inaccuracy of using an external measurement system and transferring back into the machine.

- Besides the machine and in-process measurement fundamentals themselves, there is a need for a high-precision zero-point clamping system throughout the process chain. This allows the transfer of the products through different processing steps without losing the initial adjustment, with which it matches the different machining coordinate systems. Furthermore, it allows also for external measurement, for example, measurement of aspherical curve accuracy and automatic feedback into the machine, and also alignment possibilities and defined transfer in the machining process. The accuracy of the system applied at SwissOptic lies within 2 μm of translational and 5 arcseconds of rotational repeatability.

16.5.3 Data for CNC-machined monolithic optics

16.5.3.1 Typical operation parameters (Tables 16.8, 16.9 and 16.10)

The values in Tables 16.8–16.10 are only for the central shaping process. For the technical values after polishing, please refer to the relevant processing technologies described in this book.

Table 16.8 Typical technical parameters.

Dimensional position accuracy	3 μm
Dimensional angle accuracy	5 arcseconds
Form accuracy	≤2 μm, depending on the shape itself
Machinable dimensions	2–400 mm
Machinable shapes	Flats, spherical, aspherical, free-form, and any combination

Table 16.9 Typical economical parameters.

Main lot sizes	1–1000 pcs/batch
Average lead time	5–6 weeks
Cost reduction	Dependent on the design but very often between 10 and 20% compared with a conventional multipart design.
Initial costs for a new product	Process planning, programming, and special tools depend on the specific product
Initial investment in technology	≥500,000 €

Table 16.10 Limits of the technology.

Surface quality	The surface quality directly achieved in the shaping process, so called prepolished surfaces, is sufficient for applications based on intensity, such as simple illumination systems. The surface quality is not sufficient for applications that aim for specific wavefront accuracies.
Cost efficiency for high volumes	If the quality requirements can be fulfilled by glass or plastic molding, within high-volume productions, molding is the preferred technology.

16.5.4 Conclusions

CNC-machined monolithic optics, the integration of flat, spherical, aspherical, and free-form optical surfaces as also mechanical reference and/or reference elements into one glass piece, opens up new ways in designing and making optical systems. The outlay as a monolithic design can lead to increased quality, robustness, and miniaturization, while at the same time usually offering the advantage of cost reduction. Today, the integration of refractive flat and spherical surfaces as mechanical elements is quite common. Furthermore, monolithic products including diffractive [2] and partially coated surfaces have been made, thereby opening up the possibility of an increased level of integration.

Regarding aspherical and free-form shaped optical surfaces, three backlogs are evident that need to be the focus of future developments. For high-precision aspheres with strong deviations from best-fit spheres, and especially free-form shapes, a compromise between flexibility and accuracy of the core shaping process becomes evident. Because of the bearing technologies applied in highly flexible machinery, there are frequencies of typically ~2–6 mm wavelength and 100–300 nm amplitude within the surface that do require the application of smoothing processes to equalize the surface. These smoothing processes can be done by fine-grinding on the specialized ultraprecision machinery that is generally used for diamond turning. Another way that is still unexplored is the development of special toolings that do absorb the distinct frequencies of the machine while equalizing the surface at the same time.

The state of the art of numerical controlled polishing technologies lacks flexibility; for example, one is missing systems with an automatic tool changer. This would open up the possibility of combining different processing steps within one machine setup, for example, prepolishing, local correction, and surface smoothing, as well as applying specialized toolsets for flat, spherical, aspherical, or free-form surfaces [3].

Finally, the development of in-process measurement capabilities for polished surfaces with regard to surface quality and accuracy in general, as well as flexible measurement capabilities regarding aspheres and free-form shapes with short setup times, is still challenging [4].

16.5.5 Further reading

1. B. Reiss, "Machining advances lead to multifunctional optics," *Photonics Spectra*, pp. 70–72 (2002).
2. H. Paul, "Diffraktives Optisches Element," in *Lexikon der Optik*, Bd. 1, Spektrum Akademischer Verlag GmbH, pp. 156–157 (1999).
3. A. Schwarzhans, "Subaperture Robotic Polishing," *Advanced Optics using Aspherical Elements*, chapter 13.5, pp. 234–239 (2007).
4. H.J. Tiziani, "Metrology," *Advanced Optics using Aspherical Elements*, chapter 6, pp. 59–74 (2007).

16.5.6 Links

- http://www.swissoptic.com/download/multifunction_monolith_d2.pdf
- http://www.optotech.de

Chapter 17

Editor and Author Biographies

17.1 Volume Editors

Dr. Bernhard Braunecker
Leica Research Fellow (Retired)
Braunecker Engineering GmbH
Rebstein, Switzerland

Bernhard Braunecker studied physics at the University of Erlangen-Nürnberg (Germany) from 1960 to 1965. The main topics of his diploma work (1968), PhD thesis (1972), and further research work (until 1974) were experimental studies in Nuclear Physics. From 1974 to 1982 he was active in optical information processing at the University of Erlangen, at the IBM San Jose Research Laboratory (USA), and at the University of Essen (Germany). His research focus was optical and digital image processing, holographic data storage, and image coding.

In 1982 he joined Wild Heerbrugg AG in Switzerland, later called Leica Geosystems AG. From 1992 to 2006 he was chief scientist of "Optics design & System modeling," responsible for the optics of the geodetic, photogrammetric, and defense surveying instruments. Special highlights of the R&D group were the work on the world first "Digital Level," the large airborne film and digital cameras, a UV laser printer for screen exposure (screen size 1.5 m × 7 m), several thermal imaging and night-vision systems, and about 10 large technology projects for ESA, the European Space Agency. He developed new measurement devices, for example, to calibrate the large photogrammetry cameras, and introduced new multifunctional sensors like those based on monolithic optics. He retired from Leica at the end of 2006.

He is a board member of the Swiss Society for Optics and Microscopy, SSOM, and of the Swiss Physical Society, SPS. He holds about 40 patents in the field of "Optical Engineering."

Dr. Rüdiger Hentschel
Schott AG
Mainz, Germany

Rüdiger Hentschel studied Physics at Johannes-Gutenberg University of Mainz (Germany). He received his diploma thesis in 1976 in atomic and nuclear spectroscopy. In his PhD thesis (1980) he developed a MR-method for investigation of orientational polymer structure.

In 1981 he joined the Optics Division of Schott Glaswerke in Mainz. In his professional career he started as head of Optics EDP (process data acquisition, job floor control, production master schedule, SAP implementation). From the end of the 1980s up to the present, he practiced several management functions within the present Optics Division of Schott (General Manager, Quality; General Manager, Product Group Optical Filters; Director of Schott Precision Optic Technologies in Singapore; Director, Production of Optics Division; and Director, Optics for Devices). During this period he was responsible for all basic functions of Optics: sales, production, and quality management, as well as the global business for optical filters (glass filters and coatings).

All these activities allowed him to consider all facets of modern optics, their application potential, and the characteristic technological needs for optimum performance. Due to the worldwide activities of Schott, he obtained a deep insight into the special demands of European, American, and Asian markets, their application requirements, and the best production technologies required for optical elements and systems. He holds several patents in this field.

Prof. Hans Tiziani
University of Stuttgart
Stuttgart, Germany

Hans J. Tiziani received his first degree in mechanical engineering (1959) from St. Gallen—Abendtechnikum—after his apprenticeship in optics and precision mechanics at Wild Heerbrugg. He obtained his engineering degree in optics from the Institute d'Optique, Paris/Orsay in 1962. His PhD work with H.H. Hopkins at Imperial College London was on "Some factors affecting image quality when using highly coherent light," and the habilitation at the physics department of the ETH-Zürich was on "Speckle metrology: Vibration analysis and deformation measurement."

In his professional activities he was a consultant at IBM in San Jose and head of the optics laboratory at ETH Zürich from 1968 to 1973. He was a director of the central laboratory of Wild Heerbrugg (today Leica Geosystems) from 1973 to 1978, before becoming a professor at the University of Stuttgart, from where he retired recently. At the University of Stuttgart he was a director of the Institute of Technical Optics, a dean of the mechanical engineering faculty, and president of the Great Senate for 14 years.

He published, partly with his co-workers, more than 350 papers in the world's most-important technical journals in Optics and Optical Engineering (refereed) and 16 articles in books.

He was a board member of the Swiss Society of Optics and Microscopy and the German DGaO for many years. He is a fellow of SPIE, OSA, and EOS (European Optical Society). He obtained the Gabor Award in 2001 and is an honorary member of the DGaO. Furthermore, he was president of European Electro Optics and the Optics Division of the European Physical Society for five years, and he was on the Board of Directors of SPIE for six years.

He worked in different areas of optical engineering, optical metrology and testing, image formation and speckle and speckle applications. In addition, he worked in the fields of interferometry, holography, diffractive optics, holographic optical elements and optical tweezers, as well as high-resolution microscopy, surface analysis, active and adaptive optics, micro and nanotechnology, and optical sensors for medical applications.

17.2 Contributing Experts

Dr. Jochen Alkemper
Schott AG
Mainz, Germany

Dr. Stefan Bauer
Schott AG
Mainz, Germany

Dr. Klaus-Friedrich Beckstette
Carl Zeiss AG
Oberkochen, Germany

Alois Bell
Meopta—Optika
Prerov, Czech Republic

Dr. Thomas Bergs
Aixtooling GmbH
Aachen, Germany

Massimo Biber
Leica Geosystems AG
Heerbrugg, Switzerland

Dr. Rainer Börret
Hochschule für Technik und Wirtschaft Aalen
Aalen, Germany

Dr. Ulf Brauneck
Schott AG
Mainz, Germany

Bernd Bresseler
Aixtooling GmbH
Aachen, Germany

Helwig Buchenauer
Schneider GmbH + Co. KG
Steffenberg, Germany

Dr. Bernd Dörband
Carl Zeiss AG
Oberkochen, Germany

Dr. Angela Duparrè
Fraunhofer Institut für angewandte Optik und Feinmechanik
Jena, Germany

Hildegard Ebbesmeier
Jos. Schneider Optische Werke GmbH
Bad Kreuznach, Germany

Dr. Martin Eisner
Suss MicroOptics SA
Neuchatel, Switzerland

Dr. Oliver Fähnle
Fisba Optik AG
St. Gallen, Switzerland

Michael Falz
VTD Vakuumtechnik Dresden GmbH
Dresden, Germany

Dr. Heiko Feldmann
Carl Zeiss SMT AG
Oberkochen, Germany

Dr. Edgar Fischer
Contraves Space AG
Zürich, Switzerland

Dr. Martin Forrer
Fisba Optic AG
St. Gallen, Switzerland

Steffan Gold
Viaoptic GmbH
Wetzlar, Germany

Dr. Claus Gunkel
Leica Microsystems Wetzlar GmbH
Wetzlar, Germany

Michael Haag-Pichel
Schneider GmbH & Co.KG
Steffenberg, Germany

Dr. Harro Hagedorn,
Leybold Systems GmbH
Alzenau, Germany

Dr. Peter Hartmann
Schott AG
Mainz, Germany

Dr. Bernhard Hladik
Schott AG
Mainz, Germany

Dr. Christoph Horneber
Leica Camera AG
Solms, Germany

Alfred Jacobsen
OpSys Project Consulting Application Field
Schoeffengrund, Germany

Dr. Norbert Kaiser
Fraunhofer Institut für angewandte Optik und Feinmechanik
Jena, Germany

Peter Karbe
Leica Camera AG
Solms, Germany

Christopher Klein
Schott North America
Duryea, Pennsylvania, USA

Jörg U. Korth
Korth Kristalle GmbH
Altenholz, Germany

Dzelal Kura
FISBA Optic AG
St. Gallen, Switzerland

Eckhard Langenbach
Fisba Optic AG
St. Gallen, Switzerland

Dr. Alexander Laschitsch
Degussa Röhm Service/Röhm GmbH & Co. KG
Darmstadt, Germany

Reinhold Litschel
Jos. Schneider Optische Werke GmbH
Bad Kreuznach, Germany

Dr. Hans-Jürgen Mann
Carl Zeiss SMT AG
Oberkochen, Germany

Dr. Ralf Mayer
Viaoptic GmbH
Wetzlar, Germany

Mark Meeder
Fisba Optic AG
St. Gallen, Switzerland

Dr. Rudolf Müller
Schott AG
Mainz, Germany

Dr. Ulrich Peuchert
Schott AG
Mainz, Germany

Hans K. Pulker
Institute of Ion Physics
Innsbruck, Austria

Bernd Reiss
SwissOptic AG
Heerbrugg, Switzerland

Dr. Detlev Ristau
Laser Zentrum Hannover
Hannover, Germany

Dr. Simone M. Ritter
Schott AG
Mainz, Germany

Uwe Schallenberg
mso jena GmbH
Jena, Germany

Dr. Bianca Schreder
Schott AG
Mainz, Germany

Dr. Andreas Schwarzhans
SwissOptic AG
Heerbrugg, Switzerland

Dr. Karine Seneschal-Merz
Schott AG
Mainz, Germany

Laurent Stauffer
Leica AG
Heerbrugg, Switzerland

Dr. Thomas Sure
Leica Microsystems Wetzlar GmbH
Wetzlar, Germany

Dr. Bernd Szyszka
Fraunhofer IST
Braunschweig, Germany

Hans K. Tafelmeier
Dünnschicht-Technik
Rosenheim, Germany

Uwe Tippner
Jos. Schneider Optische Werke GmbH
Bad Kreuznach, Germany

Wilhelm Ulrich
Carl Zeiss SMT AG
Oberkochen, Germany

Dr. Reinhard Völkel
SÜSS MicroOptics SA
Neuchatel, Switzerland

Andreas Wälti
Evatec Ltd.
Flums, Switzerland

Dr. Kenneth J. Weible
SÜSS MicroOptics SA
Neuchatel, Switzerland

Alain Würsch
Inopaq technologies Sar
Grandvaux, Switzerland

Dr. Jose Zimmer
Schott AG
Mainz, Germany

Acknowledgements

This book is written by experts for those who are interested in modern optics. It is the product of contributions from about 60 authors from various industries and institutions. Their inputs, based on their daily work, provide professional information about technologies at the leading edge of the value added chain of modern optics. We believe that such a profound presentation of optical engineering competence in one book is unique in the field of optics. Thus, many thanks to our co-authors for their enthusiasm, openness and for the hard work they did.

We are especially indebted to the following institutions, named in alphabetic order, where the technologies were developed and applied: Aixtooling GmbH, Berliner Glas GmbH, Carl Zeiss AG, Carl Zeiss SMT AG, Degussa Röhm GmbH & Co. KG, Dünnschicht Technik Tafelmeier, Evatec Ltd, Fisba Optik AG, Fraunhofer Institut IST, Fraunhofer Institut für angewandte Optik und Feinmechanik (IOF), Hochschule für Technik und Wirtschaft Aalen, Inopaq Technologies Sarl, Institute of Ion Physics University of Innsbruck, Jos. Schneider Optische Werke GmbH Kreuznach, Korth Kristalle GmbH, Laser Zentrum Hannover, Leica Camera AG, Leica-Geosystems AG, Leica-Microsystems AG, Leybold Systems GmbH, Meopta-Optika, MSO Jena GmbH, Oerlikon Space AG, OpSys Project Consulting Application Fields, Schott North America Inc., Schott AG; Schneider OM GmbH + Co. KG, Süss Microoptics SA, SwissOptic AG, University of Stuttgart, ViaOptic GmbH, VTD Vakuumtechnik Dresden GmbH.

Many thanks to Norbert Kaiser for his great support in coating technologies. Furthermore, we appreciate many fruitful discussions with Klaus-Friedrich Beckstette, Rainer Börret, Claus Gunkel, Susanne Lehnicke, Jose Zimmer and many others.

We are very grateful for the assistance of Mrs. Beate Schmitt at Schott AG/Mainz and Mrs. Andrea Bormann-Wicki at Leica Geosystems AG, Heerbrugg. They succeeded in organizing on time all chapters into a manageable form; a rather crucial task of coordination with so many authors.

Finally, it is a great pleasure we thank our publishers—in particular, Nick Barber of Techset Composition Ltd. and Tim Lamkins and Eric Pepper of SPIE–for their valuable guidance, suggestions, and the great care taken throughout the course of the production of this book. Without their professional help, this book would never have been finished.

Bernhard Braunecker, Rüdiger Hentschel and Hans Tiziani May 2007.

Index